Tourism and Transport

ASPECTS OF TOURISM TEXTS
Series Editors: Professor Chris Cooper, *University of Queensland, Australia*
Dr C. Michael Hall, *University of Canterbury, Christchurch, New Zealand*
Dr Dallen Timothy, *Arizona State University, Tempe, USA*

Other Books of Interest

Codes of Ethics in Tourism: Practice, Theory, Synthesis
David A. Fennell and David C. Malloy

Cultural Tourism in a Changing World: Politics, Participation and (Re)presentation
Melanie K. Smith and Mike Robinson (eds)

Festivals, Tourism and Social Change: Remaking Worlds
David Picard and Mike Robinson (eds)

Film-Induced Tourism
Sue Beeton

Histories of Tourism: Representation, Identity and Conflict
John K. Walton (ed.)

Learning the Arts of Linguistic Survival: Languaging, Tourism, Life
Alison Phipps

Music and Tourism: On the Road Again
Chris Gibson and John Connell

Nature-based Tourism in Peripheral Areas: Development or Disaster?
C. Michael Hall and Stephen Boyd (eds)

North America: A Tourism Handbook
David A. Fennell (ed.)

Shopping Tourism, Retailing and Leisure
Dallen J. Timothy

Tea and Tourism: Tourists, Traditions and Transformations
Lee Jolliffe (ed.)

The Global Nomad: Backpacker Travel in Theory and Practice
Greg Richards and Julie Wilson (eds)

The Tourism Area Life Cycle, Vol.1: Applications and Modifications
Richard W. Butler (ed.)

The Tourism Area Life Cycle, Vol.2: Conceptual and Theoretical Issues
Richard W. Butler (ed.)

Tourism Development: Issues for a Vulnerable Industry
Julio Aramberri and Richard Butler (eds)

Tourism, Recreation and Climate Change
C. Michael Hall and James Higham (eds)

Tourist Behaviour: Themes and Conceptual Schemes
Philip L. Pearce

Tourism Ethics
David A. Fennell

Tourism and International Exchange: Why Tourism Matters
Gavin Jack and Alison Phipps

Tourism in the Middle East: Continuity, Change and Transformation
Rami Farouk Daher (ed.)

Wildlife Tourism
David Newsome, Ross Dowling and Susan Moore

For more details of these or any other of our publications, please contact:
Channel View Publications, Frankfurt Lodge, Clevedon Hall,
Victoria Road, Clevedon, BS21 7HH, England
http://www.channelviewpublications.com

ASPECTS OF TOURISM TEXTS
Series Editors: Chris Cooper (*University of Queensland, Australia*),
C. Michael Hall (*University of Canterbury, New Zealand*)
and Dallen Timothy (*Arizona State University, USA*)

Tourism and Transport
Modes, Networks and Flows

David Timothy Duval

CHANNEL VIEW PUBLICATIONS
Clevedon • Buffalo • Toronto

For Madeline

Library of Congress Cataloging in Publication Data
Duval, David Timothy
Tourism and Transport: Modes, Networks and Flows/David Timothy Duval.
Aspects of Tourism Texts
Includes bibliographical references and index.
1. Transportation engineering. 2. Transportation–Passenger traffic. 3. Tourism.
I.Title. II. Series.
TA1145.D88 2007
388'.042–dc22   2007000284

British Library Cataloguing in Publication Data
A catalogue entry for this book is available from the British Library.

ISBN-13: 978-1-84541-064-3 (hbk)
ISBN-13: 978-1-84541-063-6 (pbk)

Channel View Publications
An imprint of Multilingual Matters Ltd

UK: Frankfurt Lodge, Clevedon Hall, Victoria Road, Clevedon BS21 7HH.
USA: 2250 Military Road, Tonawanda, NY 14150, USA.
Canada: 5201 Dufferin Street, North York, Ontario, Canada M3H 5T8.

The policy of Multilingual Matters/Channel View Publications is to use papers that
are natural, renewable and recyclable products, made from wood grown in
sustainable forests. In the manufacturing process of our books, and to further support
our policy, preference is given to printers that have FSC and PEFC Chain of Custody
certification. The FSC and/or PEFC logos will appear on those books where full
certification has been granted to the printer concerned.

Typeset by Saxon Graphics Ltd.
Printed and bound in Great Britain by MPG Books Ltd.

CONTENTS

ACKNOWLEDGEMENTS

Special thanks to the series editors, Mike Hall, Chris Cooper and Dallen Timothy, as well as Mike Grover at Channel View, for their support and encouragement. Mike Hall and Tim Coles were constant sources of encouragement and support – a debt of gratitude is indeed owed. Both Chris Cooper and Mike Grover provided useful feedback and advice on a first draft – the book is clearly better for it. Thanks also to, in no particular order, Ewan Wilson (founder of Kiwi Airlines International in New Zealand), Mike Swiatek, Stephen Jones and Rachel Gardiner at Freedom Air (New Zealand), Adam Weaver, Paul Peeters, Amanda Mabon, Paul Wilkinson, Bill Found, everyone in the Department of Tourism at Otago (Monica Gilmour, Diana Evans, James Higham, Richard Mitchell, Brent Lovelock, Hazel Tucker, Neil Carr, Anna Carr, Kat Blumberg), Micha Lueck, 'ntd-devsys', Sarah Todd, David Buisson, Alan MacGregor, Rosalie Rissetto and Mike Tod at Air New Zealand, Martin Montgomery, Jan Schlaefke, and Achim Munz. Thank you to everyone in the Commerce Division Office (Amanda, Kate, Bronwen, Kathie, Lauren, Clint, Kirstin, Sarah, and Karen) for making 2005 and 2006 smooth and enjoyable. Over the years, various students from my Tourism Transport Management course at the University of Otago have had a significant input into this project, whether they knew it or not. For his encouragement and insightful discussions I want to particularly thank Ayudh Nakaprasit, a former student of mine who, as at July 2006, is working as an Aviation Analyst in the Strategic Planning Division at Nok Air in Thailand.

I was fortunate to have a team of very capable Research Assistants over the course of this project, including Trudie Walters, David Purdie, David Scott, Paul MacDonald (for reading a final draft and offering excellent suggestions), Martine Baastians, Simon Rowe and Sarah Nicolson. Thanks to everyone at Channel View for their patience and support, particularly Sami Grover (good luck, mate), Sarah Williams, Ellie Robertson, Jonna Gilbert, Ken Hall, Kathryn King, and of course Mike and Marjukka Grover.

Background sounds courtesy of Andy Bell, Vince Clarke, Dave Gahan, Martin Gore, Andy Fletcher, Alan Wilder, Ralf Hütter and Florian Schneider, Eric Mouquet and Michel Sanchez, Client, William Orbit, Leftfield, Neil Tennant, Chris Lowe, FGTH, New Order, Pitch Black, John Digweed, and Marcus Lush in the evenings on Radio Live. Finally, I want to especially thank my wife and, indeed, pillar of support, Melissa, for both her presence and tolerance.

David Timothy Duval
Opoho, October 2006

PERMISSIONS

Thanks to the following individuals/organisations for granting permissions to utilise/ reproduce materials presented herein:

- David Fossett, for permission to base Figure 4.4 on his own map.

- www.cruisejunkie.com for permission to use data on cruise line environmental infringements.

- Jean-Paul Rodrigue for permission to base Figure 6.2 on his original figure (http://people.hofstra.edu/geotrans/).

- Elsevier for permission to reproduce Figure 2.2 and Figure 7.1.

- Air New Zealand (particularly Rosalie Rissetto) for Figures 8.1 and 8.2.

ABBREVIATIONS

AA	American Airlines
ACCC	Australian Competition and Consumer Commission
ACI	Airports Council International
AIF	airport improvement fee
ASA	air service agreement
ASA	Advertising Standards Authority
ASK	available seat kilometre
ASM	available seat mile
ATW	Air Transport World
BAA	British Airport Authority
BBC	British Broadcasting Corporation
BTS	Bureau of Transport Statistics (US)
BWIA	British West Indian Airways
CAA	Civil Aviation Authority (UK)
CAC	command and control
CBC	Canadian Broadcasting Association
CEV	Crew Exploration Vehicle
CFC	chloroflurocarbon
CLIA	Cruise Lines International Association
CLV	Crew Launch Vehicle
CRA	Customs and Revenue Agency
CRS	computer reservations systems
CTO	Caribbean Tourism Organization
DESA	Department of Economic and Social Affairs (UN)
DoT	Department of Transport (US)
DPRK	Democratic People's Republic of Korea
DSEC	Statistics and Census Service (Macau)
ECAA	European Common Aviation Area

EIS	Environmental impact statement
EPA	Environmental Protection Agency
ETEZ	Effective Tourism Exclusion Zone
FAA	Federal Aviation Administration (US)
FDPS	flight data processing system
FIT	free and independent travellers
FSA	full service airline
GAO	General Accounting Office (US)
GE	General Electric
GDP	gross domestic product
GDS	global distribution systems
GhG	greenhouse gas
GIS	geographical information system
HACAN	Heathrow Association for the Control of Aircraft Noise
IASA	International Aviation Safety Assessment
IATA	International Air Transport Association
ICAO	International Civil Aviation Organisation
ICCL	International Council of Cruise Lines
IDP	International Driving Permit
IEA	International Energy Agency
IMO	International Maritime Organization
IOC	Intergovernmental Oceanographic Commission
IPCC	Intergovernmental Panel on Climate Change
ISS	International Space Station
IT	information technology
ITA	Office of Travel & Tourism Industries (US)
JAO	joint airline operation
KTX	Korean Train Express
LCC	low-cost carrier
LCLF	low cost/low fare
LIAT	Leeward Islands Air Transport
MEAS	Macau Eagle Aviation Services
MEP	Maritime Environment Protection
MICE	meetings, incentives, conventions, exhibitions
MSD	marine sanitation device
NASA	National Aeronautics and Space Administration
NCN	National Cycling Network
NGO	non-governmental organisation
NSCR	North Sea Cycle Route
NZCC	New Zealand Commerce Commission

OACC	Oceanic Area Control Centre
OECD	Organization for Economic Cooperation and Development
PATA	Pacific Asia Travel Association
PERC	perchlorethylene
P&O	Peninsular and Oriental Steam Navigation Company
PNR	passenger name record
PRC	People's Republic of China
RCEP	Royal Commission on Environmental Pollution (UK)
RDC	Democratic Republic of Congo
RMS	revenue management system
RPK	revenue passenger kilometres
RPT	regular passenger transport
SAA	strategic alliance agreement
SAM	single aviation market
SAR	special administrative region
SARS	Severe Acute Respiratory Syndrome
SUV	sports utility vehicle
TSA	Transportation Security Administration
TTMRA	Trans-Tasman Mutual Recognition Arrangement
UNCCD	United Nations Convention to Combat Desertification
UNEP	United Nations Environment Programme
UNFCCC	United Nations Framework Convention on Climate Change
UNWTO	United Nations World Tourism Organization
VAA	Virgin Atlantic Airways
VBA	value-based airline
VFR	visiting friends and relatives
VLCV	very large cruise vessels
VOIP	voice over internet protocol
WMO	World Meteorological Organization
WTI	West Texas Intermediate
WTO	World Tourism Organisation
WTTC	World Travel and Tourism Council

CHAPTER 1:

INTRODUCTION: MANIFESTATIONS OF TRANSPORT AND TOURISM

LEARNING OBJECTIVES

After reading this chapter, you should be able to

1. Understand the complex nature by which transport is tied to tourism and tourism development.

2. Assess and describe the scope of transport operations worldwide, particularly as they relate to travel flows and tourism development.

3. Distinguish and draw correlations between modes, networks and flows and their role(s) in the development of transport networks.

4. Outline the top issues facing transport provision in the context of tourism.

5. Discuss the role of the mode and type of transport in the context of decision-making systems.

INTRODUCTION: A SITUATION ANALYSIS

Transport has emerged as one of the more ubiquitous and complex global economic sectors. It forms the backbone of national and international commerce by acting as a mechanism for the movement of freight and people. As a result, growth in transport systems share synergies with growth in tourism, and vice versa. The global reach of tourist activities has, in part, been facilitated by the increase in accessibility of tourist 'places' on a global scale, and the popularity of holidays in western countries that make use of personal transport surged throughout the 20th century, thus spawning consistent demand for accessibility. Importantly, externalities that affect the viability of tourism

1

at varying spatial levels (e.g. attractions, destinations, regions, global) can have flow-on effects to transport. As well, externalities that affect transport provision can impact on tourism demand and tourism development. Indeed, the events of 11 September 2001 in the United States demonstrated the fragility of the global tourism sector and associated transport industries.

Global tourism has grown significantly in the past few decades (Figure 1.1), and even over the past century. The stagnant growth in global tourism between 2001 and 2003 seems to have begun a renewal beginning in 2004 and carrying through to early 2006 (UNWTO, 2005a, 2006). The scope for international travel, according to the IATA, is positive in some regions (e.g. Asia and Middle East) despite overall setbacks experienced in 2005 due to increased costs for fuel spurned by rising oil prices (IATA, 2005). With UNWTO predicting almost 1.6 billion arrivals by 2020 (UNWTO, 2005a), and with several airlines (e.g. Qantas, Emirates) purchasing next-generation aircraft such as the Boeing 787, the Airbus A350 or the A380, the importance of transport provision becomes clear. Developments in transport can, and will, have an enormous impact on people's mobility, and tourist motivation and demand in general already has significant impacts on the *way* people travel (Hall, 2005). As Hall (2005: 37) notes, tourists have benefited from the introduction of new technologies in transportation, which have been developed as a direct consequence of the rise in demand of travel:

> The cost and time of moving commodities, services and people have dramatically reduced in recent years. The real cost of travelling internationally has fallen sharply, as has the time it takes to travel long distances…In the first decade of the twenty-first century marginal increases in the time saved may be achieved but, more significantly, the same flight will be undertaken by double-decker jumbo jets carrying almost twice as many people as the 'traditional' jumbo jet.

The Department for Transport in Great Britain established a baseline index from 1990 and plotted the movement of distance, time and number of trips using the National Travel Survey. The result is that the number of trips has decreased steadily since 1990 while distance has increased (Figure 1.2). The amount of time spent travelling (including all forms of travel, not just tourism) has slightly decreased.

A 20-year outlook for commercial air travel produced by Boeing (2005) suggests passenger traffic between 2005 and 2024 will increase by an average of 4.8% per year, and cargo traffic is predicted to increase by an average of 6.2% across the same period. Boeing also predicts the global fleet of aircraft will double by 2024 to over 35,000 commercial aircraft (including both cargo and passenger). Airbus, a competitor, produced its own market outlook report (Airbus, 2005) with similar predictions of growth. Airbus predicts that the number of passenger aircraft in operation will double from almost 11,000 to 22,000 between 2003 and 2023. Airbus is also predicting a doubling of frequencies on existing routes, but only an increase of 20% of the number seats on aircraft.

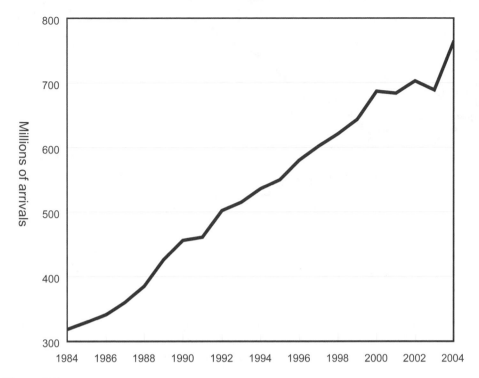

Figure 1.1 World tourist arrivals (millions)

Source: ITA (US) (2005), based on U.S. Department of Commerce, ITA, and UNWTO

ABOUT THIS BOOK: POSITIONING MODES, NETWORKS AND FLOWS

The broad purpose of this book is to map the key elements that comprise the complex relationship between transport and tourism. A framework of modes, networks and flows, as primary elements that help explain the transport/tourism relationship, is utilised. This framework has largely been adopted from the geography of transport studies literature (e.g. Hoyle & Smith, 1998), although other disciplines such as management, marketing and economics have also utlised similar approaches. Modes, network and flows can be defined as follows:

Modes: Following conventional definitions used in the business management literature with respect to categorising transport operations (see, for example, Gubbins, 2004), transport modes are manifested in three ways: ground transport, air transport and marine transport (a future mode, space transport, is considered in Chapter 9). A particular transport 'type' shall refer to the actual means of mobility realised within a particular mode. Thus, cruise tourism can be considered a type of transport that would fall under the marine mode of transport, and low-cost airlines, charter carriers and 'legacy carriers', to name a few, can be classified as a type of air transport (differentiated from other carriers

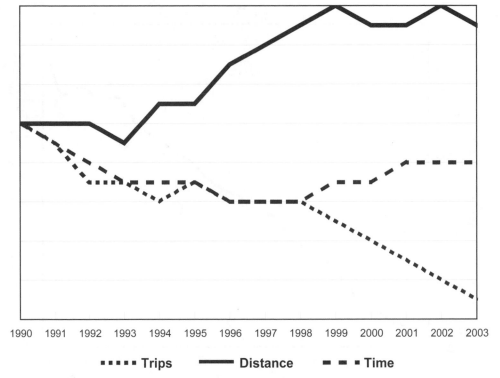

1990 1991 1992 1993 1994 1995 1996 1997 1998 1999 2000 2001 2002 2003

▪▪▪▪▪ Trips   ▬▬▬ Distance   ▬ ▬ ▪ Time

Figure 1.2 Relative change of time, distance and number of trips (all types) in Great Britain using a baseline starting point of 1990

Source: Department for Transport (2005) based on the National Travel Survey

on the basis of the business model and network served). While somewhat autocratic and rigid in its function, this classification system allows for consideration of the importance of tourism to each and to showcase useful examples of integration and importance.

Networks: If modes broadly represent the means of travel, then network structure underpins the ability of a mode or type of transport to profitably provide service and facilitate mobility. Hoyle and Smith (1998: 14) summarise the importance of networks succinctly:

> A pattern of links and nodes produces a network, a physical arrangement of trans-
> port facilities; and the design, development and management of that network requires
> a multifaceted transport system, which is ultimately both a response to demand and
> an expression of technological capability and economic resources.

Flows: Understandably, the profitability of networks depends on parameters of demand, externalities and competition. Networks are integral, therefore, in positioning modes and flows in the context of tourism such that they help explain how each work together to shape international (and regional/national) tourism. Traffic flows across networks represent the tangible measures of accessibility; they are captured in arrival statistics, load factors and demand models, and are governed by ability of modes and types of transport to service demand for traffic flow. Flows are therefore influenced by factors such as motivation and demand (incorporating economic and social variables) as well as supply.

When viewed as a wider system of variables that influence the transport/tourism relationship, modes, networks and flows can be seen to have a substantial impact on the structure of global tourism. Indeed, it can be argued that there is a significant degree of positive correlation and dependence between each concept:

1. *Networks determine flows and flows justify networks*: The spatial layout, and subsequent linkages established, of transport networks govern the flow of passengers. Thus, the ability of an airline to offer services to a particular destination, for example, plays an integral role in tourism development as it is the vector by which some tourists will arrive. This also means that, if flows are hindered by externalities such as market economics or simple demand, the scope and size of networks can change. Operators of modes of transport are therefore constantly reviewing the viability, in financial terms, of their networks. Rail providers may elect to cease services where demand is muted due to the introduction of air services, and airlines may alter their own network structure in response to decreasing demand along one or more network segments. Not surprisingly, significant investment in capital and infrastructure is required in order to maintain global networks as conduits of flows. What this can often mean is that the cost of entry can be prohibitive to the point that provision can often be concentrated in the hands of a few providers. For example, as of 2004 the top five airlines in terms of operating fleet size account for over 2400 aircraft (Table 1.1).

Table 1.1 World's top five airlines by operating fleet size (2004)

	Airline	*Number of operating aircraft*
1	American Airlines	709
2	Delta Airlines	495
3	Northwest Airlines	433
4	Southwest airlines	424
5	United Airlines	418

Source: ATW (2005a)

It is interesting to note that three of the top five airlines (American, United and Delta) have experienced fluctuating profitability problems since 2000 (see Chapter 6). Thus, while the scale of transport, particularly air transport, is massive, the fragility of the provision of services cannot be underestimated. The servicing of networks is perilous and subject to numerous externalities, including, but certainly not limited to, war, terrorist, natural disasters and economic malaise. These consequently affect cost, demand and supply. Equally important to note is that networks are not only global in scale. Regional and even national networks (for example highways or rail transport) can also influence movement and mobility at that scale, but even these can be subject to global and regional economic conditions.

2. *Patterns and intensities of flows determine the viability of networks.* Tourism is a fickle economic sector in that it relies upon the management of image, yet at the same time is vulnerable to similar externalities as transport (Hall, 2005). When image and perception change to the extent that demand is reduced, the viability of a transport provider's networks may be threatened if the pattern and/or intensity of flows is diminished. The correlation between patterns and intensities of traffic flows and the financial viability of network is therefore positive. It is for this reason that transport providers often have a financial interest in ensuring a destination is marketable and thus attractive to tourists.

3. *Regulations govern modal operations.* Despite movement toward deregulation of some transport modes (e.g. passenger air transport), transport remains a highly regulated economic industry. According to Forsyth (2006: 3), liberalisation of some air transport sectors has helped to fuel tourism growth: 'Tourism demand is quite price elastic, and aviation liberalisation has brought down fares, thus increasing tourism overall, and often, altering patterns of tourism." Despite liberalisation, however, the provision of transport services is still governed by policies and laws relating to safety, operations and competition (see Chapter 7).

4. *Transport networks play a key role in the development of destinations, especially in the context of accessibility and connectivity.* The pattern and scope of tourism is ultimately governed by the degree of accessibility and connectivity within a transport network (e.g. Butler, 1997). New Zealand as a tourism destination, for example, benefits from long-haul, non-stop air services to the United States, one of its key markets. The Caribbean has historically received a significant proportion of its overseas visitors from the United States because of non-stop flights from major urban areas such as New York and Washington DC. Likewise, rural areas popular with second home owners depend on suitable road access. As a result, the importance of accessibility is such that the ability of a destination to attract tourists is largely contingent on the availability and efficiency of transport needed to travel to that destination.

5. *Growth in tourism and transport is bi-directional and reasonably symbiotic.* In some cases this may be true, but Bieger and Wittmer (2006) rightly note that transport growth is not the only determining factor in tourism development. They argue that favourable conditions of demand and supply in the origin and destination must also be present, with transport providing the vector by which each may be satisfied. The variability in tourism amenities, attractions and new forms of mobility, such as visiting friends and relatives (Duval, 2003), second homes (Hall & Müller, 2004) and return migration (Duval, 2002) have contributed to global and regional transport demand. Growth in both tourism and transport, of course, is not universally welcomed. For example, while the economic importance of tourism for the economy is critical (e.g. Domroes, 1999), the rapid development of tourism in the Maldives has brought with it several concerns over the impact of tourism on the environment and local populations (thus raising issues of tourism as a new form of dependency [see Bastin, 1984, for example]). According to official traffic statistics at Malé International Airport, passengers disembarkations more than doubled from 1986 to 1997 (Table 1.2).

Table 1.2 Passenger movements to Malé International Airport, Maldives

	Inbound	*Outbound*
1986	124,622	123,578
1987	144,254	123,578
1988	172,119	171,561
1989	179,488	178,994
1990	217,114	216,538
1991	220,720	220,450
1992	273,982	279,645
1993	305,071	299,626
1994	348,312	345,753
1995	371,055	373,368
1996	400,300	403,645
1997	447,823	443,311
1998	468,766	460,119
1999	505,919	513,010
2000	548,518	535,658
2001	538,576	533,985
2002	557,459	552,311

Source: http://www.airports.com.mv/pastyearspf.asp (accessed 26 September 2005)

The Maldives example is by no means unique. Several trends have emerged that have influenced the manner in which transport and tourism co-exist (Table 1.3). Some of these relate to operations, such as how transport firms manage the provision of transport relating to tourism (and leisure, for that matter), while others are associated specifically with markets. As well, the scope and scale at which transport providers operate has a strong bearing on their exposure to externalities and vulnerabilities. Externalities such as the price of crude oil can impact on transport providers at smaller spatial scales (e.g. regions) just as much as those which operate on larger scales (e.g. globally).

Table 1.3 Major trends in transport and tourism

Markets

- (Continued) tailoring of services and equipment for specific tourism-related needs and market demand (e.g., business-class only airlines, niche cruising, heritage rail packages)
- International transactions and flows are increasingly becoming easier through expansion of nodes and modes as well as deregulation of global distribution systems and networks
- Consolidation of service offerings, often through vertical and horizontal integration (e.g., charter airlines, rail package tours, cruise packages)
- Some modal services (e.g., airlines) moving towards service provisions that are truly global

Tourist motivation and demand

- Continued year-on-year growth of international arrivals since 2003, following a period of negative or stagnant growth in 2001 and 2002 (UNWTO, 2006)
- A general trend of increasing discretionary travel, both international and domestic
- Demand for unique, alternative tourist experiences (e.g., geotourism [Dowling & Newsome, 2005])

Barriers to mobility

- Economic pressures (unfavourable exchange rates) that can govern flows, and to which transport modes must adapt
- Political strife (either in the origin or destination)
- Terrorist activities (either actual or the threat thereof)

- War or conflict (origin or destination)

Transport supply

- Move towards direct selling of product/service by providers (e.g., airlines, rail)
- Consolidation in air, rail and personal transport modes, with the aim of streamlining service options and availability
- Technological advances designed to cut costs and improve service offerings
- Global alliances among major service providers with the aim of streamlining services and improving market access

Operations management

- Operations linked closely with marketing and sales
- Leasing (as opposed to outright ownership) of equipment for provision of services
- Information-driven organisational structures
- Sophisticated uses of yield management to set appropriate pricing levels to match fluctuating and complex demand

Government policies and regulation

- Deregulation continues to govern government-based policy on service provision in some, but not all, jurisdictions
- Increase in multilateral (or 'plurilateral' [Holloway, 2003]), bilateral and open skies service agreements worldwide governs air access and, by extension, tourist flows
- Increase in regulation for non-economic activities such as environmental compliance and safety/security measures
- Privatisation of infrastructure associated with air and rail networks and services, but at the same time increasing tensions between airlines and airport companies of the financing of new terminal developments

Externalities

- Volatility in fuel prices, largely due to related volatility in crude oil prices that affect transport operators' ability to hedge fuel costs
- Safety and security issues not in the control of transport providers and stemming from war or terrorism activities (see above)
- Environmental (negative) externalities and the consequential pressure to limit emissions

Source: Adapted, revised and expanded from Coyle *et al.* (1994)

INNOVATION AND TRANSPORT DEVELOPMENT

Historically, technological innovation in transport is directly associated with the scope and intensity of tourism and leisure activities (Table 1.4; Figures 1.1 and 1.2). According to Butler (1980), there have been five ways in which innovation in transport has affected tourism:

1. time reduction, where a reduction in travel time, if perceived as a cost, is a reduction in cost (although this is not always the case given yield management, demand and route structures, as demonstrated in Chapter 6 with respect to air travel);

2. a reduction in financial cost, especially on a per capita basis, of travel;

3. improvement in comfort and safety for passengers;

4. increased convenience (generally through increases in connectivity of destinations);

5. increased accessibility of destinations, and as Butler (1997: 40) notes: 'The innovation often becomes a type of tourism itself; for example, ocean cruising is both a means of transportation and a form of tourism, and so too are flying, driving for pleasure and sailing.'

Table 1.4 Technological innovations and their impact on tourism

Technological innovation	*Tourism relationship*
Wheeled wagons, roading systems (Minoan and Mycenaean civilisations, Greeks)	Spread of economic activities and associated travel; increased speed of travel
Introduction of extensive paved roads (Romans)	Annual holidays, itineraries, seaside resorts; network formation
Railway development, motorised vehicles (late 19th century, early 20th century)	Modern resort developments, organised tours (see Cocks, 2001)
Air travel, cruise tourism	Mass tourism, introduction of alternative and niche forms of tourism requiring lower intensity transport modes

Source: Adapted from Prideaux (2000b)

The demand for transport has fuelled innovations in transport design and provision, and likewise transport innovation has played an integral role in tourism development in many destinations. In the same way that rail travel opened up destinations in the 19th century (see Prideaux, 2000a, 2000b), the impact of low-cost air carriers has been similar in the past decade, with increased flows of travellers who otherwise may have not travelled due to price elasticities. Worldwide, there are several examples (e.g. China, India, the United States, Europe, Australia, New Zealand and Canada) where new low-cost air access has been beneficial to tourism overall. While LCCs have proliferated, some network or 'legacy' carriers have not fared as well. In the United States, where both domestic and international routes are offered across complex networks, some network carriers have shifted their focus to international routes since 11 September 2001 and reduced costs by culling employment levels (BTS, 2005a).

While the relationship between transport and tourism is, in one sense, practical and simple in principle, the reality is that transport is manifested both within and for tourism, as well as other forms of temporary mobility. As complex as 'tourism' as a phenomenon or mobile activity is, transport is equally diverse and multifarious. Air transport, because it is truly a global phenomenon and thus almost directly responsible for global growth in tourist arrivals, is given slightly more emphasis than other transport means. In other words, where rail and steam travel revolutionised tourism and leisure travel in the 19th century (see Figure 1.3), and personal automobiles achieved a similar impact beginning in the early 20th century, air travel has, since the 1950s and 1960s (Page, 1999), revolutionised global tourism in terms of frequency and volume previously unseen.

There are three critical aspects that frame the transport/tourism relationship:

1. *Tourists ultimately travel to and from a destination (or several destinations).* Given the most oft-cited definition of a tourist from the UNWTO (WTO, 1991) is one who 'travels to a country other than that in which he/she has his/her usual residence for at least one night but not more than one year, and whose main purpose of visit is other than the exercise of an activity remunerated from within the country visited", it is integral to understand how the travel portion of this process is manifested in order to understand the wider travel and tourism system. This realisation is, of course, not new, as most introductory tourist textbooks highlight how travel and transport fit within the wider tourism system (Page, 1999). Beyond wider scoping questions, however, questions arise such as: How and why is transport provided (raising issues of competition, government intervention, supply and demand)? To what extent can it be suggested that transport (or at least some modes of transport in certain situations) is almost entirely dependent on tourist use? Likewise, to what extent does tourism itself, as a phenomenon, rely on transport? How might one characterise tourist forms of transport versus non-tourist forms of transport, and does this distinction help or hinder a full understanding the relationship between transport and tourism?

2. *Understanding supply and demand characteristics in relation to tourism is critical in the planning and management of transport infrastructure.* Tourism is pervasive in numerous different environments, ranging from constructed or built environments such as urban cities (see Page &

DR. ALBERT SCHWEITZER, philosopher—musician—surgeon, and winner of the *Nobel Prize for Peace*, says:

"Friends I used to know only by mail, I now meet face to face!"

● Take a look at a map of Africa. —Along the line where the Equator cuts across the west coast of the "Dark Continent" you would hardly expect to find a hospital caring for 5,000 patients a year . . .

But there at Lambaréné, about 150 miles up from the mouth of the Ogowe River, a man dedicated to the service of other men, has established just such a healing mission.

This great man, Dr. Albert Schweitzer, was born in Alsace in 1875, and awarded the Nobel Prize for Peace in 1953 . . . In 1954, Pan American received a

letter, written in longhand, from Dr. Schweitzer. In it he said (translation from the French)—

"I now have the opportunity to receive friends at Lambaréné who are interested in my work. Many friends from America have arrived by Pan American Airways.

"They have been delighted with the way in which these planes have accomplished this long journey, and especially with the solicitude with which they have been received by the

company and lodged upon their arrival in Léopoldville until they continue their journey to Lambaréné.

"I am grateful to the Company Pan American Airways. Thanks to this company, it has been possible for me to meet personally friends from America, with whom I would otherwise have been in contact only by correspondence. With best wishes,

Your devoted,

Albert Schweitzer

The *only* U. S. airline flying the length of Africa is—

PAN AMERICAN
WORLD'S MOST EXPERIENCED AIRLINE

Figure 1.3 Historical snapshot of tourism/transport relationship: Pan American Airlines print advertisement, 1954

Hall, 2003) to natural environments. If it is recognised and accepted that there are clear (and often not so clear, as discussed in Chapter 2) distinctions between tourist forms of transport and non-tourist forms of transport, then it is vital to identify and measure the nature of demand for tourism-related experiences in order to allow for accurate and meaningful forecasting of transport demand. Further, shifts in preferences can have immediate and long-term impacts on transport provision, and such shifts are critical for transport planners to recognise.

3. *The dynamic nature of transport can have significant impact on tourism in a destination.* Because transport is an integral part of the tourist system, disruptions and changes to route structures (in the case of, for example, air access or shifts in cruise ship itineraries) can have a significant impact upon some destinations. Disruptions can be political (in the case of two or more countries negotiating access), natural (e.g. weather-related events that temporarily 'cut off' destinations to international or domestic traffic) or business- or operations-related (increased competition forcing some operations to close permanently or perhaps adopt new business models).

Kaul (1985, in Prideaux, 2000a) outlined several arguments that highlight how transport links with tourism (Table 1.5). Importantly, these postulates ultimately serve as the basis by which one is able to assess the *extent* to which transport has (and will likely continue to have) links with tourism, thus turning away from the question of *whether* this is indeed the case.

Table 1.5 Kaul's postulates

The evolution of tourism is greatly influenced by and is a function of the development of the means of transport.
Tourism is a mass phenomenon as well as an individual activity, which needs and calls for transport and other facilities suitable for each category.
Transport facilities are an initial and integral need for tourism and operate both as an expanding as well as a delimiting factor for traffic flows; the quality of transport services offered also influences the type of tourist flow.
The planned development, maintenance and operation of transport infrastructure under a well-conceived overall transport policy, to meet the present and future technology and demand requirements, is the key to the success of the transport system contributing to the growth of tourism.
Transport prices influence elasticity of demand for traffic and diversification of price structure, and competition has encouraged price reduction and qualitative improvements amongst modes of transport much to the benefit of tourism.
The integration of domestic and international transport systems, and parallel coordination with other countries, contributes to the ease of tourism flow and growth of domestic and international tourism.

Transport technological developments would exercise a deep influence on the means and patterns of transport in both developing and developed societies, with the result that a more efficient, faster and safer transport system, beneficial to the growth and expansion of tourism, would emerge and evolve.

Accommodation, as an essential ingredient of tourism development and success, must maintain comparative growth to meet the increasing and diverse demands and transport expansion.

The satisfactory development and equipping of terminal and en-route facilities, the systematic improvement in infrastructure, the absorption and adoption of new technology and appropriate mass marketing techniques in transport would have a pervasive impact in the continued growth of future world tourism.

Source: Kaul (1985) adapted from Prideaux (2000a: 55 [Table 1])

TRANSPORT IN THE TOURISM SYSTEM

The UNWTO (2005b) published a news release on 14 November 2005 suggesting that, despite record high oil prices, global tourism has not been affected significantly. This is perhaps the most telling example of how tourism is inextricably allied with transport. First, it suggests a relative price inelasticity of global tourism overall, although what the UNWTO statement does not indicate is the extent to which any shifts in travel behaviour and subsequent transport usage had been affected. Second, it also demonstrates the potential fragility of global tourist flows: high oil prices in 2005 and 2006 were expected to contribute to reduce demand for travel flows, and the question was raised as to how robust global tourism actually is in the face of increased costs for transport provision. In other words, what are the consequences for tourism when the cost of transport becomes too expensive for those travellers who have previously enjoyed lower transport costs and increased choice in destinations?

Positioning transport within the tourism system is useful in that it demonstrates how transport modes, nodes and networks interact to *facilitate* tourism (Page, 1999). As indicated earlier in this chapter, various external and internal elements can be identified in order to identify any potential shifts or changes. Figure 1.4 is a simplified graphical representation of a tourism system adapted from Page (1999) in which key areas of significant transport link-ages are emphasised. For the sake of argument, it can be assumed that the users of this simplified system are tourists, even though, as discussed below (and more thoroughly in Chapter 2), many forms of transport utilised by tourists are also utilised by non-tourists. There are a number of salient features and assumptions within this model:

1. It suggests that specific flows of tourists utilise networks and routings that link origin and destinations (or nodes). The flows between these nodes are made possible by various

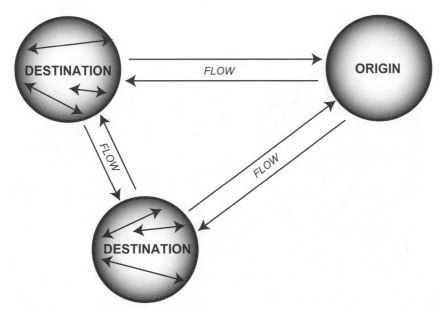

Figure 1.4 Model of the tourism/transport system

Source: Adapted from Page (1999)

modes of transport, such as commercial (or otherwise) airlines, rail or other forms of land-based transport, or even water-based transport. The operation of these flows are naturally variable and depend on numerous factors, including:

(a) the availability of specific modes of transport;

(b) cost-effectiveness of utilising various modes (discussed in Chapter 2 with respect to supply and demand) (see Figure 1.5);

(c) the motivation for travel, which is an important consideration for determining the feasibility of establishing new routes and networks or enhancing existing ones; and

(d) the time budgets associated with the potential users (or tourists) of these networks (Hall, 2005).

2. The model suggests that tourists might utilise various modes of transport within a particular destination, such as taxis, public transport, rail and air. What is important to remember, therefore, is that tourism occurs at a variety of spatial scales: air transport may be used to travel from origin to destination; rail travel may be utilised to travel within a particular country; bus services or taxis may be used to get around, for example, urban areas. Thus, the constraints and influences associated with the propensity to utilise various modes and transport can be influenced by several factors:

Figure 1.5 Reduced transport cost as a catalyst for tourism? Billboard at Exeter International Airport, UK, August 2004

(a) time-budgets whilst on holiday, such that the extent to which certain attractions or services are utilised could have an impact on the use of intra-destination modes of transport, and vice versa; and

(b) cost-effectiveness of specific modes of transport as governed by desire or motivation to undertake specific activities. One example is paying a cost premium for the opportunity to be transported by a more expensive mode that is rationalised by other motivations.

3. Finally, the model suggests that more than one destination is involved in any particular trip, and this ultimately has an impact on the utilisation of various modes of transport. For example, a family embarking upon a two-week holiday using a motor home will likely stop at numerous destinations and attractions. The flows of their travel will be influenced by the availability of infrastructure (i.e. highways and other roads), type of trip in the context of motivations (i.e. a family may elect to concentrate on destinations or attractions that are more oriented towards outdoor recreation) and the length of time available to them.

TRANSPORT SELECTION

Some studies (e.g. Ritchie, 1998) have examined the underlying motivation to travel in the context of transport, arguing that such information will allow planners and managers of tourism-related services to 'fine-tune' and 'tweak' services and products. An extension to this considers whether it is possible to suggest that different motivations to travel might have some impact on the particular mode of transport utilised. One might be inclined to think that common modes of transport are utilised concurrently by different 'types' of tourists. For example, a 747-400 travelling from Los Angeles to Sydney might have on-board numerous 'types' of tourists as described in the tourism literature. To some extent, then, the model depicted in Figure 1.4 certainly argues that the networks and flows of international tourism are inherently subject to transport provision (as argued earlier), but it inadvertently simplifies the matter in that it assigns tourism-related transport simply to an origin-destination pairing. As discussed above, transport in relation to tourism can be implemented on a number of spatial levels (i.e. within the destination, between destinations), and the extent to which transport may or may not function as part of the overall experience needs to be taken into consideration. Of equal importance, perhaps, is how the mode (or type) of transport is selected.

Mill and Morrison (1985) provide a model (roughly based on Sheth, 1975) that outlines the various elements and choices involved in transport mode selection (Figure 1.6). The importance of this model is that it captures elements of trip purpose, motivation, psychological and sociological characteristics (such as 'lifestyle') in the decision to select a particular mode or type of transport. At the same time, however, it also suggests that these variables connect into various characteristics (what Mill and Morrison call 'utilities') that ultimately govern the transport mode/type decision.

Mill and Morrison's model is important because it suggests that existing studies of travel/tourist motivations for visiting destinations or attractions need to be taken into consideration in establishing the role of transport in the wider tourism system. Perhaps what is missing, however, is a feedback mechanism whereby deficiencies in one variable can be compensated for by another. For example, less net income may result in the desire to acquire significant savings with the explicit purpose of, for example, using a particular form of transport whilst on holiday (e.g. flying on the Concorde or taking an expensive cruise). Similarly, mode accessibility may be sacrificed for mode design, such that an individual may elect to utilise a specific mode of transport because of, for example, the aesthetic qualities that it offers, even though that particular mode operates out of specific nodes that are, comparatively, inaccessible or uneconomical.

STRUCTURE OF THE BOOK

Structuring a book on transport and tourism presents several challenges. First, one is inclined to allow tourism to take 'centre stage' and weave transport into conceptualisations of the tourist system. Second, because of the complexity of the relationship (discussed in more detail in Chapter 2), it is difficult to organise the centralities and commonalities between different modes and tourism. Consequently, this book generally allows

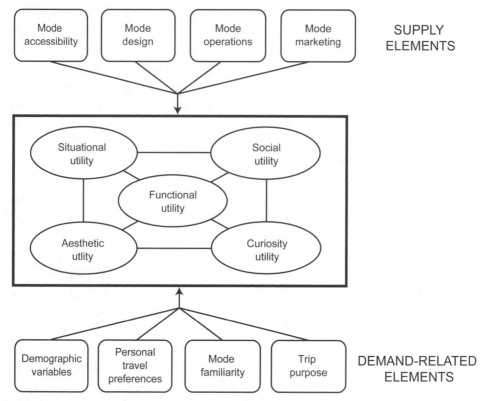

Figure 1.6 Transport mode selection model

Source: Adapted from Mill and Morrison (1985) and from Sheth (1975)

the nature of transport to take centre stage and it is therefore organised largely around the various modes and types that permeate tourism systems today. Third, while tourism as a field of enquiry is inherently multidisciplinary or transdisciplinary in nature, topics of interest to tourism development beg consideration from multiple disciplinary perspectives. In this case, the transport literature often takes centre stage throughout this book in an attempt to marry it with consideration for tourism (as well as, in some places, recreation and leisure).

Chapter 2 intentionally leaves many questions for the reader to consider. The intent of the chapter is to introduce the reader to the principles that guide this relationship. As such, the blurry distinctions between transport and tourism are outlined, with the suggestion that, quite often, delineating between tourist-focused transport and non-tourist-focused transport is problematic. Of course, much of this blurriness is a consequence of the definitional endeavour of characterising tourists. To this end, consideration is made of not only tourists but also recreation and leisure activities. Rather than enter into the debate of whether tourism is a subset of leisure (or vice versa), this book shall adopt the view that some modes and types of transport can be used in different contexts (leisure for some, tourism for others). In some ways, this amplifies the blurry distinctions discussed in Chapter 2. Chapter 2 also considers how transport is manifested within the tourist system, utilising the work of Mill and Morrison (1985) in the premise that touristic activities, and the provision for movement and mobility to facilitate these activities, are embedded within a complex system. The role of government is also discussed, primarily because the policy and planning measures and decisions rendered by national and local governments (and wider NGOs that span the global space) determine the extent of the relationship between tourism and transport. Thus, the extent to which government is involved (e.g. regulation and planning) is discussed. This gives rise to supply and demand issues and an important synopsis of models of elasticities.

As the book engages in a transport-centred focus, the third chapter reflects this in its discussion of the spatial aspects of transport networks. This is an important point because it means consideration is given to the physical manifestation of transport and, consequently, what that may mean for tourism at a variety of spatial levels (i.e. global, regional, local). Transport, quite simply, is rarely plotted outside of prescribed routes that are dependent on infrastructure. Passengers (and goods, for that matter) follow specific routes, most of which are pre-determined. Chapters 4 through 6 outline trends in each of the three main modes of transport (ground, marine, air). Each chapter examines critical issues associated with these modes and briefly interrogates the notion of sustainable modes of transport, a topic which is visited more broadly in Chapter 9. Chapter 7 highlights three critical areas of management with respect to transport management: nodal frameworks (using the example of airports), yield management, and safety and security issues. The chapter is inherently business-oriented in its presentation, but the three areas considered are critical in understanding (1) how networks govern tourism and mobility, and (2) how management considerations in tourism and transport need to consider auxiliary nodal elements (e.g. airports). Chapter 8 considers the marketing of transport in relation to tourism, and the nature of the 'product versus service' element of transport is considered. As well, various strategic marketing initiatives are discussed, including market penetration, market development and product development. Alliances are examined within a marketing context and a discussion on relationship marketing demonstrates how some transport providers use loyalty systems such as frequent flyer miles to understand their markets. Finally, Chapter 9 consider future trends in transport in two broad contexts: sustainability (particularly emissions control and peak oil) and space tourism, which may represent a fourth mode of tourism transport. Several future trends are discussed next with a view towards highlighting potential areas for further exploration and evaluation.

CHAPTER SUMMARY

Transport is tied to tourism and tourism development in a variety of ways. It functions as a means by which tourist flows are manifested, but it also plays a role in determining the success of tourism development and the financial viability of existing tourism services. This chapter has argued that, using modes, networks and flows as the central framework, transport and tourism relationships can be explored more fully. Importantly, these relationships are complex and in an almost constant state of flux. Transport provision worldwide is manifested on a variety of spatial scales using a variety of modes and types. As well, transport networks can range from being simple to complex. Importantly, the close relationship that is shared between transport and tourism ensures that neither is particularly immune from the externalities that each faces.

SELF-REVIEW QUESTIONS

1. Define mode, network and flow.

2. What is the general trend of world tourist arrivals and what does this mean for transport?

3. What are the key elements of a tourism system?

ESSAY QUESTIONS

1. Why are modes, networks and flows important in understanding the relationship between transport and tourism?

2. How has innovation in transport impacted upon tourist flows?

3. Tourism and transport have a uniquely symbiotic relationship. Write an essay that discusses this conclusion and provide examples to support your arguments.

KEY SOURCES

Page, S. (2005) *Transport and Tourism: Global Perspectives*. Harlow: Pearson Education.

A seminal work on transport and tourism, Page has succinctly woven issues of supply, demand and management together in an accessible text. The second edition (published in 2005) features an exhaustive reference list, useful for most students embarking upon assessments or theses.

Hall, C.M. (2005) *Tourism: Rethinking the Social Science of Mobility*. Harlow: Pearson Education/Prentice Hall.

Hall's book is an excellent introduction and background to some of the concepts introduced in this volume. The book is well situated to provide an introduction to such topics as mobilities, globalisation, production and consumption, and spatial interaction models. Hall also explores other critical issues of relevance to transport and tourism, including environmental change, security issues, geopolitics and tourism development.

CHAPTER 2:

TOURISM-TRANSPORT RELATIONSHIPS

LEARNING OBJECTIVES

After reading this chapter, you should be able to

1. Recognise and explain the blurry distinctions when investigating transport and tourism relationships.

2. Develop your own example of where transport and tourism have a symbiotic relationship.

3. Assess and differentiate transport development as a catalyst for tourism development from the development of tourism as a catalyst for transport initiatives.

4. Position and identify the role of government in transport and tourism relationships.

5. Understand the nature of supply and demand in relation to transport and tourism.

INTRODUCTION

The purpose of this chapter is to outline some of the salient issues that emerge when exploring the linkages between tourism and transport. These linkages, or 'interfaces' (Lumsdon & Page, 2004), are a useful point of departure because their sheer complexity would suggest that identifying and measuring a direct correlation between transport and tourism services and industries can often be difficult. Transport systems encompass numerous modes and types of transport. They feature complicated and dynamic routes determined by governmental policy and planning as well as underlying political factors (e.g. Nayar, 1995). Other factors include economic aspects, turbulent business operating environments and external issues of demand. It is, therefore, somewhat difficult to arrive at a final model or equation that definitely explains how transport fits within tourism and, conversely, how tourism fits within transport. Nonetheless, some top-line issues that highlight the tourism/transport relationship can be discussed, which is the intent of this chapter.

The shape of the chapter centres around three themes: 1) defining the transport/tourism relationship; 2) integration of transport within tourism systems; and 3) the nature of supply and demand. The chapter begins by considering the 'blurry distinctions' that characterise tourism and transport relationships, which roughly correlate to the perceived (and actual) difficulty in associating or establishing concrete relationships between modal forms/types of tourism and modes of transport. Following this, the integration of transport within the wider tourism system is considered, with a special emphasis on identifying transport as a particular (and often critical) element within the tourism system. Finally, consideration of supply and demand in the context of transport is examined.

BLURRY DISTINCTIONS

While the importance of transport to tourist flows (at a variety of spatial levels – e.g. international, domestic, local) was highlighted in the previous chapter, what remains to be considered is the *manifestation* of this relationship. In many ways, *how* transport and tourism are co-dependent is not always apparent. As such, it can be argued that there is a blurriness in the way transport and tourism co-exist. There are three reasons why the linkages and subsequent relationships between transport and tourism can be said to be blurry. First, a particular mode of transport can be both a facilitator of mobility (by tourists and/ or recreationalists) as well as an attraction (Lumsdon & Page, 2004). Second, the provision of transport explicitly for tourist use is inherently rare (Hall, 1999), aside from perhaps explicit examples where the mode of transport is either the key attraction or the means by which an attraction is consumed or experienced (e.g. boat cruise around the fjords of Iceland). Thus, as Halsall (1992) points out, there is often very little to distinguish between tourist and non-tourist use of transport services. Third, the extent to which transport plays a role in tourism development, particularly in the development of destinations in international, national and local contexts, is often not entirely clear. Put another way, the question becomes as follows: does the presence of transport facilitate tourism growth in a destination, or must a destination be suitably developed in order to attract new transport provisions or enhance existing ones?

Transport as a facilitator and an attraction

For many tourists and recreationalists, transport is seen as merely utilitarian or functional. As an example, a business traveller flying from Singapore to London Heathrow may be interested primarily in securing a flight that is within his or her budget and immediately accessible at a particular time. A family taking a holiday from London to Singapore may use similar decision-making criteria. As a result, the decision-making criteria may be focused on the cost and efficiency of this particular mode of transport. A cruise passenger, on the other hand, certainly uses a large ship for transport to and from certain destinations or ports of call, but the ship itself may often be an attraction in its own right. To this end, many cruise companies are increasingly offering on-board shopping and activities, for a variety of reasons that will be discussed in Chapter 5. In these two examples, the functional aspect transport element is shared, but its use is quite different as are the motivations and key elements used by the traveller in ultimately deciding on the particular mode or type of transport. For some, transport facilitates travel/tourism/leisure experiences. In other situations, transport is deliberately part of (or is engineered to be) the experience.

BOX 2.1 The world's longest flight? EWR-SIN on SQ

The introduction of non-stop services between Newark, New York (IATA: EWR) and Singapore (IATA: SIN) by Singapore Airlines (using an Airbus A340-500 aircraft) in late 2004 is an example that illustrates how a mode of transport can also be the attraction. Billed as the world's longest non-stop flight (at 16,603km), the inaugural flight attracted a great deal of attention from aviation enthusiasts and frequent flyers (and has been accompanied by its own website, www.nonstop2newyork.com), with some purposely booking seats on the flight in order to quite literally be a part of history. One posting on an online discussion forum for travellers epitomises this desire: 'For me the chance to be on the historic first flight was reason enough – the miles, PPS sector credits and upgrade vouchers earnt [sic] along the way are just the icing on the cake!' If one was to profile those who took part in the inaugural flight from EWR to SIN, it would likely be possible to apply several leisure and tourism motivation theories to this kind of activity. Beard and Ragheb's (1983) competency-mastery component might go some way to explaining the desire of frequently flyers (some of whom regularly make 'mileage runs' for the explicit purpose of building up their frequent flyer account balances) to embark upon such a trip. For them, travelling such a distance on the inaugural flight might well be an achievement, and may serve to elevate status in an online community/discussion fora where issues relating to aviation, frequent flyer programmes and airlines are discussed. For others, the flight could be merely a novelty and offer a sense of prestige (Crompton, 1979; Dann, 1981). Finally, if Iso-Ahola's (1980, 1982) matrix of intrinsic rewards and interpersonal environments is adopted, it might highlight individuals who are keen to enhance their own ego but at the same time acquaint themselves with others (whom they may perceive as being similar to themselves) in interpersonal environments.

Questions for consideration

1. How might such long flights be marketed to potential travellers?

2. How might long distance travel, with fewer stops, affect global flows of people and some tourism destinations?

Tourist versus non-tourist use of transport services

Hall (1999: 182) suggests that there are three categories of transport users:

a) the host community (i.e. residents not directly involved in the tourism sector);
b) employees from the tourism sector;
c) tourists or excursionists.

Outside of the contested definition of a tourist and the relationship between travel and types of mobility, it can be suggested that there is a blurry distinction between non-tourist use and tourist use of certain modes and types of transport (see also Lumsdon & Page, 2004). There are few examples of a particular mode or type of transport reserved for the exclusive use of tourists. One could argue that examples such as light rail transit within a theme park or cruise ships might be designed exclusively for tourist use, but they may also be used by other people as well (e.g. employees or management executives associated with the operation itself), albeit in smaller proportions. The Rottnest Ferry in Western Australia, for example, carries both tourists and island employees to and from the city of Perth and Rottnest Island (Figure 2.1). But what of other modes of transport such as air travel or personal vehicles? It could be argued that these modes of transport may not be exclusively designed for and utilised by tourists. A country's road network, for example, is used by both residents (in day-to-day activities, or as excursionists and/or tourists), tourists and for commercial activities. Likewise, air travel is used by a variety of types of mobile individuals: tourists, migrants, excursionists, commuters, to name a few examples.

In recognising that there are, in fact, many different users of most modes of transport, it then becomes difficult for planners, policy-makers and, indeed, local and national governments to accurately measure the level of tourist use of these modes. Some destinations or regions, for example, may be inclined to charge a road toll for the use of specific road networks and, more often than not, these charges are designed to supplement declining funding for infrastructure renewal in the face of bourgeoning tourism activities. For example, a proposal was considered in 2002 by the Queenstown-Lakes District Council

Figure 2.1 Blurry distinctions: the Rottnest Express Ferry (Perth, Western Australia) carries both tourists and employees working on the island

in New Zealand, where specific road tolls would be implemented on scenic routes, thus directly targeting tourists and excursionists. This raises several questions: which tourists are being targeted? Domestic tourists might see this as a 'tax grab' and argue that their taxes already fund the road network, including upgrades, repairs and new developments. If targeted at international tourists, the question then becomes: to what extent, if any, will such tolls have a detrimental impact on visitation and visitor flows? Is it possible that the implementation of tolls might have a negative impact on demand? In 2004, the city of Edinburgh announced plans to implement road tolls on two key cordons in the city, but various members of the local tourism industry warned that such measures could result in tourists bypassing the city completely whilst visiting Scotland (Edinburgh News.com, 2004). What is important, then, is to recognise that transport is utilised by both tourists and non-tourists, yet within the 'tourist group', to use Hall's (1999) characterisation, multiple segments can be identified and it is not unreasonable to suggest that each have different needs with respect to transport.

Tourism for transport development or transport for tourism development?

Gauthier's (1970) review of the relationship between transport and economic development led him to characterise three ways in which the tourism development/transport relationship can be manifested: positive, permissive and negative (Table 2.1).

Table 2.1 Gauthier's (1970) transport/development characterisations as applied to tourism and transport

	Gauthier's economic development/transport relationship	*The resulting tourism/transport relationship*
Positive	Where transport has a direct and positive impact on economic development (i.e. providing transport directly influences economic development)	The use of transport to facilitate domestic and international mobility, linking origin and potentially several destinations to the benefit of the destinations in question, although it is recognised that this may also contribute to the economic well-being of the origin point (through positive balance of payment on services and retail sales associated with mobile residents). The management of transport industries in concert with tourism destinations, then, would seem to be critical. Partnerships and agreements should be established for the purpose of recognising the importance of gateways (e.g. Palhares, 2003).
Permissive	Where transport does not directly influence economic development (i.e. the provision of transport alone is not enough to influence economic development, thus it is relegated to a passive, but supportive role)	Transport itself may not be a significant contributing factor to tourism development, but other factors (e.g. shifting market trends and consumer preferences) may, in aggregate, shift demand and, ultimately, mobility flows, thus spawning a reaction from transport mode service providers.

Negative	Where the provision of transport is fuelled by investment to the detriment of other economic development alternatives (i.e. transport hinders economic growth and productivity)	The overcapitalisation of transport flows and networks between multiple nodes usually results in the fragility of some operations. If and/when such operations become financially unviable, the reduction of routes along networks can have a negative impact, perhaps in the short term, for economies that may rely upon these modes of transport to bring tourists to them.

As useful as Gauthier's (1970) characterisations are, it is necessary to consider the geographic scale of analysis in determining whether tourism has more of an impact on transport development than transport has on tourism development. There is no question that new forms of technology, as applied to transport services, have resulted in a 'globalising' effect of tourism. The road systems established by the Romans, for example, were used not only for military campaigns and maintenance of the empire but also for facilitating recreation and tourism by Roman citizens across modern Europe (Prideaux, 2000a). Further still, and as argued in the last chapter, it is evident that technological developments in transport have resulted in faster travel over further distances, perhaps even at a greater frequency (all other things being equal) and, as Prideaux (2000a) notes, in greater comfort. Thus, on a global scale, there is no question that transport has facilitated tourism development by 'opening up' new destinations or enhanced existing ones (Page, 1999).

BOX 2.2 Case study – Tourism development in the West Indies: Transport as a catalyst for tourism?

Tourism, it seems, may be one of the more significant economic sectors upon which stability in the West Indies is afforded. For centuries, agriculture (sugar and bananas, both of which are under threat because larger producers in other countries have argued to the WTO that the preferential markets for both contravene natural global trade practices) has been the mainstay of the region's economy. There are signs, however, that this is changing. St Lucia, for example, has all but abandoned sugar production in favour of tourism; indeed, tourists now ride the rail system that used to carry cane sugar. The wider West Indies region, comprised of a series of island states (and dependencies of varying size and stages of development), is often regarded as synonymous with conventional mass tourism. The region has hosted visitors from Europe and North America for centuries, but it was not until the development of the jet engine aircraft that tourism became an important social and economic force in some countries in the region. In the end, people from North America could get to the Caribbean and back again without spending significant portions of their vacations in transit. The result is that, starting in the 1950s, North America became the most important market for the region as a whole (Duval, 2004a).

While transport has led to significant development of tourism in the region, it has also contributed to specific forms of tourism. Cruise tourism, as discussed in Chapter 5, acts as both the means of transport and the attraction in the region, with 43% (2000 data) of all visitors to region arriving on pre-packaged cruise ships (CTO, 2002). The region is served by numerous international airlines (e.g. Air Canada, British Airways, Air France, United Airlines, US Airways, AA), although the recent financial crises surrounding US Airways, United Airlines and Air Canada highlight the rather fragile nature of existing routes and passenger mobility from North America. Of great concern when considering the United States market is the financial difficulties that have plagued AA since 2003. AA controls upwards of 70% of the market to the Caribbean from the United States (Smith, 2002). While regional carrier LIAT and BWIA continue to operate in the region, the reality is that adequate air access has recently become more dependent upon government involvement, either financially or logistically (Duval, 2004b).

Duval (2004b: 294) notes that 'it is likely that a formal emphasis on curtailing substantial competition from numerous, small airlines and subsequently shaping a regional transport plan of one or two strong RPT (Regular Passenger Transport) providers will transpire within the next five years'. Indeed, more recently BWIA may be re-branded as Caribbean Airlines in early 2007 and LIAT and Caribbean Star are exploring merger options.

As air transport was critical in helping the region develop into a premier tourism destination, it seems that the future of the region's tourism sector is still quite reliant on air access. In fact, Bahamian Premier Perry Christie suggested in 2002 that 'there can be no rejuvenation of Caribbean tourism without the simultaneous reinvention of Caribbean air transportation' (Smith, 2002). What is needed, according to Duval (2004b), is regional control over airlift, if only to secure the provision of transport in the hands of the countries that would benefit most. One step towards this was the 1996 Caribbean Community Air Services Agreement, which not only saw the liberalisation of air travel and access throughout the region, but also allowed for multi-nation investments in air transport operations (OECD, 1999). Although this solidified intra-regional travel, international access to the region has been, and still is for the most part, provided by foreign carriers, although the exception would be BWIA, which services Miami, New York, Washington DC, Toronto, Georgetown, Paramaribo, London and Manchester with nine aircraft. Recently, several Caribbean hotel owners have suggested that ownership of Caribbean-based airlines should perhaps be held across the region to address competitiveness and the importance of tourism overall to the region (Jamaica Gleaner, 2005).

Questions for consideration

1. Are there other areas of the world that have developed in a similar manner to the Caribbean with respect to transport?

2. Why would politicians and governments in general have a vested interest in ensuring that adequate airlift is maintained through the Caribbean?

On a more micro scale, however, there is room to suggest that tourism development, or more properly, the demand for particular services and/or attractions by tourists, can play a substantial role in facilitating or creating demand for either new modes of transport or enhancements to existing modes. A few examples:

1. *Firms offering specific modal services may elect to offer new services on routes where they perceive significant demand.* Air New Zealand, for example, announced in early 2004 the intention to offer direct flights between Auckland and Shanghai, largely because of increases in the number of Chinese tourists visiting New Zealand (New Zealand Herald, 2004), but also because of increasing migration-related connections between the two countries, such as permanent or temporary student migration. Similarly, a proposed high-speed rail line between Las Vegas and Los Angeles, at an estimated cost of US$1 billion, is meant to not only link two significant tourism centres but also stimulate travel between them (Las Vegas Sun, 2004; see also www.transrapid.de).

2. *There exists a synergy between new transport services and increases in demand.* The supply of transport can have some impact on demand (either in the positive or negative) through attempts to generate market interest or turn latent demand into actual demand. One example is the introduction of Kiwi Air International and Freedom Air in New Zealand, both of which began operating on routes between New Zealand and Australia in the mid-1990s. Consequently, a rise in the number of New Zealand residents visiting Australia was recorded (Table 2.2). Granted, the affordable fares on offer from both airlines (among other variables, of course) could very well have had some impact on encouraging travel, but it was the introduction of these new services that was the proximate cause of the 'fare war' (Duval, 2005c).

Table 2.2 Resident departures from New Zealand to Australia, 1993–2001

Year	Resident departures	Annual change	Percent change
1993	418,738	30,596	+7.9
1994	407,408	−11,330	−2.7
1995	460,266	52,858	+13.0
1996	587,488	127,222	+27.6
1997	598,612	11,124	+1.9
1998	616,743	18,131	+3.0
1999	620,027	3284	+0.5
2000	684,934	64,907	+10.5
2001	676,047	−8887	−1.3

Source: Statistics New Zealand (2002)

3. *Integration between transport providers and other tourism-related operations targets specific markets with the intent of stimulating latent demand.* The integration of transport and non-transport services allows for the possibility of numerous strategic management decisions, including the introduction of new products, enhancing market share or even facilitating market entry, reducing costs and increasing profitability (see, for example, Oum *et al.,* 2004). Although *horizontal* integration generally involves harmonising similar products across two or more firms (and airline alliances, in this regard, are discussed extensively in Chapter 6 as examples of integration), of particular interest is *vertical* integration, where transport can be linked and/or merged with tourism-related operations.

Vertical integration has the ability of generating efficiencies at the competitive level. By integrating product and service offerings across a value chain, future operations are more certain and the likelihood of competitor entries is somewhat diminished (Lafferty & van Fossen, 2001). Several airlines have interests in firms and operations outside of aircraft operations. For example, it can be said that Air Canada and Qantas have financial and operational extensions of their primary operations that include interests in hotels and tour companies, thus enabling each to sell products and services in conjunction with their respective air services (Lafferty & van Fossen, 2001: Table 1). Likewise, Disney holds ownership and control over Castaway Cay in the Bahamas, where its own ships dock. Acknowledging and understanding vertical integration that features a transit or transport element is important for two reasons: 1) to control the sales and distribution of the means of travel and other associated services; and 2) to stimulate latent demand for profitable routes by capturing market interest in package holidays.

INTEGRATING TRANSPORT IN TOURISM DEVELOPMENT

Transport development, as a process, can occur at a variety of spatial levels and at varying degrees of integration with tourism development:

1. Global perspectives of tourism development generally examine the rate of tourist flows, north–south distinctions in the speed of development, pro-poor tourism development designed to empower less developed countries through tourist activities (and the associated debates that this brings), and wider consideration of the concept of mobility of which tourism is but one example. At the global level, transport is largely functional in the context of facilitating flows via prescribed networks, but there is some room to consider how some modes of transport can be utilised in branding and marketing exercises.

2. At the destination level, development includes, among other considerations, planning and policy measures including issues of implementation and monitoring; resource management and governance; physical landscape changes and any resulting impacts; indigenous peoples' and cultural impacts of tourism; and place promotion and marketing. Here, transport can be used not only as a facilitator (within smaller networks) but also an attraction (e.g. historic rail package tours). Transport should be integrated into destination development, planning and marketing decisions.

3. At the level of the attraction, micro-level business models of investment (sometimes foreign direct investment), financial viability of operations, seasonality and specific lifestyle factors, and organisational structures and behaviour are areas that have been investigated. Transport may not have an overt degree of integration at this level, but its importance cannot be overstated. The enterprise operator would undoubtedly have a strong interest in a transport operator's network development plans, marketing and overall successes (or otherwise).

More generally, the degree of transport integration can cross many spatial fields and involve multiple elements. These can include the availability of quality facilities relating to transport both within and outside the destination (Hall, 2004), appropriate routing to facilitate ease of flow (Halsall, 1992) and provision of adequate infrastructure for network development. For example, air transport on a global level is carefully monitored not only in the form of regulation and bilateral trade (see Chapter 7), but also destinations understandably form strong relationships with transport providers for the purpose of ensuring viable markets and a continuous (at least as much as possible) flow of tourists. Through destination marketing organisations, destinations will therefore work to ensure that they are accessible via one or more modes or types of transport.

An ideal model of the role of transport integration within tourism development would see three key stakeholders: tourists or travellers, destinations (including local stakeholders such as government and the community) and various relevant transport industries. Destinations have an interest in securing and maintaining market share. Transport providers will thus strive to ensure that their own market share and financial viability is maintained and will thus take great interest in the manner in which the destination undertakes development and marketing efforts. At the same time, destinations may attempt to entice transport providers to service particular places or locations, thus embarking on marketing to a different audience. Tourists and travellers, of course, form the primary user group and thus have expectations of the quality of the experience for both the mode of transport (and whether this is between origin and destination, or intra- or inter-destination transport) and within the destination itself.

The path of development, then, requires the considered coordination of how transport will fit within the tourist experience at two levels: a) the journey to and from the destination (or within the destination); and b) the extent to which transport provision can be made profitable through partnerships with destinations. Three areas of integration (although they are more) can be examined: the role of government, regulation and deregulation, and planning.

THE ROLE OF GOVERNMENT IN TOURISM/TRANSPORT RELATIONSHIPS

The tourism literature offers examples of the use of tourism by many governments as a form of economic restructuring (see, for example, Page and Hall [2003], in the context of urban tourism). The extent to which transport is positioned within such restructuring is, however, variable. As Hall (1993: 207–208) points out, there are several conditions that characterise how transport, tourism and economic restructuring are intertwined, some of which include:

(a) The national economic context: notably the availability of capital investment to develop the necessary infrastructure, not least transport and communications.

(b) Scale: small economies are more likely to depend upon imports.

(c) Level of development: lower levels of development, including that of domestic tourism, produce poor economies of scale for suppliers.

(d) The organisation of capital: the extent of the penetration of international capital may be critical in assisting development and/or leading to leakages of income abroad through payment of royalties, profits and dividends which the economy may ill afford.

The roles of foreign tour companies, carriers (notably airlines) and transport manufacturers (of, for example, luxury tourist coaches), are significant here.

(e) Nature of tourism and background of tourists: tourists' contribution to economic development may vary considerably in terms of per capita spending power,
infrastructural demands and the forms of tourism in which they participate.

Hall's observations are important for several reasons:

1. Broadly speaking, the extent to which governments are able to internalise transport
provision for the benefit of tourism are vast, and can include (a) the creation of favourable economic conditions or environments that encourage investment in transport services;
(b) direct subsidies designed to offset unprofitable routes or operations (e.g. rail services
in the United States – see Chapter 4); (c) direct regulation of services (discussed below
and in Chapter 7); and (d) the establishment of policies and guidelines relating to operational procedures.

2. The type(s) of tourism present (or under consideration) at a particular destination
requires consideration of whether to invest or encourage private development of specific
modes of transport. For example, Australia may be considered a long-haul destination for
several of its non-Asian markets. Despite recent moves towards domestic liberalisation of
air transport routes, the Australian government is likely keen to ensure the survival of
Qantas in a crowded and competitive international environment as it represents at least
some control of tourist flows to and from the country itself.

3. The varying size of destinations, and perhaps by extension the varying sizes of countries and governments within which such destinations are located, can be critical when
examining their ability to control how transport fits within economic restructuring.
Smaller countries may not be able to afford to handle some aspects of tourism-related
transport, and thus leave such operations to foreign-owned conglomerates or firms. The
risk here, then, is that tourist mobility, most importantly to and from a destination,
becomes an externality that is driven almost solely by profit as opposed to serving the
greater economic good of the destination.

Given the complex relationship between transport and tourism and the economic framework within which both operate on a local, national and regional level, the wider role of
government in the provision of transport as it relates to tourism can essentially be captured in two broad ways: the regulation or deregulation of services, including the
monitoring of competition, and the issuing of transport planning measures.

REGULATION AND DEREGULATION

A regulated transport environment refers to those instances where price and market access are tightly controlled by governments in the form of nationalised firms or local-level firms (Hoyle & Smith, 1998; Morrison & Winston, 1989) or through specific policy measures designed to influence terms of trade. In a deregulated environment, direct governmental controls are relaxed in favour of allowing market conditions to dictate operational success. While regulation and deregulation are often discussed in the context of aviation, a more general consideration would include issues relating to the facilitation (or otherwise) of competition and legislating through policy frameworks that speak to the overall economic importance of transport to national economies. Worldwide, governments must therefore decide whether to:

1. formally regulate the provision of transport services and access, including, of course, those blurry areas that could ultimately have some degree of impact on tourist flows. These actions would thus exert some control over routes and flows, pricing and frequency of access;

2. allow whole or partial market access to privately owned firms that may still need to meet standards of varying degrees with respect to routes and/or price; or

3. allow unfettered access to local or national markets without restriction.

As Button (1993: 244–245) outlines, paraphrased here, there are several arguments in favour of transport regulation:

1. Markets are, for the most part, somewhat imperfect and prone to failure, which could lead to high fares and dangerous service practices. In fact, the concern over safety and congestion became two of several reasons why, shortly after the 1978 deregulation of the United States aviation industry, some people were calling for the re-regulation of airline services (Morrison & Winston, 1989; see also Tretheway & Waters, 1998).

2. Regulation stifles what may otherwise be monopolistic operations within imperfect market structures.

3. Unregulated services could result in degradation of service quality through focused efforts on profitable route structures as opposed to a more general offering of services to multiple markets; to a large extent, this explains why some airlines which are government-owned offer services across networks which, if privately-owned, would likely be abandoned due to unprofitability.

4. Regulation is necessary in those instances where transport can be perceived as a public good; this may partially explain why the United States government chose to financially

protect and support many major airlines following the 11 September 2001 attacks in the United States as it was argued that the national infrastructure, as well as the flows of goods and people, must not be interrupted.

5. Government involvement is necessary in many forms of transport where the underlying infrastructure cost is high. Thus, the argument is that private-sector firms are less likely to be able (or willing) to direct significant capital into risky ventures where the bottom-line return on investment is years away.

6. A regulated market may result in certain externalities (such as environmental compliance measures) falling under government control and mandate, whereas such externalities may be ignored in a deregulated environment due to excessive cost and perceived lower returns-on-investment.

From the perspective of integration with tourism, deregulation can force destinations to become more competitive because the market is less restricted, and thus marketing and overall development plans become critical. It could even be argued that, in a deregulated environment, and particularly with respect to international air transport, tourism destinations must market to transport providers in much the same way as they advertise and promote their image(s) to tourists.

The benefits to passengers in a deregulated environment, especially with respect to air transport, are not necessarily a given. Morrison and Winston (1989) argue that the benefits of deregulation to the traveller should be matched by policies that encourage competition and efficient airport use, especially when some of the initial consequences of deregulation in the United States market (e.g. airport congestion, alliances) were found to be less beneficial to passengers. Further, McHardy and Trotter (2006: 87) argue that deregulation as it applies to air services may not, itself, provide passengers with benefits when other elements of the transport system (e.g. airports) may not be subjected to similar levels of competition. This element of competition is critical in a deregulated environment, where transport is based on the theory of contestable markets (Levine, 1987), although the context of such a deregulated environment itself technically pre-dates the introduction of this theory. The basis of the theory of contestable markets in the context of transport is that the free entry, or even the threat of entry, is enough to encourage efficient transport operations that would be adaptive in pricing to supply and demand vagaries. Put another way:

> Contestability theory was used to underpin the ideological moves to deregulate and privat e transport, which started in the USA and Great Britain at the end of the 1970s. This process then spread rapidly throughout the world and included many less-developed countries often at the instigation of the World Bank and the International Monetary Fund (Bell and Cloke, 1990). (Knowles & Hall, 1998: 76)

Under contestable markets, 'sunk costs, which a firm incurs in order to produce and which would not be recuperable if the firm left the industry, are not significant' (Sinclair & Stabler, 1997: 61). In such an environment, firms operate under the assumption that (continued) competition is inevitable as the barriers to entry are negligible. There are several examples of transport modes operating under deregulation that ultimately have some bearing on the shape of tourism and tourist flows:

1. The United States Airline Deregulation Act of 1978 facilitated the introduction of competitive air services, thus helping to establish a competitive market and subsequent growth in domestic travel (Adler, 2001).

2. Starting in the 1980s, the British government embarked upon a programme to privatise several transport-related services and operations, including Sealink Ferries, British Airways, the BAA, the National Bus Company and the Scottish Bus Group. The result is a fiercely competitive environment, especially in the context of air travel with the introduction of several new LCCs beginning in the 1990s (see Knowles & Hall, 1998 and Chapter 6).

3. Transport services serving third world countries (especially international air transport) were often affected by government reforms designed to deregulate the market in their own countries. Thus, with an increasing privatised transport market, coupled with the desire by many third world countries to encourage increases in visitation and, by extension, visitor expenditures, the deregulation or loosening of state controls over competition and transport operations in these countries has become somewhat inevitable. However, this can bring significant problems, not the least of which is the inability of third world-based operations to compete with large-scale services originating from developed countries. In the case of Air Zimbabwe, for example, Turton (2004: 75) notes that deregulation outside of the country in other jurisdictions was a factor in the restructuring of its own operations:

> In the 1990s Zimbabwe, in common with other developing states, embarked upon a program of economic structural adjustment. Many of the state undertakings, such as Air Zimbabwe, were subjected to a reappraisal by independent advisers and the report by a European consultancy recommended a policy of liberalisation for the airline and the loosening of constraints upon competition for traffic from other airlines. Within this competitive framework, long haul flights into Harare were introduced by Air Frances, KLM and Sabena and the national airline was ill-equipped to counter this with its limited financial resources and lack of aircraft suited to its various routes. However, the restructuring of Air Zimbabwe, with a loosening of state control, was viewed by the government as an opportunity for the airline to improve its performance on all routes and to compete more effectively in the tourism market.

In the past five years, Air Zimbabwe has faced rising operational costs and shrinking revenue. In September 2006, the airline was largely insolvent, despite injections of capital from the Reserve Bank of Zimbabwe designed to prevent a full collapse of the airline.

Regulation and deregulation policies may not be as 'clear cut' as they may seem. For example, a deregulated environment does not automatically mean the absence of regulatory structure. In the many countries, safety and operational procedures (such as pilot training) are heavily regulated, as are systems such as air traffic control. Road transport is regulated (largely in relation to speed) in most countries, as are the terms and conditions governing, for example, rental car contracts (i.e. mandatory third-party accident insurance, a critical issue in countries where tourists in rented vehicles can be responsible for a significant number of accidents). Ultimately, the deregulation of modes of transport has more to do with market access by competing firms than it does with operational standards and assessments. As a result, government role in transport planning, even in deregulated environments, can be critical.

TRANSPORT PLANNING

If the distinction between deregulation and regulation with respect to transport services as it relates to tourism can be considered a critical issue in transport policy, what remains is the *extent* to which government involvement is prevalent. With transport operations deregulated in many countries around the world (in many developing countries, some transport is heavily protected by government interests), it would be incorrect to assume that governments no longer have a financial or operational interest in the provision of transport services, both for tourists and non-tourists. Many national governments around the world have developed detailed transport plans that target areas of growth and shifts in demand. Banister (2002) outlines several agendas implicit in the relationship between markets and governments with respect to transport planning, two of which are highlighted here as they are integral in the specific context of tourism and transport planning:

1. 'There is a need for *intervention in the market* with new forms of regulation' (Banister, 2002: 229; italics in original). The issue here is the provision of a safe and stable business environment within which transport companies (and consortiums, if applicable) can operate. Destinations (on a wider spatial scale) and attractions (on a more micro level) are thus interested in how governments approach planning and, by extension, how it will impact on tourist flows and mobility patterns.

2. '*Environmental issues* must now be seen as integral to all transport policy decisions and investment proposals' (Banister, 2002: 229; italics in original). Clearly the sustainable operation of transport, from an environmental or biophysical perspective, is a moral and ethical idea (see Chapter 9), but this must be couched in the reality of externalities and cost-benefits. Thus, by engaging in transport planning governments must provide the appropriate incentives to encourage the incorporation of appropriate environmental objectives without jeopardising the financial health of transport firms or the social benefits they provide.

While industries involved in the provision of tourism services, such as hospitality, may not have a direct and frequent impact on transport planning, government or state involvement in transport planning can have a significant impact on their performance. The question then remains: to what extent do governments recognise the impact of transport planning and policy-making on the tourism sector? The United States DOT recently released its 2003–2008 Strategic Plan (entitled 'Safer, Simpler, Smarter Transport Solutions') that is decidedly sector-wide in its focus. Interestingly, tourism is only mentioned twice in the entire Plan. It initially appears in a section in which future trends are discussed in relation to international transportation:

> Goal: To develop, coordinate and implement DOT's international transportation and trade policies and ensure that the U.S. transportation system supports the competitiveness of the U.S. transportation industry, and rapidly expanding global trade and *tourism*. (United States DOT, 2003; emphasis added)

This goal is linked to two desired strategic outcomes that are desired, namely 'Enhanced international competitiveness of US transport providers and manufacturers' and 'Harmon ed and standard ed regulatory and facilitation requirements'. The other mention of tourism in the Plan is in the context of domestic tourism, specifically:

> Investment in domestic and international transportation systems is central to survival in the global marketplace. Given the important role that transportation plays in commerce and *tourism*, if there is not greater private sector investment and improved coordination of public–private sector investment in domestic and international intermodal transportation connections, U.S. businesses will not be competitive in the global marketplace. (United States DOT, 2003; emphasis added)

Transport planning is closely tied to understanding the market for transport services. For example, the relationship between transport and domestic tourism is an important one as the ratio of domestic to international tourism can often be quite high. With existing constraints such as age, gender, income and social factors such as consumer culture and cultural context, the location of leisure and tourism activities can vary significantly. Georggi and Pendyala (2001), for example, note that older age segments and groups in lower income brackets are less likely to undertake long-distance travel when compared to younger population segments. In Canada, Statistics Canada estimates that almost 84 million non-business domestic trips were taken in 2003. A substantial number were same-day trips taken by automobile, thus suggesting that, spatially, many peoples' activity areas for same-day domestic tourism and recreational activities is very much local or regional. Table 2.3 shows the breakdown of domestic travel by mode of transport in Canada for 2003.

Table 2.3 Domestic travel in Canada by mode/type of transport (in thousands)

	Total	*Same-day*	*Non-business*	*Business*
			OVERNIGHT/TRIP PURPOSE	
Automobile	173,400	89,664	77,036	6700
Air	6807	515	3561	2731
Bus	4763	1717	2792	254
Rail	1479	240	938	301
Boat	589	135	416	38

Source: Adapted from Transport Canada (2004)

Despite the obvious size and economic importance of domestic tourism in Canada, a review of Transport Canada's 2001–2004 Business Plan (titled 'Looking to the New Millennium') reveals no mention of tourism, although it does discuss briefly safety and security issues. Similarly, Transport Canada's *Straight Ahead – A Vision for Transportation in Canada* (released in early 2004) makes no mention of the linkages between tourism and transport. The affiliated document, *Creating a Transportation Blueprint for the Next Decade and Beyond*, makes a somewhat generalist statement on the importance of transport in general: '…attention is…required to connect Canada's vast territory to our principal gateways by supporting corridors that are important to international trade and tourism' (Transport Canada, 2004: 8). It also makes reference to transport and rural areas: 'Rural and remote communities are often dependent on exporting resources to world markets, or on tourism, and transportation is a critical factor in the competitiveness of these products and services (Transport Canada, 2004: 10)'.

DEMAND AND SUPPLY

Tourism and transport has thus far been positioned as somewhat co-dependent, but what remains to be considered is how the relationships discussed above transfer to the actual use of transport by tourists. In other words, given either the deregulated or regulated environment of transport operations, what is the process by which people decide to travel on a certain mode or type of transport as tourists? The answer to this requires examination of transport demand and supply. A number of elements or variables effectively govern the extent to which modes of transport are selected. These can include:

1. *Family structure.* The role of family structures (which is associated with other lifestyle variables) in the selection of destinations is paramount, but it also raises the issue of the particular life-course of individuals utilising various transport services. For example, Thorton *et al.*'s (1997) travel diary study of tourists staying in Cornwall suggests that adults-only travel parties spent more time sightseeing by car or coach than travel parties that included children (see also Giuliano, 1997a). Such variables may also exert some influences on the choice of a particular transport mode or type. This has been raised in the transport geography literature (e.g. Giuliano, 1997a, 1997b), and the effect ageing populations will have some impact not only on the destination but also the choice of transport.

2. *Social and behavioural considerations.* These govern the ability to undertake travel (e.g. Wang, 2001). Cohen and Harris (1998), for example, suggest that British VFR travellers chose a particular mode based solely on economic reasons. Further, visitors to national parks in Britain who tend to utilise their own vehicle for transport purposes may be persuaded to use public transport in order to keep pace with sustainable management of the parks system (which is growing) and, in theory, enhance the value of the experience (Eaton & Holding, 1996; see also Crabtree, 2000).

3. *The total cost of the trip.* The selection of a mode of transport needs to be considered in the wider context of the availability of the total trip budget. Thus, transport is one element within a series of variables that can include activities, accommodation and other expenditures. Prideaux (2000a) suggests that the demand for transport can be expressed as follows:

$$E_i = f\left(D_i, A_i, T_i\right)$$

where

E_i is the total holiday expenditure for destination i
D_i is the discretionary spending
A_i is the accommodation costs
T_i is the transport access costs

Ortuzar and Willumsen (2001) suggest that time, cost, reliability, convenience and comfort that impact upon modal choice, with time and cost generally dominating in importance (cf. Crocket & Hounsell, 2005).

DEMAND FOR TRANSPORT

Although there are numerous variables involved, it is possible to generate models that illustrate the relationship they have with transport demand. Prideaux's (2004) transport cost model (Figure 2.2) not only neatly outlines how expenditure on transport cost is

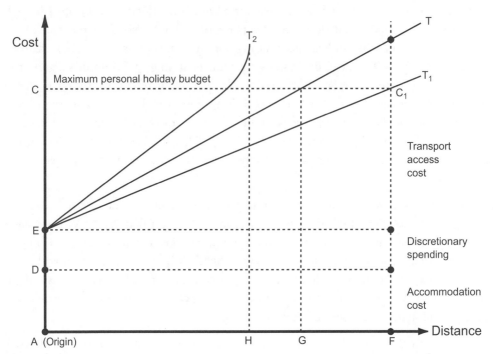

Figure 2.2 Prideaux's transport cost model

Source: From Prideaux (2004); reproduced with permission from Elsevier

affected by the function of distance, but it also highlights how other variables are also considered when selecting a destination. Prideaux's model suggests that trade-offs become necessary as an individual considers vacation destinations (H,G,F) further away from their point of origin (A). What is important to consider is the maximum personal holiday budget (CC_1) and the access cost for each destination depending on the mode of transport. For example, accessing destination H using transport mode/type T_1 is clearly within budget (given AD is accommodation cost, DE is discretionary spending and EC is the transport access cost). In fact, travel using mode T is also within budget. However, travel using mode T_2 is not within personal holiday budget. As destination G is further away from the point of origin, the model suggests that the cost of T and T_1 modes of transport will be higher. Selecting G as the destination will still be within budget, although note that selection mode T for destination G will result in maximising the total amount of money reserved for the entire trip.

A few issues with respect to how this model can be operationalised deserve consideration:

1. As Prideaux (2004: 83) notes, what appears as fixed costs for both accommodation (AD) and discretionary spending (DE) are, in reality, anything but fixed. Travellers may, for example, opt to forfeit the cost of accommodation in favour of more expensive transport costs because of perceived value or comfort. Similarly, discretionary spending may also be adjusted in favour of the variable transport access cost for similar reasons. Much of this depends on the perceived value of the mode or type of transport, which is discussed more fully below.

2. There are situations where the transport access cost for some destination does in fact increase the further away from the point of origin one travels. Prideaux (2004: 83–85) uses this model to outline how peripheral regions can remain competitive in the face of increasing access costs (which he breaks into three categories: fare cost, time spent travelling from origin to destination, and the 'cost of comfort'). In urban and peri-urban situations, for example, modern road systems (e.g. motorways) may provide quick access to destinations (or attractions) located along these routes (or zones of interaction, as Hall [2005] refers to them). Attractions or destinations that are not as easily accessible may suffer from higher access costs because of the increased time (due to the lack of motorways for transit) and the lack of value placed in the time taken to access. However, on a much larger scale, the issues of supply, demand, competition and pricing structures (among other elements) need to be taken into consideration. For example, to travel from Dunedin, New Zealand to London is generally cheaper than to travel from Dunedin to Toronto, which is shorter in both time and distance. Much of this has to do with the supply and demand and the resulting pricing structures of airlines in both New Zealand and North America, but it illustrates how distance is not always a factor in determining access cost.

3. In some respects, the model treats transport as '*for*' tourism as opposed to '*as*' tourism, and thus speaks to the earlier discussion on the blurry distinction between transport as a facilitator and transport as the attraction. In fact, Lumsdon and Page (2004: 5) make this distinction clear when they speak of 'transport for tourism' and 'transport as tourism'.

While models of transport expenditure, such as Prideaux's model, are useful for understanding the transactional nature of the holiday decision-making process, it is important to understand how demand can be understood with respect to transportation. Transport geographers and economists often distinguish between demand at the aggregate level (i.e. examining flows and patterns in a particular zone) or the disaggregate level (i.e. at the level of the individual). In the context of transport and tourism, the most useful manner in which to approach demand for particular modes of transport is at the individual or disaggregate level, largely because it affords a more explanatory framework of analysis as opposed to a generally normative one (Fischer, 1993). Below, a microeconomic approach to analysing demand for tourism transport is discussed, as are general models relating to decision-making processes that use the individual as their unit of analysis. Hoyle and

Knowles (1998) suggest that, as well, there are numerous elements that influence the provision and overall assessment of transport (Table 2.4).

Table 2.4 Factors relating to transport demand, provision and assessment

Demand	Demographic conditions	Structure, density/distribution/accessibility/connectivity, mobility
	Economic structures	Resource planning, land use, employment, other sectors
	Trade relations	Structure/flows, markets and market access, controls, regulations
Provision	Political structures	Planning, investment, regulations, partnerships
	Environmental constraints	Management, conservation
	Finance	Cost/revenue, investment, regulation
	Technology	Cost, distribution
Assessment	Usage	Cost/behaviour, trends, modal split
	Scales	Local, national/regional, international, global
	Dimensions	Spatial, temporal, structural
	Influences	Development, land use, marketing

Source: Adapted from Hoyle and Knowles (1998)

Generally, the demand for a particular good can be expressed as follows (Button, 1993: 54):

$$D_a = f\left(P_a P_1, P_2 ... P_w Y\right)$$

where

D = demand
P_a = price of the particular good
$P_1 ... P_w$ = price of other goods
Y = income

However, as Button (1993) points out, applying such a model to transport may not be entirely appropriate, largely because the factors involved, and as represented, tend to obscure the complex and important relationships between the factors themselves. The earlier discussion on the blurry distinctions of transport and tourism plays an important role here. Where the mode of transport is merely functional, price may play a pivotal role in measuring or assessing demand, yet when the mode of transport is also the attraction, such as a cruise experience, the price of the cruise is weighed up against the other benefits and services on offer. As a result, Button (1993) raises caution as to what product or service is actually under demand. Once this is recognised, the balance of price versus the price of other goods, as well as the degree to which other complex factors are involved, can be established.

A further point by Button (1993), with respect to the problem with applying a generalised model of demand for transport, relates to the costs incurred in order to utilise the service itself. Button (1993) argues that, for example, the cost of time whilst utilising a particular mode of transport needs to be taken into consideration (see also Preston, 2001). Thus, a particular mode or type of transport under consideration also plays a factor in deciding price versus the cost of time. For example, a business traveller deciding between taking the train or taking a plane may include in his or her decision the cost of parking or other forms of transport that are necessary. In essence, the argument here is that there is an opportunity cost that needs to be considered when a particular mode of transport is evaluated if the primary variable is price. For example, while the cost of transport mode A might be significantly lower than transport mode B, the amount of time spent travelling if one were to use mode B may be significantly less. Thus, for some travellers the added cost may be a justifiable expense and thus select mode B. Quinet and Vickerman (2004: 97) model this as follows:

$$C_f^i = p_f + h_i t_f$$

and

$$C_a^i = p_a + h_i t_a$$

where

C_f^i, C_a^i are the generalised costs of two modes of transport (e.g. commercial air carrier and rail, respectively)
p_a and p_f are the prices of the two modes
t_a and t_f are the travel times
h_i is the value of the time of the user

Thus, user i will likely choose to utilise a commercial air carrier (C_f^i) if $C_f^i \leq C_a^i$. Another way of representing this is as follows (Quinet & Vickerman, 2004: 97):

$$h_i \leq \frac{p_a - p_f}{t_f - t_a}$$

What this suggests is that a traveller could make a decision on the mode of transport on the basis of the amount of time necessary when using one mode of transport versus the other, but only if the value of time to that particular traveller warrants the additional cost incurred.

Transport economists recognise that the cost of transport and the time utilised 'in transit' are not the sole variables to consider in the demand for transport. There are numerous additional variables that have some bearing on the demand for a particular mode of transport. A few of the more obvious examples might include:

1. the cost of transport to and from the modal transport point. A good example is the cost to take a taxi to and from the airport from a place of residence or an office;

2. the amount of time necessary for transport from a place of residence or an office to the node (i.e. traffic congestion); and

3. the level of service provided by a particular mode of transport. For example, some travellers may be members of frequent flyer programmes that provide access to lounges and/or preferred check-in procedures.

By way of an example, and in order to illustrate the complex number and variety of variables involved in the demand for transport associated with tourism, assume that a couple are interested in a week-long holiday to Las Vegas. The city in which they live, Los Angeles, is a hub for a variety of transport options, not the least of which includes ground coach and commercial air carriers. They also have the option of using their own means of transport (i.e. a personal vehicle). All three modes of transport differ in terms of travel time, with air being the quickest and, for illustrative purposes in this example, the cheapest. In adopting a standard model of demand on the basis of price, the couple might be expected to be interested in pursuing this option solely because of cost. If, however, the couple live quite a distance away from the nearest airport that services their destination, the cost of travelling to the airport (both in terms of time and actual expenditures) may be seen to be too high to justify flying. Further, the couple may also not wish to subject themselves to the stress of proceeding through check-in and security, which has been substantially revamped and heightened as a result of the 11 September 2001 terrorist attacks in New York and Washington and again in August 2006 following the threats to aircraft originating in the United Kingdom but bound for the United States. The result is that the couple may decide to opt for one of the alternative transport modes (i.e. their own personal vehicle or a coach) to reach their destination, or they may decide not to travel to Las Vegas and, instead, embark upon another form of leisure or recreation closer to home.

This last point is critical and serves as a reminder as to why it is important to understand demand for transport and the numerous variables involved in selecting a particular mode of transport. Destinations are certainly interested in hosting visitors and ensuring their stay is enjoyable, but the argument here is that there are factors that are often beyond their control that govern whether they, as a destination, are even selected by potential tourists. In some cases, tourists may decide against visiting a particular destination when non-cost-related variables (e.g. security queues) are simply not strong enough to trump cost-related variables. Further, they may decide to travel to a destination at a further distance simply because the time and money required are potentially no greater than if they decided to travel to a destination closer to home.

Elasticity of demand

Elasticity of demand (or price elasticity) refers to the degree to which customers or users respond to shifts in prices. As a ratio, it is the 'percent change in quantity demanded to percent change in price' (Coyle *et al.*, 1994). As formula, elasticity is expressed as follows (Doganis, 2002):

$$\text{Price elasticity} = \frac{\%\ \text{change in quantity}}{\%\ \text{change in price}}$$

The relationship between price and quality of a good or product is generally referred to as either elastic or inelastic. If the demand for a product is elastic, the amount demanded is more than the total price change amount (thus resulting in a positive elasticity coefficient). If the demand for a product is inelastic, then the amount of product demanded is less than that the total change in price (thus resulting in a negative elasticity coefficient). An inelastic product is said, therefore, to be insensitive to price change. Understanding elasticity in relation to transport and tourism is important for two reasons:

1. It is not entirely clear in all cases whether the price of transport in relation to tourism is elastic or inelastic. In examining shits in demand and price, however, one also needs to take into consideration the length of the observation. Economists have recognised for quite some time that the cost of petrol at service stations is relatively inelastic because, quite simply, most consumers either genuinely need their personal vehicle (for example, to commute to and from their place of employment) or have no desire to give up the 'luxury' of personalised motor transport. Thus, over the long term, the price of petrol tends to be rather inelastic as shifts in consumption of other goods (e.g. dining out) may be affected in the first instance. In the short term, however, people may opt to not purchase petrol or reduce the frequency and duration of non-essential motorised travel. As a consequence, when examining the demand elasticity of travel, it may be necessary to measure the relatively elasticity over longer periods of time.

2. Where price is not a discriminating factor in either the selection of a particular mode or type of transport, or where competition exists to the point where price differentiation is negligible, then elasticity of service becomes important (Coyle *et al.*, 1994). In this case, what distinguishes one transport mode from another (or even two similar transport modes) ultimately becomes the more intangible service-related items as opposed to price (see Chapter 8). Competition between two modes of transport serving a particular route may, for example, result in very little difference in price. As a result, the competition for increased market share becomes based on levels of service. As Coyle *et al.* (1994: 33) point out:

> Assuming no price changes, the modal or specific carrier demand is much more sensitive to changes in service levels provided. Many air passengers monitor the 'on-time' service levels of the various air carriers and, when possible, will select the air carrier that provides the best 'on-time' service.

In relation to tourism, Doganis (2002: 203) argues that airlines (but this also applies to other providers of tourism-related transport) 'need to have a feel for the price elasticity of the various market segments on the route or routes they are dealing with. Without such a feel, they may make major planning and pricing errors.' In other words, different market segments will be more or less elastic/inelastic based on historic characteristics. Leisure travel, for example, may be perceived as more elastic than business travel or travel for emergency purposes (e.g. to a funeral). As a result, transport providers need to carefully consider the nature of the market when adopting pricing schemes for their product. As Doganis (2002: 204) points out, a market that is elastic may respond to a fare decrease with stronger demand, resulting in more people travelling, and thus higher revenues despite the fare decrease. Conversely, a market that is inelastic may respond to a fare increase with less demand, and the increase in fare in this case may not be enough to cover the drop in demand. A further discussion on pricing structures is considered in Chapter 7 when yield management (or price discrimination) is discussed.

ISSUES OF SUPPLY IN TOURISM AND TRANSPORT

The supply side of transport in relation to tourism is generally concerned with the ability of firms (sometimes engaged in public–private operational endeavours) to provide adequate levels of service to support nodes and affiliated networks. Supply issues can be examined from a macro perspective, where the structure of the market within which a mode of transport operates, or from a micro perspective where distribution characteristics are considered in the context of matching or meeting demand.

Macro determinant perspectives

The wider political environment within which transport is supplied for the purposes of tourism necessarily needs to consider issues of bilateral agreements (Chapter 7) and regulations of operations. As discussed, regulation and deregulation are efforts on the part of governments to exert some degree of control over market structures, either directly through regulation or the removal of control through deregulation. Whereas the previous discussion was concerned with the role of government in market structures or forms, further consideration is due regarding how deregulated market structures in transport have materialised and what impact this has on tourism. The shape of the macro operating environment is largely determined by the presence or absence of significant competitors. Various market forms exist that determine supply chains and distribution channels, but these largely exist as either imperfect or perfect competition models. Imperfect competition forms include pure monopolies, where a single firm is the only seller of a product. Contestable markets were emphasised earlier in the context of deregulation, but in some cases even deregulated environments can produce unbalanced supply of services. According to Sinclair and Stabler (1997: 81), deregulation of the air transport market has introduced competition where previously regulated environments would produce state-controlled or supported monopolies or oligopolies: 'In the international market some routes are competitive, being served by many carriers. Most of the others are served by at least two carriers, indicating an oligopolistic market, although a few routes are served by a single carrier which may be tempted to exercise monopoly powers.'

In a regional or national context, Graham (1997) and Goetz (2002) note that oligopolistic market structures in passenger air transport may be largely relegated to main trunk routes, perhaps to the detriment of regional feeder routes. In some markets, rail transport can exhibit market structures that are similar, with as few as two firms exerting oligopoly control (e.g. Knowles, 1998). Understanding macro-level market structures is critical for tourism because these can form supply landscapes to which demand elements are matched. The shape of tourism flows depends largely on the competitiveness of transport conveyance. Market structures that promote competition benefit the traveller, but quite often such structures can come at the expense of firms unable to compete. As Eccles and Costa (1996: 48) note, deregulation in the European market has led to considerable upheaval that may ultimately be of some concern to future tourist flows:

> ... many small European carriers will be swallowed up by the stronger, existing privat ed European airlines. If this is the case the consumer will have fewer airlines to choose from, the exact situation that has occurred in the USA. As airlines continue to strive for European dominance, the Asian carriers are seeking to link into the European network, and, to gain access, require strategic alliances, in particular with those carriers that are currently looking to control European aviation.

Micro determinant perspectives

Micro determinants in transport and tourism supply considerations can be said to focus on the transaction levied between the transport provider and the customer or user. To this end, the focus is on the product or service that is sold (see Chapter 8 for a discussion on products/services). In recent decades there has been a marked trend towards the use of information technology to facilitate this process, and often at the expense of more traditional distribution and supply chains, such as travel agents. In fact, many transport companies are utilising technology to target existing markets in an effort to consolidate distribution and acquire greater market penetration through innovation (Jarach, 2002). For Jarach (2002), a distinct difference exists between e-commerce (which focuses primarily on the manner by which outputs, or products, are sold) and e-business, which he classifies as broader value relationships. For airlines, Jarach (2002: 120) notes that '[t]his means satisfying the needs of consumers, as well as attracting, fascinating and tying the consumer to the airline in a creative and entertaining way'. Buhalis (2004) found that the airline industry utilises information technology to improve distribution efforts while at the same time reducing costs. While LCCs have generally pioneered the use of online sales, Buhalis (2004: 819) notes that some larger carriers have recently adopted similar distribution mechanisms in order to 'demonstrate the extra value they offer for comparable prices'. Many cruise companies now feature online booking, following a trend introduced by airlines. Royal Caribbean launched an online advertising campaign (utilising banners on web pages) that featured the tag-line: 'You can wait for warm weather or go find it. Get started now' (Van der Pool, 2004). Of interest here is what this means for the traditional distribution channels, such as the use of travel agents for ticket booking and sales.

When a passenger books a flight on a major carrier through a travel agent, it is not uncommon for the agent to use some form of CRS (e.g. Galileo, Sabre) to determine fare and routing options and pass along the options to the customer. In the 1970s and 1980s, many individual airlines operated their own CRS. CRS are designed to enhance information provision, such as offering instant choice on sales options for a particular carrier. However, that choice (or lack thereof) in product has not been without controversy:

> There has been concern about the market power given to operators of particular CRS and the partiality of systems both in the exclusion – rationalised on technical grounds – of other rival systems and the influence owners can exert of travel agents and other airlines … Allegations that six American airlines were using a CRS as a means of establishing a price fixing ring have been denied by following Department of Justice [US] pressure the practice has been discontinued (Tomkins, 1994) … Other forms of alleged abuse include the selective display of some airline flights to the detriment of alternatives. Initially CRS owners arranged for their own flights to be displayed first on possible flight listings and systems have also showed a display bias in favour of online over inter-line connections. (Driver, 1999: 135, 136)

Where this becomes significant is in human behaviour at the point of purchase. As Driver (1999) points out, customers may elect to refrain from searching for alternatives if their perception of the offer in front of them is deemed reasonable. Thus, an individual airline's CRS, coupled with a travel agent who is incentivised to book passengers with a particular carrier that offers higher commission rates, is a powerful tool. In the past few decades GDS, which are CRS affiliated with airlines (Buhalis & Licata, 2002; Page, 1999:

186) have become popular. These systems display several pricing offers from a variety of airlines, including schedules, availability, passenger information and fare rules (Page, 1999: 187). As efficient as these systems are, airlines are still paying travel agents commission on the sales they run through when utilising GDS. As many airlines have, for the past few decades, explored areas in which to cut costs as a means of boosting revenue, the commission paid to travel agents came under close scrutiny. For many, this means a fundamental re-think of the nature of the manner in which their product/service is distributed. Buhalis and Licata (2002) note that future distribution channels (which they call ePlatforms) will take three forms: online (internet) sales, interactive digital television and mobile devices. Online sales and e-ticketing are expected to be the dominant form of distribution for air travel by 2007 (O'Toole, 2002, referenced in Buhalis & Licata, 2002), yet the emergence of web portals (e.g. Yahoo) and online travel agents (e.g. Orbitz) has meant new challenges, particularly over the control of these systems by larger conglomerates (Field, 2005). Airlines are, of course, not the only mode to utilise online sales. Many railway operations and cruise companies offer the ability to book and pay for tickets online.

What do these new forms of distribution mean for traditional intermediaries such as travel agents? In 1998, the Guild of European Business Travel Agents ran a conference at which Kieron Brennan, a consultant, urged agents to develop 'journey management' businesses, where such intangible services such as tracking flight alerts, providing recommendation on routes, and utilising databases on customers' past purchases to, in the future, provide tailored options for travel (Travel Trade Gazetta Europa, 1998). The reduction in travel agent commission as a result of shifting distribution methods and new ePlatforms (Buhalis & Lacata, 2002) for generated sales is interesting for several reasons. First, at the same time many airlines were adopting online purchasing of tickets as a means of providing a direct service to their customers. While establishing online e-commerce activities came at a cost, airlines wanted to effectively set the price at which their seats were sold. At the same time, by adopting a model where direct purchase was possible, it achieved a degree of what has been termed by one industry observer as 'disintermediation' (Kennedy, 1996), where airlines were keen to cut out the middleperson in the distribution of travel sales and thus introduce a new distribution chain. Much of this started in the mid- to late 1990s when the global economy was relatively strong. However, following economic recessions (especially in the United States) caused some companies, airlines included, to re-think cost structures. Although the period since then has witnessed substantial fluctuations, many agents struggle to recover their costs in markets where direct purchasing (either online or over the telephone) is strong (although not at all dominant in most environments).

Second, many airlines also started to establish a strong online presence with booking capabilities as a means of getting closer to their customer. In fact, Ryanair is often regarded as one of the more successful carriers to implement an easy-to-use and highly successful online booking engine, which followed on the heels of a large telephone-based reservation system (which was subsequently dismantled for the purpose of cutting costs). By operating online booking engines themselves, airlines could track usage and itineraries as a means to help its yield management system in the future. Despite the simplicity of booking online, and thus an airline's ability to provide direct sales to customers, the systems are not without problems. Booking a series of flights that begin with domestic flights, follow through to international flights, and then back on domestic flights is problematic for most, if not all, airline online booking and reservation systems. In those cases, customers must either telephone the airline directly or utilise the services of a travel agent. Also, with the advent of cheaper fares by LCCs, many mistakes are made by customers when booking and purchasing online. When combined with the fact that the most inexpensive seats are often non-refundable, those customers who make mistakes often have little, if any, redress.

CHAPTER SUMMARY

This chapter has outlined some of the core elements of the tourism/transport relationship. It considered the blurry distinction between how tourism transport is actually conceptualised (especially with respect to tourist versus non-tourist use of transport services) and examined how transport can actually become the attraction rather than (or sometimes in addition to) the destination or attraction itself. The wider tourism system was also considered, and the nature of transport's position within this system was discussed. The model proposed (based on and adapted from Page, 1999) that various aspects of the tourism/transport relationship benefit from a systems approach, namely the flows of tourists, the mode of travel, the motivation to travel and the available time for travel. It also highlighted the fact that transport permeates many levels of the tourism experience. For example, many trips that involve the use of various modes of transport do so at various spatial levels. Transport therefore plays an integral role in tourist flows from origin to destination, from destination to destination, and within destinations.

The role of government is a critical area in which the tourism/transport relationship is manifested. Key among these is regulation and deregulation (or privatisation) and what impacts they have on tourist flows and the ability of modes to even offer services to tourists. Similarly, the planning for transport and whether tourism is a factor were considered, and it was found that tourism is often only given cursory consideration. This could be detrimental in the long run because it fails to consider the close relationship between economic development and tourism, and the important role that transport plays in facilitating tourism development at a variety of levels (both locally, regionally and nationally).

Issues of demand and supply were also considered, and it was noted that traditional models of demand for goods may not be adequate. As demand for transport is influenced by a number of factors (e.g. time, access, price), more comprehensive models are necessary in order to capture the nuances embedded within this reality. The cost of transport, for example, may not simply be the cost for the actual mode of transport, but also the cost in order access that particular mode. As well, cost is traditionally expressed in monetary figures, whereas the cost of time (e.g. time spent utilising and/or accessing a particular mode) must also be considered. The elasticity of transport demand was also considered in order to demonstrate the variability of such a concept as applied to tourism-related transport. In other words, demand elasticity for tourism transport may initially be somewhat elastic because new innovations or pricing structures may spawn increased travel. In the long run, however, it is not entirely clear whether this elasticity is maintained. To a large extent, it depends on the level of competition and the overall demand for the product or service. As a result, in a highly competitive environment, price elasticity may be replaced by service elasticity as more potential consumers opt to decide on travel based on variables other than price.

BOX 2.3 Case study – The Channel Tunnel: The right idea at the wrong time?

To a large extent, the rationale behind the construction of the Channel Tunnel was to provide fast, reliable and comfortable transit between two nodes, namely Britain and France. Generally, politicians in both countries argued that the overall economic benefit of going forward with a tunnel (the plans for which date back, in some form or another, to the 1960s [Gibb, 1994]) would bring significant economic benefits to both countries. What is significant for the development of tourism, as Essex (1994) points out, is the fact that 30 million people live within three hours travel of the Tunnel. Essex (1994) also suggested that the presence of the tunnel will have definitive impacts on tourism:

1. The Tunnel will 'remove the physical and psychological barrier of the Channel for some travellers, who perceive visits across the Channel as arduous and requiring careful organisation' (Essex, 1994: 81). This means that rail travel can now link continental Europe with the UK.

2. The Tunnel will substantially increase the carrying capacity between the two countries, and certainly from continental Europe to the UK.

3. The presence of the Tunnel, according to Essex (1994: 81), results in 'competition for the existing modes of passenger travel and so create greater choice and flexibility for the customer'. Essex uses the example of the London–Paris link: with the Tunnel, this would be a 3-hour service, compared to 7 hours by ferry and slightly less than 4 hours by air.

The Tunnel represents one of the largest privately funded infrastructure projects in the world (Tran, 2004), but the cost of development has been steep and the problems facing the operation have been numerous. First, and largely a function of high interest rates at the time of peak construction (early 1990s) as well as a total development cost of £10 billion, the Tunnel has struggled to encourage traffic (both passenger and freight) to a level where costs can be recovered. Second, Tran (2004) reports that shareholders are, as of early 2004, not entirely happy with how the Tunnel is managed. Third, competition from low-cost airlines operating throughout Europe (which may not necessarily have been foreseen during planning stages) have inhibited passenger travel using the Tunnel, especially given that Essex (1994) notes that price will be the deciding factor for most people when deciding on whether the Tunnel represents a suitable mode of transport. Overall, the future of the Tunnel seems somewhat turbulent, and it may be quite some time before the initial projections of use are realised. In recent years, the Tunnel's operators, Eurotunnel, have encountered financial problems due to lacklustre performance and problems with shedding costs. As of July 2006, Eurotunnel was actively seeking protection from creditors after negotiations to restructure the companies massive debt (some £6.2 billion) failed.

Questions for consideration

1. Why was the Channel Tunnel a good idea from the perspective of securing access between two countries?

2. Why are LCCs threatening the financial viability of the Tunnel?

SELF-REVIEW QUESTIONS

1. What are the three blurry distinctions that arise when describing tourism/transport relationships?

2. What are the key factors that influence demand for transport?

3. What is elasticity and how does it affect transport and, consequently, tourism?

ESSAY QUESTIONS

1. Why is it important to understand the principles of supply and demand when discussing the relationship between tourism and transport?

2. Compare and contrast two examples of where transport would fit within the tourism system.

3. Find an example of a local transport plan where you live and write an essay that evaluates it from the perspective of support for, and facilitation of, the local tourism sector.

KEY SOURCES

Lumsdon, L. and Page, SJ (eds) (2004) *Tourism and Transport: Issues and Agenda for the New Millennium*. Amsterdam: Elsevier.

The Lumsdon and Page volume explores significant issues revolving around the subject. Notable contributions are an update to Prideaux's transport cost model, Hall's discussion on sustainability and equity, and Freathy's discussion on future directions for airports. The lead chapter by the editors is also useful to capture a wider perspective on a variety of issues, and the notion of the transport/tourism 'interface' is addressed here.

Travel Industry Association of America – www.tia.org

Bureau of Transportation Statistics (US) – www.bts.gov

Department for Transport (UK) – www.dft.gov.uk

These three sites will generally offer recently available statistics and general measures relating to transport operations and use. In some instances, full Excel tables may be obtained.

Key journals that explore tourism and transport relationships include most of the existing tourism journals, but other transport industry-specific journals such as the *Journal of Air Transport Management, Journal of Transport Geography* and *Transportation Research* (Parts A, B and D) often feature articles relating to tourism and/or leisure and recreation.

CHAPTER 3:

TRANSPORT NETWORKS AND FLOWS

LEARNING OBJECTIVES

After reading this chapter, you should be able to

1. Describe and critically evaluate the importance of connectivity and accessibility of transport networks and their role in tourism development.

2. Recognise the important of networks in business-related decisions relating to transport provision.

3. Interpret the approach(es) of government versus the private sector with reference to connectivity and accessibility.

4. Describe various models of spatial interaction in transport and what this means for tourism flows and development.

5. Generate your own examples of connectivity and accessibility models and explain their structure.

INTRODUCTION

In Chapter 1 several of the key concepts that underpin much of the discussion in this book were outlined. Modes (and types), networks and flows were positioned within a wider tourism/transport framework, and it was suggested that these three elements, when combined, effectively determine the nature of tourist demand for transport products and services (although the exact distinction between transport products and services will be discussed in more detail in Chapter 8). By extension, much of the second chapter was economic (or at least business, management and politics) in format and presentation; it not only considered issues of supply and demand, but it also considered how governments are involved in transport planning. This chapter investigates the physical manifestation and spatial interactions surrounding tourism and transport. Here, the intent is to discuss the role of space and place, from a spatial geography perspective, in characterising the various *networks* of transport (Figure 3.1). In the context of tourism, the issue is the degree of complementarity between destinations and origins and, further, how transport flows and networks work to embed destinations or nodes and origins within the wider tourism system. This complementarity is related closely to demand characteristics and is essentially a form of 'time–space coordination' (Nutley, 1998). The purpose of this chapter is to suggest that the *mapping* of networks and tourists flows is integral to understanding how transport works with tourism. The central argument is that, through understanding the spatial relationships and spatial interaction within and across networks and flows, any analysis of supply, demand and government planning and policy-making becomes more meaningful. It is important to keep in mind, however, that the graphical representations of transport networks presented in this chapter are, in fact, heavily influenced by management, planning, marketing, demand and other external variables. In other words, this chapter is meant to introduce and position spatial concepts as complementary to the other forms of analysis (e.g. supply/demand, regulatory systems, environment, marketing, management) by which transport and tourism can be assessed.

Spatial processes have been adopted in recreation and leisure studies (e.g. Wolfe, 1952, 1966), but they have only recently been applied to tourism (e.g. Hall, 2005; although see van der Knaap, 1999 and McAdam, 1999 for an overview of GIS applications in tourism). Thus, in contrast with the chapters that follow, rather than focus on specific modes and types of transport, this chapter examines structures and activities associated with the flows of tourists and the networks of transport they utilise. More importantly, of interest here is determining *how* and *why* such networks are important to understand.

MOBILITY, SPACE AND PLACE: A BRIEF OVERVIEW

How do space and place, from a non-human/social geography perspective, relate to transport? Put simply, transport is the vector by which movement and mobility is facilitated. It represents the means by which people are shuttled from place to place, but more importantly it allows for some places to become accessible and connected across networks. Accessibility is perhaps the most critical aspect of understanding transport networks in the context of the layering tourist flows (e.g. multiple modes, multiple destinations, multiple routes). Aside from a few exceptions, there is usually more than one mode and associated route serving a particular destination, and most destinations are connected to large networks of modal and multi-modal transport services. As a result, understanding how accessible a destination is from the perspective of potential transport options is critical for destination planning and management.

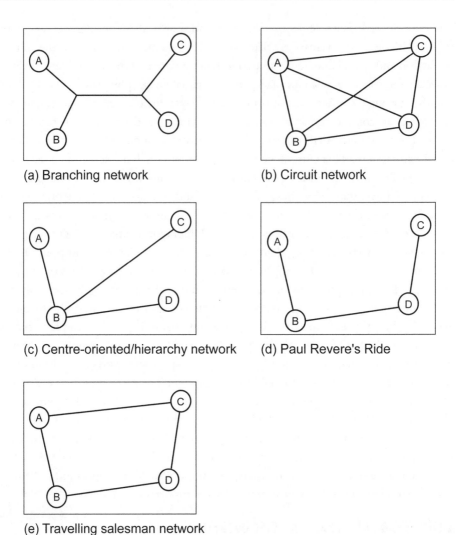

(a) Branching network

(b) Circuit network

(c) Centre-oriented/hierarchy network

(d) Paul Revere's Ride

(e) Travelling salesman network

Figure 3.1 A typology of networks

Source: Adapted from Fellman *et al.* (1992); Lowe and Moryadas (1975)

The fact that tourists and travellers are able to be whisked around the world in little more than 24 hours means that there are more places open and available to potential travellers, and thus represents a manifestation of globalisation (Hall, 2005). This is important because it points to what has been termed in the geography literature as time–space compression (Harvey, 1990) or time–space convergence (Janelle, 1969). At the beginning of the 21st century, increasing numbers of people have the ability to 1) travel further; 2) travel faster, given technological advances; and 3) experience and, most importantly, conceptualise more spaces and places. In short, mobility and the linking of places or nodes and spaces hinges on the provision of transport.

The importance of transport as it exists in conceptions of mobility and the extent to which individuals are mobile is not always apparent. It can be said, therefore, that transport provision is paramount for understanding mobility for five reasons:

1. Transport generally acts as the vector by which persons enter existing, or even non-existing (i.e. free form), networks and flows on a variety of spatial scales.

2. By extension, the spatial scales in which transport permeates is wide, ranging from local movement, within a relatively smaller geographic scale, to international, utilising modes such as air- and water-based transport to traverse large distances. In spatial geography, the extent to which people are mobile within a given area is referred to as their 'activity space' (Hägerstrand, 1970; see also Hall, 2005).

3. The range of transport modes available that facilitate mobility, and the fact that each generally features differing levels of speed for covering distances, also means that important decisions must be made with respect to personal time-budgets. Time-budgets refer to the amount of time an individual is willing to, or must, allocate to certain activities, in this case transport (see Chapin, 1966 and Hall, 2005 for a useful discussion on time-budget in relation to tourism). Building on the discussion from the previous chapter that considered travellers' choice of transport and its relationship to the value of time, the existence of multiple transport modes and corresponding types would suggest that decisions need to be made on whether a particular mode of transport is worth utilising given the often restricted amount of time available for leisure pursuits, including tourism.

4. Mobility is largely dependent on existing transport and related infrastructure in so far as it is allied with (a) the actual propensity for mobility (i.e. availability of existing networks, irrespective of demand in some cases), (b) the amount of time needed to travel from origin to destination (i.e. inaccessible destinations may require more time to travel to, such that sacrifices to other trip activities need to be made), and (c) the level of existing competition serving one or more destinations from a point of origin (i.e. the presence of competition can often result in lower costs to utilise transport, thus facilitating demand and ultimately the propensity for mobility).

5. In the context of transport and tourism, understanding mobility and the underlying structures of time–space utilisation, a complete picture of discretionary activities reflecting choice in participation, as opposed to activities that are more obligatory (Parkes & Wallis, 1978), is afforded. As such, time-budgets and activity spaces determine the spatial extent of mobility and accessibility (Ross, 2000) and may play an important role in determining tourist flows and flows for non-tourist purposes. Families, for example, may elect to go on holiday to locations that are convenient in a geographic sense and thus minimise travel time or expense (and stay confined to a particular activity space), but obligatory travel, such as visiting a sick relative, may not leave such choices open to alternatives.

Before discussing the mathematical foundations of spatial networks, it is important to consider a few of the wider social implications of movement and how it relates to 'space' and 'place'. The issue of how destinations function as places to be enjoyed and 'consumed' has been well researched both within the tourism literature (e.g. Meethan, 2001) and within other social science disciplines (e.g. Urry, 1995). Geographers have been interested in the spatial elements of places, but particularly in how place comes to be represented and juxtaposed with other places (see, for example, Hubbard *et al.*, 2004 for an excellent overview). Characterisations of space and place range from positivist constructions from the quantitative revolution in geography (e.g. Hägerstrand, 1967, 1970; Haggett *et al.*, 1977) to more behaviouralist approaches to spatial behaviour (e.g. Golledge, 1981) to the 'cultural turn' in human geography (e.g. Jackson, 1989).

The study of mobilities from a social perspective has only recently been appropriated by tourism studies for the purpose of cataloguing and explaining both individual and aggregate movements through time and space (e.g. Coles *et al.*, 2004, Hall, 2005). Characterising movement and mobility is, however, difficult and at times rather confusing. This is largely because of the numerous classifications of movement and the variability of the individuals undertaking that movement. The WTO, for example, recently started to classify day trips as a form of tourism, but what this highlights is the blurring of the distinction between tourism (defined as movement away from home for a particular period of time, but involving an overnight stay), leisure and even recreation. As Coles *et al.* (2004) argue, tourism should perhaps best be seen as a leisure-oriented form of mobility. Thinking of a range of mobilities, then, tourism is but one, and each of these manifestations of mobilities carries with them significant transport considerations (e.g. Box 3.1):

> Tourism [can be] seen as a leisure-oriented component of a continuum of mobilities that stretch from commuting and shopping through to what is usually categorised as migration. Such a representation of tourism clearly seeks to explicitly connect tourism not only with other discussions of mobility in the social sciences but also to integrate macro and micro scale understandings of mobility. Moreover, such a conceptualisation assists in further integrating research on the leisure dimensions of other forms of mobility such as second homes (e.g. Coppock, 1977; Hall and Müller, 2004), mobility of the highly skilled (OECD, 2002), travel for overseas work experience (Mason, 2001), and educational travel (Kraft *et al.*, 1994) with that of tourism mobility. (Coles *et al.*, 2004: 465–466)

BOX 3.1 Bottlenecks in connectivity and accessibility: Cottage country and weekend transport

The use of cottages and cabins (or holiday homes) can put enormous strain on highway networks during weekends (generally Friday afternoons/evenings and Sunday afternoon/evenings) and during extended holidays, especially when areas of cottage recreation (embedded within 'cottage landscapes' of rural recreation, Halseth, 2004) are within driving distance of major urban areas. Kremarik (2002) notes that approximately 823,000 Canadian households owned second homes or cottages, and approximately three-quarters of these were located within Canada. Svenson's (2004) research on cottage lifestyles shows that, while cottage owner-ship generally disfavours families, there is room to suggest that cottage use (as opposed to outright ownership) may be more family-oriented and thus used by various members of a particular family: 'The cottage, for many Canadians, is a place where extended family and friends gather together, where work is meaning-ful, where there is time for leisure and contact with nature, where community feels present' (Svenson, 2004: 73).

While studies of cottage use among urban residents has been conducted, little attention has been paid to the role of transport in these activities. One area where cottage use, as second homes (e.g. Hall & Müller, 2004), in relation to urban environments is significant is the metropolitan area of Toronto. The major highway (Highway 400) leading to an area of substantial cottage development north from Toronto (Muskoka) often becomes extremely congested on Friday afternoons in the summer as weekend recreationalists drive north to their cot-tages (see also Halseth, 2004). Anecdotal evidence from Toronto suggests that traffic congestion is even having an impact on decisions when to travel. Many opt to leave on either Thursday evening (thus taking Fridays as annual leave days over the summer until their leave is used up) or early on Saturday morning. In this case, transport has a strong impact on mobility patterns. It may also have an impact on how holidays are utilised; for example, occupants of cottages/cabins/second homes may elect to have multiple long weekends rather than one block of two or three weeks of annual leave. Svenson (2004) points out that driving times might be having some impact on decisions to purchase cottages in some areas of Canada. What this suggests is that access to cottage areas may be having significant impacts on second home recreational activities, although these assump-tions are currently in need of empirical validation:

1. Extensive traffic congestion may inhibit the enjoyment of second home/cottage properties if work/leisure flexibility is not realised (i.e. if an individual or family is unable to travel during a period of time where congestion is less onerous).

2. As Svenson (2004) notes, the time needed to travel from urban areas to areas of second home/cottage developments may ultimately have an impact on the overall attractiveness of the cottage as a suitable weekend-oriented recreational activity. Thus, in this case accessibility may in fact be responsible for altering recreational patterns.

3. Increasingly, accessibility may have an impact on the demographic profile of cottagers, such that cottage may cease to become weekend second home getaways and instead become locations of longer stays.

Questions for consideration

1. How has the congestion associated with major metropolitan areas such as Toronto influenced or changed mobility patterns of those with cottages?

2. What might the impact be of rising petrol/gas prices in a situation like the one described above?

SPATIAL ASSESSMENTS IN TOURISM AND TRANSPORT: CONNECTIVITY AND ACCESSIBILITY

With transport acting as a vector of mobility, there is a need to consider connectivity and accessibility between nodes, which are often referred to as 'vertices' in the spatial geography literature, as these two concepts underpin the ability of tourism destinations to exist. Broadly, there are several reasons for examining connectivity and accessibility in relation to transport and tourism:

1. for the purpose of establishing the role of government in public access to transport (including those modes of transport used by tourists);

2. assessing the priorities of private firms wishing to establish or expand current transport operations across a network; and

3. for use by other industries involved (to varying degrees) in the provision of tourism services in order to plan future expansion and/or diversification efforts.

As Chapman (1979: 209) notes, '[c]onnectivity indices provide useful aggregate measures of the spatial structure of a network'. Spatial analysis exercises in which connectivity and accessibility are measured can achieve several outcomes:

1. The ability to examine the relationship between different modes of transport and the wider networks within which they function. In other words, how well does a particular mode or type of transport service a particular region?

2. The ability to compare, between two places or nodes, the relative strength of a particular mode of transport to another mode of transport (i.e. the degree of connectivity or accessibility of air versus rail) and even by type. For example, in some heritage tourism destinations, this could mean comparing the utility and experiential benefits of certain rail services, such as an historic tram (e.g. Pearce 2001a, 2001b), with a guided bus trip.

3. The evolution of transport networks over time can be measured, including consideration for whether government policies designed to enhance or encourage transport innovation and service have been beneficial or detrimental. Fan (2006) found a remarkable increase in inter-city connectivity between the UK and Ireland and continental Europe between 1996 and 2004, largely attributable to the expanding networks created by LCCs (see also Dobruszkes, 2006).

The power of spatial assessments can even be used in making business decisions. For example, suppose a newly formed national coach company is interested in developing services across a specific network, incorporating several destinations. In the course of its planning, decisions need to be made with respect to which destinations could be served and where it might be best to establish an operational base (where routine maintenance and large repairs are carried out, and where a significant population base exists such that it can use the population base as its key market to service outlying destinations; this is commonly known as the hub-and-spoke concept and is discussed in Chapter 6 in relation to air transport). When determining which destinations are to be included, the company must carefully evaluate the overall value of servicing a particular node. This would be done not only in reference to the direct routes that it would bring, but also the integration of one node into the wider network and whether this would generate subsequent demand and a subsequent positive return on investment. By using connectivity and accessibility analysis, the company might be able to understand the true spatial value of the proposed network and, more importantly, institute pricing schemes or consider alternatives. Further, if certain routes along a proposed network (or the entire network itself) needed government approval (perhaps in order to satisfy that it was not being anti-competitive), the company could demonstrate the economic and social benefits of the proposed network structure by the fact that it is proposing to link destinations that may have otherwise been under-serviced or totally disassociated from connections to other nodes.

PUBLIC SERVICE CONSIDERATIONS: THE ROLE OF GOVERNMENT

From a government perspective, planning priorities can be established if demand for transport is known and the network along which that demand may be manifested is established and mapped (Werner, 1985). Governments have long been responsible for the provision of transport for the greater good of society, as Werner (1985: 12) notes:

> In a highly organised and diversified society such as ours, adequate transportation requires enormous resources (on the order of 20 percent of the gross national product) and sophisticated organisation and management. Such expenditures call for intelligent planning, which in turn requires that we understand the principles that govern transportation phenomena, their causes, their dynamics, their distribution, and their impact on space and time.

Yet governments are often also tasked with ensuring that transport networks are adequate in order to service tourists, both domestic and international. For example, most tourism destinations have multiple networks associated with them, largely because more than one mode of transport can be used for access. Various levels of government may therefore use information on the overall pattern of transport networks serving destinations in order to make decisions regarding, for example, incentives for upstart transport providers or gauging the extent to which public support for transport infrastructure is necessary. If a particular mode of transport is not adequately increasing the degree of connectivity between nodes, a government may wish to focus on supporting alternative means of transport through direct policies or subsidies. A good example is the global air services network, which is managed and regulated through a series of regulatory agreements and treaties (see Chapter 7). In some cases, governments may elect to liberalise access to its key nodes in order to enhance or increase the economic contribution that additional traffic may bring.

LOCATION, LOCATION, LOCATION: THE VIEW FROM THE PRIVATE SECTOR

From a private-firm perspective, potential destinations or nodes can be examined and assessed as to whether the existing population base is enough to warrant new service development given the additional costs needed in developing an expanded network (not to mention subsequent marketing costs). Analysing location is important as it allows for consideration of potential costs and risks for competitiveness (Daskin, 1995; Kidokoro, 2004). Examining accessibility and connectivity in these situations could help to address several questions:

1. Which routes would a private transport firm ideally select (i.e. what is the existing demand for services across a wide geographic area, such that specific profitable places can be identified)?

2. Given demand considerations, what are the key criteria involved in deciding whether to provide transport service to these places (i.e. cost of service, potential competition, future demand shifts and access to ancillary services would all enter into the equation)?

3. With respect to potential competition, private firms offering services across a network may also utilise connectivity and accessibility information to determine barriers to entry. For example, the coach company profiled above may decide to limit services along a specific route (or edge) and concentrate on a new destination in order to facilitate flow-on effects to other destinations. Alternatively, the same company may decide to operate services to a remote destination knowing that travellers will continue along the network using high-profit routes. In effect, incorporating some destinations may be financially detrimental to the overall operation, but necessary in that they demonstrate market presence and, as a result, cause potential competitors to closely evaluate the cost of entry. These kinds of decisions are common in the aviation industry. An airline, therefore, may service an intermediate destination solely for the purpose of getting customers integrated into their network and thus on more profitable routes. Some international airlines, such as Air Canada and Qantas, operate in a 'two tier' manner servicing both smaller domestic nodes and larger international hubs, and thus carefully monitor the extent to which their smaller domestic networks are able to feed their larger international networks. Other carriers, such as Virgin Atlantic, Singapore Airlines and Emirates, selectively fly between major hub airports and are not generally concerned with feeder traffic as these are handled by other airlines.

As tourism in general relies heavily on accessibility, other industries within the tourism sector can benefit from the knowledge gained from understanding connectivity and accessibility across multiple networks pertaining to specific modes and types of transport (see Hodge, 1997). For example, theme parks may decide to locate either immediately within or just outside of a particular vertex or destination. Such a decision may be made possible by examining the potential flows across a network of, for example, inter-regional motorways.

CONSTRUCTING CONNECTIVITY AND ACCESSIBILITY MATRICES

One of the more common means by which connectivity within a network is measured is graph theory. As Taaffe *et al.* (1996: 250; italics in original) note:

> Graph theory, a branch of topology, deals with abstract configurations of points and lines, or nodes and linkages. It does have considerable potential real-world usefulness, however, since it can provide empirical measures of the structural properties of any system once that system is translated as a set of nodes connected by a set of linkages. In graph theory terminology, *nodes* or points are usually referred to as *vertices*, and *linkages* or line segments are usually referred to as *edges*.

An example of a simple connectivity network is presented in Figure 3.2. In this particular network, five vertices (or nodes) (A, B, C, D and E) and six edges can be identified. The first step in examining this particular network is to determine its *overall* connectivity. This is important because it gives an indication (and numeric value) of the robustness of existing networks. A gamma index is calculated as one way of measuring overall connectivity (Scott *et al.*, 2006):

$$\gamma = \frac{e}{e_{max}}$$

where

e represents the total number of existing links in a network
e_{max} is the maximum (or potential) number of links in a network
e_{max} is calculated thus:

$$e_{max} = 3(v - 2)$$

where

v represents the total number of vertices or nodes in a network.

In this hypothetical simple network from Figure 3.2, the value is as follows:

$$\gamma = \frac{6}{9} \text{ and thus } \gamma = 0.67$$

A gamma index will range from 0 to 1. Progressively higher gamma indices indicate more connected networks. The purpose of the index is to arrive at an aggregate measure, but one that is reasonably indicative of the overall connectivity of a network. Another measure of connectivity is the beta index, which is calculated by dividing the number of edges (e) by the number of vertices (v):

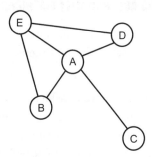

Figure 3.2 A simple connectivity network

$$\beta = \frac{e}{v}$$

In this hypothetical simple network from Figure 3.2, the value is as follows:

$$\beta = \frac{6}{5} \text{ and thus } \beta = 1.2$$

A higher beta index value means a network that is, overall, more connected. According to Chapman (1979: 206), it 'expresses the number of edges present in relation to the number of vertices to be connected and therefore may be regarded as indicating the average number of links leading into or out of each node'. If one were to remove one vertex and three edges, for example, the overall connectivity of the network would be reduced:

$$\gamma = \frac{3}{4} \text{ and thus } \gamma = 0.75$$

Once the overall connectivity value is known, it is possible to utilise other spatial analytical tools to help consider its connectivity and accessibility in relation to both the vertices and edges. For example, in considering the specific vertices within the sample network from Figure 3.2, it might be concluded that, upon visual inspection, vertex A seems to be the most connected as it has edges that connect it to vertices B, C, D and E. Vertex C, however, is only connected to vertex A, and is thus is not nearly as well connected as vertex A. To get a more precise measure of which vertex is the most *directly* connected within a particular network, the degree of connectivity can be plotted as a matrix (Figure 3.3[a]).

In Figure 3.3(a) matrix, any direct connection (that is, a connection where it is not necessary to pass through another vertex on an edge) receives a '1' in the corresponding cell. If no direct connection exists, a '0' is recorded. Thus, while there is a direct connection between A and B through a single edge, there is no direct connection between C and E using one single edge only, but it is certainly possible to reach E from C by going through A. This is referred to as a single step, and will be discussed below in the context of examining multiple steps. Initially, however, the idea is to establish the nature of the *direct* connections within the network. To arrive at a final numeric value that will help determine which vertex is the most connected, the rows for each vertex are summed. As a result, vertex A (with a score of 4) is the most *directly* connected as it has the highest number of direct connections (Figure 3.3[a]). The dispersion index is a relative measure of the overall connectivity of a network. It utilises the number of steps from each vertex to another vertex in the network. Thus lower dispersion indices reflect lower overall connectivity (Figure 3.3[b]).

As a result, the coach company may decide to set up its operations base at vertex A. In order to further help the company with this decision, and even to help schedule fleet

(a) Connectivity matrix

	A	B	C	D	E	
A	0	1	1	1	1	4
B	1	0	0	0	1	2
C	1	0	0	0	0	1
D	1	0	0	0	1	2
E	1	1	0	1	0	3

(b) Shortest path matrix

	A	B	C	D	E	
A	—	1	1	1	1	4
B	1	—	2	2	1	6
C	1	2	—	2	2	7
D	1	2	2	—	1	6
E	1	1	2	1	—	5

Dispersion index = 28
Mean dispersion index = 5.6

Figure 3.3 Connectivity and shortest path matrices

assignments, a further measure of accessibility can be measured. Figure 3.4 depicts the existing network with actual distances between the vertices indicated.

From this, it is possible to calculate an accessibility matrix that takes into consideration the *distances* (as opposed to the direct connectivity by presence/absence of a connection) throughout the company's network. To calculate, the required distances between vertices across the network are plotted in a matrix (Figure 3.5). Then, totals are summed and divided by the total number of vertices in the network. The result is that vertex A is spatially (by distance) central within the network, further supporting the decision by the coach company to locate its base of operations there (Figure 3.5).

From a simple spatial perspective, these measures may be useful, but there are issues with this kind of analysis that need to be considered before business decisions are considered:

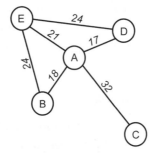

Figure 3.4 Connectivity network showing distance between vertices

	A	B	C	D	E	
A	0	18	32	17	21	17.6
B	18	0	50	35	24	25.4
C	32	50	0	49	53	36.8
D	17	35	49	0	24	25
E	21	24	53	24	0	24.4
	17.6	25.4	36.8	25	24.4	129.2

Figure 3.5 Measuring accessibility

1. In the case of land-based transport, topography is not considered when utilising connectivity and accessibility matrices. For example, in some small Caribbean islands, it can often take up to two hours to travel between two points that are, seemingly, geographically proximate to one another and thus appear to be relatively accessible. The reasons for this are the condition of the roads and the variable topography. Ultimately, these considerations can become diluted if they exist throughout the entire network, but the larger the spatial scale under consideration the more likely that these variables will have some degree of impact.

2. In measuring the connectivity of an overall network or the accessibility associated, individual vertices may be conceptually meaningful, but the *propensity* to utilise the network is not entirely clear. For example, two or three vertices may in fact be well connected from a transport perspective, but the demand to visit and thus utilise the network may not be present. Barriers to accessibility should be measured, as argued by Geurs and van Wee (2004), at the level of individual constraints in relation to time and distance. Somewhat related, and as discussed in Chapter 2, while transport links may indeed be present, the tourism product (or products) must be substantial enough to warrant travel and convert latent demand to actual demand. As a result, van Wee *et al.*'s (2001) research into the nature of competition across networks (in relation to employment markets and competition for jobs) can be applied here because different destinations will have different marketing programmes, stakeholder planning networks (Hall, 2005) and levels of development that may influence choice.

MODELLING ACCESSIBLITY AND DEMAND

In Chapter 2, issues relating to supply and demand in relation to transport were highlighted. There, the strength of transport provision in relation to tourism was expressed primarily in economic terms, and it was noted that shifts in either supply or demand can be modelled and thus help inform propensity for travel. Building on this, other factors can be considered. Janelle (1969), for example, examined the function of transport innovation with respect to time–space affordability in his concept of location utility. Location utility

> is a measure of the utility of specific places or areas, which in this case is defined by the aggregate time-expenditure (cost or effort) in transport required for that place or area to satisfy its needs. Operational need refers to those natural and human resource requirements which permit the place or area to fulfil its functional roles in the larger spatial systems of places and areas.
>
> (Janelle, 1969: 349)

Put simply, a place experiences an increase in locational utility when transport innovations are captured positively in an individual's time-budget. Thus, the place has more value. Janelle's work was not tourism-specific, but it does introduce a central concept to this discussion: that of the perception of value as determined by cost, effort and time in relation to transport. Thus, innovation in transport allows for greater accessibility: 'As man speeds up his means of movement, it becomes possible for him to travel further in a given time, to increase his access to a larger area and, possibly, to more and better resources' (Janelle, 1969: 352).

Rather than focus on the propensity for travel, as Chapter 2 outlined, an alternative approach can be offered: spatial interaction modelling. The premise of this type of modelling is that space and the relationships between places can inform the *potential* for mobility. In other words, outside of the supply/demand arguments considered in the previous chapter, it is argued that a spatial consideration can be employed to help map potential flows and, indeed, may influence demand. Once a spatial system is understood, it is possible to arrive at a more coherent view of the relationship of places to each other and what bearing that may have on transport provision (see, for example, Werner, 1985).

There are several models of traffic growth that emerge from the spatial transport literature that are worthy of attention. The Growth Factor Model, for example, established by Martin *et al.* (1961) suggests that the growth of transport/traffic or movement between two locations will be roughly equivalent to growth across an entire area. This may be ideal in theory, but it does not consider future growth where transport does not already exist (Werner, 1985). Somewhat related, intervening opportunity models attempt to capture multiple choice assessment made on the basis of existing opportunities. Thus, they rely on the notion of Central Place Theory (see Christaller, 1972) where locations of services (i.e. tourism attractions or destinations) are established based on the nearest opportunity.

GRAVITY MODELS

Gravity models are commonly used in measuring the extent of spatial interaction (Taaffe *et al.*, 1996). A gravity model is spatial approximation that, in reality, can be directly corre-lated to demand, but in a more substantial and meaningful manner because it incorporates two key elements that support demand: population and distance (Taaffe *et al.*, 1996; see also Lowe & Moryadas, 1975s). The gravity model can be traced back to Carey (1858: 42), who stated: 'Man tends, of necessity, to gravitate towards his fellow man. Of all animals he is the most gregarious, and the greater in number collected in a given space the greater is the attractive force there exerted.' A gravity model is used to capture two characteristics (Haynes & Fotheringham, 1984: 11):

1) scale impacts: for example, cities with larger populations tend to generate and attract more activities than cities with small populations; and
2) distance impacts: for example, the farther places, people, or activities are apart, the less they interact.

Further, Chapman (1979: 196) notes that the gravity model allows for investigations of 'notions of complementarity, transferability, and intervening opportunity' with reference to spatial interaction.

A simplified equation of the model is as follows (Chapman, 1979: 196):

$$I_{ij} = \frac{P_i P_j}{d_{ij}^e}$$

where

I_{ij} is the interaction between place i and place j
P_i and P_j represents the population of places i and j
d_{ij} is the distance between i and j
e is a distance-decay function.

Expressed verbally, Chapman (1979: 196) suggests that the model 'implies that the amount of interaction between any two places will be directly proportional to the products of their populations and inversely proportional to some power of the distance between them'. In this sense, distance-decay can be interpreted as the relative power of a place

71

being inversely proportional to its distance from an origin. As tourism occurs between origin and (often multiple) destinations, the relationship in terms of distance between these points out to be of concern, especially if the model is suggesting the distance plays a critical role in determining the power of a destination (or destinations) to attract people from an origin rests with distance.

A few points of interest with respect to this model warrant elaboration. The first is that the distance between i and j as represented in the model is a strictly *geographic* distance. As discussed below, this can be a significant limitation to the gravity model if applied to tourism and transport relationships.

The second point is the distance-decay effect. Oppermann (1995) recognised the effect of distance in pleasure travel, and geographers with an interest in transport often focus on the effect of distance on the movement and mobility of goods. In other words, the volume of movement declines the further one moves away from their point of origin (Chapman, 1979). Expressed as an equation, the rate of decay of movement (e) further from an origin may be calculated using:

$$ Q_{ij} = \frac{1}{d^{e}_{ij}} $$

where

Q_{ij} represents the quantity of movement between i and j
d_{ij} represents the distance between i and j.

From this equation, it can be seen that the smaller the value of e, the less distance-decay is experienced. As an integral part of the gravity model, it helps to model movement in the context of space and, more accurately, with causation related to distance. Problematic with the gravity model, however, is the extent to which it can be accurately applied to tourist movements between two places/nodes/destinations, or even touristic movements across multiple places (or destinations).

Figure 3.6 outlines an evolutionary model using the size and distance relationship and the potential flows of movement between four hypothetical urban areas using a gravity model function (with the thicker lines denoting larger flows and, thus, the strength of the relationship between two vertices). In this example, (re-drawn from Chapman, 1979: Figure 8.10), travel between A and C (Figure 3.6[b]) is substantial not only because of the relative sizes of their populations (which suggests that the greater the population, the higher the incidence for travel) but also because the distance between them is advantageous and supportive of substantial interactions. This explains the relatively smaller amount of interaction afforded between City A and City B. However, this stochastic model makes several assumptions that may not always be applicable in the context of transport and tourism:

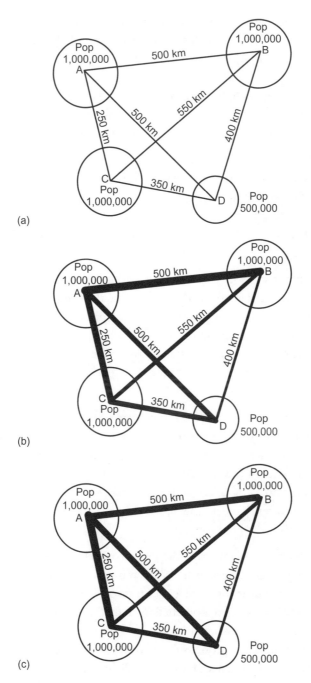

Figure 3.6 Size and distance relationships (a) and predicted movement flows (b) using gravity model

Source: Adapted from Chapman (1979)

1. It assumes that transport initiatives would not eventually be developed, thus enhancing travel between A and B, potentially to the detriment of A and C (assuming a finite amount of demand at City A) (Figure 3.6[c]). For example, B may be twice a far from A as C, but new transport infrastructure can effectively negate the affect of distance. Twinned motorways, for example, might reduce the travel time considerably. New airline upstarts may offer services that, in reality, significantly cut the time of travel between A and B to the point where it is quicker to get to B than C from A (see Figure 3.6[c]).

2. Somewhat related, the gravity model somewhat overstates the effect of distance (Taafe *et al.*, 1996) by assuming that the travel distance is roughly equal in perception between A and B and A and C. In other words, the perception might be that B is more than twice as far from A as C when in fact it is not if the effect of travel time is taken into account. Haynes and Fotheringham (1984: 12) suggest that this is one of the key modifications that are necessary to the basic gravity model:

> For example, the cost per mile of traveling may decrease with distance, as in air travel. Obviously the operational effect of distance would therefore not be directly proportional to airline miles and the negative aspect of distance would need to be reduced or dampened so that the model properly reflects its effects. On the other hand the effect of distance may be underestimated by mileage because the opportunity to know people in cities far away may be reduced by language, culture, and information.

What this means is that the *effects* of distance may not always be proportionate to distances travelled, especially with the introduction of transport services that may minimise such distances (as discussed in Chapter 1). As well, the perception of travel distance may be prone to the frequency of travel: 'frequent travelers to a certain destination are expected to perceive a destination as more accessible than those who do not have the chances to visit the destination often, even if the actual distance they have to travel is the same' (Hwang *et al.*, 2002: 53).

3. The gravity model also assumes that the type of travel between A and B and A and C is similar enough to warrant comparison. When measuring commodity flows, this may be entirely appropriate as similar forms of cargo transport tend to be used (e.g. truck, rail), but the nature of tourism and development is such that product development in two locations is rarely the same. As a result, markets can be substantially different, even if they are from a similar origin, thus the propensity for travel is considerably more complex than just being associated with distance. Heggie (1969) noted this deficiency in the gravity model concept by arguing that there is no limitation in supply, and no formulaic function to account for differentiation in supply. Further, the assumption made by the model is that demand is directly dependent on distance, thus ignoring economic realities, in this case, in either A, B or C (Heggie, 1969; see also Gutiérrez, 2001) and access to individual means of transport (Lanzendorf, 2000).

Given these limitations, a revised gravity model that would take into account issues of the amount of time necessary to travel from place to place can be formulated as follows:

$$I_{ij} = \frac{P_i P_j}{t_{ij}^e}$$

where

I_{ij} is the interaction between place i and place j
P_i and P_j represents the population of places i and j
t_{ij} is the normal travel time between i and j
e is a distance-decay function.

In this equation, the function of the distance-decay variable is related directly to the amount of time taken to travel between two places. In other words, decay becomes purely a function of time as opposed to distance. While this avoids some of the frictional effects traditionally measured in gravity models where distance is sometimes found to not be a strong factor in determining flows (Lösch, 1954; see also Huff & Jenks, 1968), it does not entirely dismiss the fact that distance plays some role. This relationship can be more accurately modelled, however, by taking accessibility measures into consideration. The result is as follows:

$$I_{ij} = \frac{a_{ij}}{t_{ij}^e}$$

where

a_{ij} is the degree of accessibility between i and j
t_{ij}^e is the distance-decay function in relation to the length of time it takes to travel from i to j

As such, it is possible to use the same four cities used in Figure 3.6 to demonstrate how new innovations in transport effectively reduce the distance-decay of time. Figure 3.7 represents another evolutionary depiction of movement. Figure 3.7(a) shows the existing travel times using ground transport (and for the sake of argument, personal automobiles are used for modelling purposes in this instance). In Figure 3.7(b) travel from B to C is longer, assuming the presence of natural barriers (e.g. mountains). Similarly, travel between A and C is generally quick because the mode of transport used facilitates rapid transit times. If, however, changes were made to the infrastructure (for example, twinning the motorway) such that travel from A to B was effectively halved, one could expect, all other things being equal, more flows between the two cities as a result (Figure 3.7[c]), and potentially to the detriment of flows between A and C. Similarly, if the travel time between

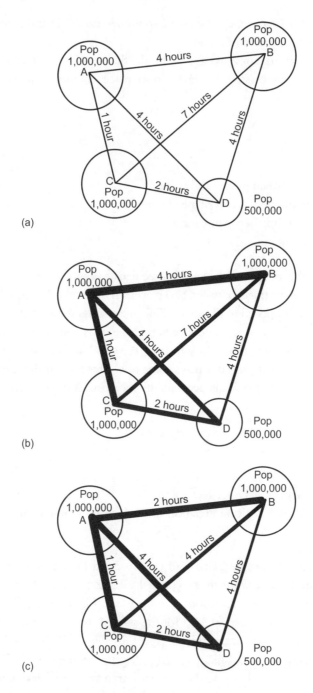

Figure 3.7 Gravity model application with length of time dependency

C and B were to be reduced through similar infrastructural development (or perhaps through the introduction of air transport), then one might expect to measure increased movement between them (Figure 3.7[c]).

The practicalities of examining accessibility and distance-decay with reference to tourism have been examined by McKercher and Lew (2003), who discuss the relationship between distance-decay and what they call ETEZ. These zones are areas where factors such as spatial incongruities, geomorphology and even political realities combine to limit and promote flows along certain routes and in certain directions: 'Collectively, these factors produce ETEZs, areas where little or no tourism activity occurs that is relevant to the source market while demand is concentrated at certain peak locations. In this way, ETEZs exert a distorting effect on the standard distance-decay curve' (McKercher & Lew, 2003: 161). The result of this is a revised distance-decay curve (Figure 3.8), where the standard decay exhibited as a result of diminishing distance is replaced by a secondary peak:

> Demand still peaks relatively close to the origin and declines rapidly with distance. However, the distance decay curve is shaped by expanses of ETEZs, secondary peaks

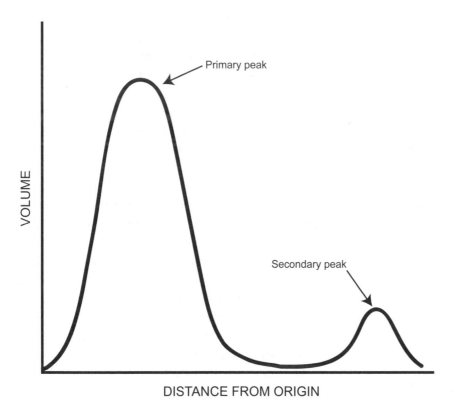

Figure 3.8 Primary and secondary peaks in tourism-related distance-decay curve

Source: Adapted and re-drawn from McKercher and Lew (2003)

are likely to occur at great distances from the origin, where a compelling pull of exceptional attractions overcomes the friction of travel. (McKercher & Lew, 2003: 161)

McKercher and Lew (2003) argue that there are several important factors that affect the flow of international tourism. Certainly the availability of transport modes to and from international destinations will affect flows, and will therefore affect the distance-decay curve indicated. For example, despite the fact that seasoned travellers may think nothing of embarking upon a 24-hour journey between New Zealand and the United Kingdom, some travellers may elect to travel to destinations that are closer to their point of origin. Thus, certain types of travellers may elect to ignore distance-decay effects. An important point made by McKercher and Lew (2004) is that the primary peak may, in fact, be extended in distance. Put another way, the initial peak may in fact be represented across a wider distance (as a plateau) and thus not be represented a narrow point as depicted in Figure 3.8. This may occur in situations where the draw of destinations or attractions along a route may be strong enough to warrant demand irrespective of distance. At some point, however, that demand would drop off as distance becomes a factor in the decision-making process.

Of course, and as mentioned briefly above, factors of distance and time are only part of a wider equation of demand for particular products/services/experiences in tourism. Another factor to be considered is the type of tourist. For example, using Cohen's (1974) typology of travellers, the explorer may be expected to visit destinations that are decidedly outside of more common transport routes or networks. Similarly, families may elect to travel to destinations that are closer, thus favouring more time at the destination as opposed to being in transit. Clearly, then, the distance-decay effect in relation to time spent travelling is an important consideration, but it is also one that may depend on numerous other variables. Also important is the degree of marketing that is undertaken by destinations around the world. As McKercher and Lew (2003) emphasise, no two destinations are alike and marketing efforts, combined with transport options that are available that support such efforts, can effectively serve to increase latent demand for their products and services (see, for example, Box 2.2).

The incorporation of distance-decay effects, in combination with accessibility and connectivity considerations, is, on the one hand, useful in order to obtain a snapshot of the position of destinations with networks of varying scales (international, national, local, for example), but, on the other hand, the dynamic nature of tourism and the complexity involved in developing tourism destinations can also suggest that modelling travel flows is difficult. It is possible, however, to construct a model that takes accessibility into consideration when attempting to profile the relationship between place, time and space (Figure 3.9). The model suggests that the number of trips may not always correlate precisely with high degrees of accessibility because such trips may be taken out of necessity or obligation. When thinking about the nature of mobility and its relation to transport, it

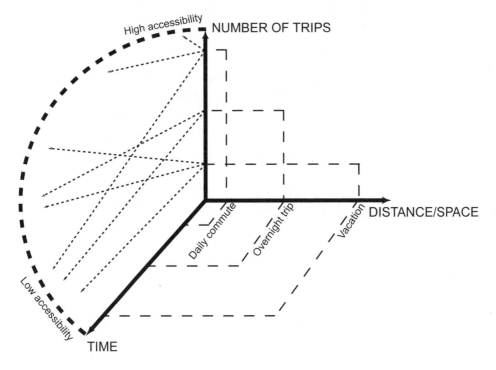

Figure 3.9 Time–space optimisation model in the context of accessibility

is not only the type of trip that needs to be considered, but also the frequency and duration of the mode of transport. As such, whilst numerically small, vacations and holidays
may incorporate both accessible (qualitatively evaluated or otherwise) and inaccessible.
Much of this accessibility depends on the availability of transport networks and thus their
ability to satisfy latent demand for a particular route. Similarly, daily commutes may be frequent, but the path followed from home to work may be inaccessible due to limitations in
the transport network (i.e. infrastructural limitations or temporary blockages). The fundamental concept underlying the model in Figure 3.9 is that accessibility is conditional and
continuous. While it can be measured in terms of actual modelling as discussed above, it
is also important to consider that perceived accessibility also plays a critical role. Thus,
whilst some individuals may consider a two-hour commute in a larger urban megacity to
be perfectly acceptable, others may feel otherwise.

Table 3.1 Application of spatial techniques to hypothetical business development options in transport and tourism

Scenario	Procedures	Output
Introduction of new airline to a region, ostensibly servicing three or more nodes	Model accessibility and demand using gravity model	Measure of potential accessibility; central node of operations can potential be designated
New public transit route carrying passengers to and from airport and key drop-off points throughout an urban area	1) Potential accessibility if choice of nodes needed to determine logical, practical and profitable drop-off points 2) Beta index to determine potential routes 3) Connectivity matrix to show route plans	1) Measure of demand in relation to location (i.e. potential accessibility given presence of tourists any/all sites) 2) Optimal routing 3) Optimal scheduling based on distance needed, but other factors (i.e. traffic) need to be considered

Table 3.1 outlines two scenarios of when the tools outlined in this chapter might be put to use. These situations reflect generic representations of real-world business opportunities, and while it is important to keep in mind that business decisions largely incorporate many other elements beyond that which is presented, it is argued that these issues should be considered initially before other considerations are given attention.

CHAPTER SUMMARY

This chapter has attempted to provide something of a spatial toolkit for the analysis of transport demand in relation to locations and places. It has shown how issues of connectivity and accessibility need to be carefully considered in establishing both the initial and long-term viability of transport operations. While spatial approaches in geography and transport studies have traditionally examined such issues in greater detail than what has been provided here, the intention is to gain a better understanding of the relationship between places and the potential interactions between them.

While this may seem somewhat out of place in a book on transport and tourism, understanding networks is important for several reasons:

1. Networks are inherently bounded by particular nodes and routes, and tracking movement through space and time allows for a clear understanding of how well positioned (or otherwise) a particular node is within a wider network (Figure 3.1). In the context of tourism and transport, this means that some destinations can spatially analyse the degree of connectivity and accessibility before examining market-based forces such as competition and regulation of routes and networks.

2. Furthermore, plotting accessibility and connectivity networks allows for a simplified view of accessibility and connectivity; in other words, the degree of accessibility and connectivity can ultimately play a key role in the success of a particular destination, especially when considerations of elasticity are involved.

3. Measuring connectivity along a network on a variety of levels can be a first step in determining a variety of business and operational decisions in transport, primarily those relating to barriers to entry and pricing/demand.

4. Above all else, the mobility of travellers (with tourists included in that characterisation) ultimately dictates the use of and demand for transport, and thus plays a significant role in destination development.

BOX 3.2 Case study – Practical examples of connectivity and accessibility concepts

The purpose of this case study is to demonstrate how spatial methods alone may not necessarily be adequate in understanding the relationship between tourism and transport. Several examples are shown. The first examines drive tourism in South Africa, while the second compares the route networks of jetstar Asia (based in Singapore) and Nok Air (based in Bangkok).

Drive tourism in South Africa
Although the concept of drive tourism has been addressed in the literature (e.g. Hardy, 2003; Laws & Scott, 2003; Olsen, 2003; Prideaux & Carson, 2003), what has yet to be fully assessed is the extent to which tourists make decisions regarding destinations during a self-drive holiday using spatial information. More importantly, the question remains as to how the tools discussed in this chapter might be used by those tourism operations with an interest in self-drive holidays (e.g. rental companies, attractions, accommodations, to name a few). In other words, how might strategic decisions regarding marketing and development make use of such information, and what problems are introduced in doing so?

Independent tourists represent a significant proportion of South Africa's travel market. Once in the country, many tourists opt to embark upon self-drive excursions. The impact of the internet in the distribution of tourist experiences in South Africa has been documented (Wynne *et al.*, 2001). www.drivesouthsfrica. co.za was established as a portal for visitors to rent a variety of personal vehicles (including campers and cars, as well as 4×4s), design tours, book accommodation and learn about festivals and events. South Africa is a sizeable country, so the extent to which it is connected comes into question. Table 3.2 outlines various distances of several key destinations in South Africa.

The modern roadway system assures almost full connectivity to most major centres, so generating a simple connectivity matrix may not be the most useful starting point. Instead, a full connectivity matrix with associated driving distances may provide some insight into the extent to which certain centres or places are more proximal to other areas. What is immediately apparent is that Kimberley, located close to the border of Northern Cape and Free State, is located the shortest distance within the network of locations and places in Table 3.2. This might imply that Kimberley should be the base of operations for any tourism business that features at its core the rental of vehicles for tourists on self-drive holidays. The problem with this conclusion is that, as indicated earlier in the chapter, this type of analysis ignores topography and actual travel time. Thus, while Kimberley may be spatially proximal to most other destinations, the actual travel time from Kimberley to other locations may be hindered by external variables such as poor roading (including surface type and speed limits), varying topography that can influence overall time irrespective of spatial distance. Thus, while connectivity matrices can be useful in establishing a theoretical spatial structure, they must be treated as the starting point only in terms of understanding how and why a network is constructed and how this information can be used in business decisions.

Comparing two route networks – jetstar Asia versus Nok Air

Based on the route network featured on jetstar's website (www.jetstar.com), a full connectivity matrix can be calculated (Table 3.3). This matrix is based on the ability of the passenger to book a complete ticket to the destinations listed. In many situations, where connectivity is not implied (e.g. an '0' is indicated) it is entirely possible for the passenger to travel between the two places but not on one single ticket.

Table 3.2 Distances (kms) between selected South Africa locations

	George	Bloemfontein	Cape Town	Durban	East London	Grahamstown	Johannesburg	Kimberley	Ladysmith	Mafikeng	Port Elizabeth	Umtata	Welkom	Messina	
George	0	773	438	1319	645	465	1171	762	1183	1203	335	880	926	1701	**842.9**
Bloemfontein	773	0	1004	634	584	601	398	177	410	464	677	570	153	928	**526.6**
Cape Town	438	1004	0	1753	1079	899	1402	962	1431	1343	769	1314	1156	1932	**1105.9**
Durban	1319	634	1753	0	674	854	578	811	236	821	984	439	564	1118	**770.4**
East London	645	584	1079	674	0	180	982	780	752	1048	310	235	737	1512	**679.9**
Grahamstown	465	601	899	854	180	0	999	667	932	1065	130	415	754	1529	**677.9**
Johannesburg	1171	398	1402	578	982	999	0	472	356	287	1075	869	258	530	**669.8**
Kimberley	762	177	962	811	780	667	472	0	587	380	743	747	294	1002	**598.9**
Ladysmith	1183	410	1431	236	752	932	356	587	0	597	1062	517	340	894	**664.1**
Mafikeng	1203	464	1343	821	1048	1065	287	380	597	0	1548	1003	451	808	**787.0**
Port Elizabeth	335	677	769	984	310	130	1075	743	1062	1548	0	545	830	1605	**758.1**
Umtata	880	570	1314	439	235	415	869	747	517	1003	545	0	718	1403	**689.6**
Welkom	926	153	1156	564	737	754	258	294	340	451	830	718	0	788	**569.2**
Messina	1701	928	1932	1118	1512	1529	530	1002	894	808	1605	1403	788	0	**1125.0**
	842.9	526.6	1105.9	770.4	679.9	677.9	669.8	598.9	664.1	787.0	758.1	689.6	569.2	1125.0	10465.3

Source: From www.places.co.za

Table 3.3 jetstar Asia connectivity matrix

	Singapore	Bangkok	Bangalore	Jakarta	Denpasar	Hong Kong	Phuket	Manila	Phnom Penh	Siem Reap	Yangon	Surabaya	Taipei	
Singapore	0	1	1	1	1	1	1	1	1	1	1	1	1	12
Bangkok	1	0	2	2	0	0	0	0	0	0	0	0	0	5
Bangalore	1	2	0	0	0	0	0	0	0	0	0	0	0	3
Jakarta	1	2	0	0	0	0	0	0	0	0	0	0	0	3
Denpasar	1	0	0	0	0	0	0	0	0	0	0	0	0	1
Hong Kong	1	0	2	0	0	0	0	0	0	0	0	0	0	3
Phuket	1	0	0	0	0	0	0	2	0	0	0	0	0	3
Manila	1	0	0	2	0	0	2	0	0	0	0	0	0	5
Phnom Penh	1	0	0	0	0	0	0	0	0	0	0	0	0	1
Siem Reap	1	0	0	0	0	0	0	0	0	0	0	0	2	3
Yangon	1	0	0	0	0	0	0	0	0	0	0	0	0	1
Surabaya	1	0	0	0	0	0	0	0	0	0	0	0	0	1
Taipei	1	0	0	0	0	0	2	0	0	2	0	0	0	5
	12	5	5	5	1	1	5	3	1	3	1	1	3	
													Dispersion	46
													Mean Dispersion	3.5

As shown, the jetstar Asia network is not particularly dense from a spatial perspective, but the power of the hub (see Chapter 6) in Singapore can be seen. As well, the airline uses Singapore as a stopover for several key routes, including Jakarta, Bangalore, Phuket and Taipei. The introduction of these routes is purely strategic, focusing on satisfying demand for travel to these destinations from origins such as Hong Kong, Manila and Phuket.

Compare this network with that of Nok Air (Table 3.4), operating out of Bangkok. Again, this matrix reflects the ability of a passenger to book a ticket from a place of origin to the destination indicated. Like the jetstar Asia example above, travel between two places is possible where connectivity is implied to be non-existent, but this would require two separate tickets.

Table 3.4 Nok Air connectivity matrix

	Bangkok	Udon Thani	Loei	Chiang Mai	Mae Hong Son	Phuket	Trang	Nakhon Si Thammarat	Hat Yai	
Bangkok	0	1	1	1	0	1	1	1	1	7
Udon Thani	1	0	1	1	0	0	0	0	0	3
Loei	1	1	0	0	0	0	0	0	0	2
Chiang Mai	1	1	0	0	1	0	0	0	0	3
Mae Hong Son	0	0	0	1	0	0	0	0	0	1
Phuket	1	0	0	0	0	0	0	0	0	1
Trang	1	0	0	0	0	0	0	0	0	1
Nakhon Si Thammarat	1	0	0	0	0	0	0	0	0	1
Hat Yai	1	0	0	0	0	0	0	0	0	1
	7	3	2	3	1	1	1	1	1	
								Dispersion		20
								Mean Dispersion		2.2

What is evident from the Nok Air connectivity matrix is that, first, the network itself is comprised of fewer vertices or nodes (9 versus 13 for jetstar Asia). The second feature of the matrix is that two distinct sub-networks are apparent. The first involves destinations that are north of Bangkok, including Mae Hong Son, Chiang Mai, Loei and Udon Thani. Among these destinations, including Bangkok, a considerably more connected network has been established when compared to the second sub-network, consisting of the remaining four destinations (Phuket, Trang, Nakhon Si Thammarat and Hat Yai). The second sub-group, all of which are located south of Bangkok, are considerably less connected among themselves and thus contribute to the lower dispersion and mean dispersion indicators for the entire network. To illustrate this, a graphical representation of the network appears in Figure 3.10.

The corresponding connectivity matrix, taking into account the distance depicted in Figure 3.10, is presented in Table 3.5.

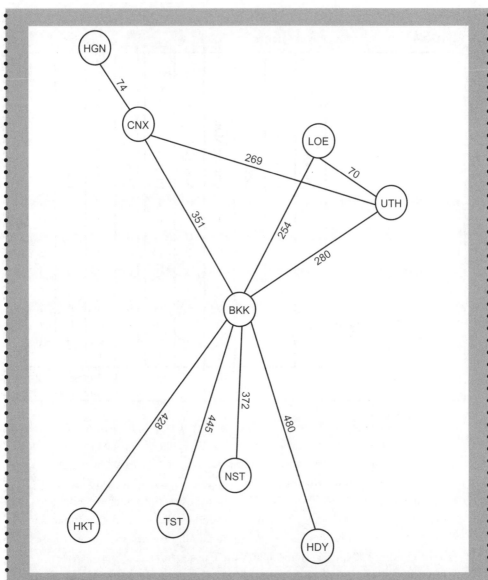

Figure 3.10 Nok Air network

Table 3.5 Nok Air connectivity matrix with distances

	Bangkok	Udon Thani	Loei	Chiang Mai	Mae Hong Son	Phuket	Trang	Nakhon Si Thammarat	Hat Yai	
Bangkok	0	280	254	351	0	428	445	381	480	**290.0**
Udon Thani	280	0	70	269	0	0	0	0	0	**68.8**
Loei	254	70	0	0	0	0	0	0	0	**36.0**
Chiang Mai	351	269	0	0	74	0	0	0	0	**77.1**
Mae Hong Son	0	0	0	74	0	0	0	0	0	**8.2**
Phuket	428	0	0	0	0	0	0	0	0	**47.6**
Trang	445	0	0	0	0	0	0	0	0	**49.4**
Nakhon Si Thammarat	372	0	0	0	0	0	0	0	0	**42.3**
Hat Yai	480	0	0	0	0	0	0	0	0	**53.3**
	290.0	**68.8**	**36.0**	**77.1**	**8.2**	**47.6**	**49.4**	**42.3**	**53.3**	**672.7**

Once again, this connectivity matrix has been constructed on the basis on how Nok Air actually offers sales of flights. If, however, the airline allowed for one ticket to cover, for example, Mae Hong Son through to Phuket or from Hat Yai through to Trang (via Bangkok), a very different matrix could be constructed (Table 3.6)

The result is that Bangkok continues to feature as the central hub, but there is room for the airline to consider future enhancements to allow the network to be more complete. For one, travel between Phuket, Trang, Nakhon Si Thammarat and Hat Yai requires a stop in Bangkok. Granted, an airline might be inclined to suggest that expanding the network to include flights between these four destinations would be beneficial to the traveller given that the stop in Bangkok would no longer be necessary, but the larger question is whether such routes would be profitable. In this particular case, Nok Air has decided that the opportunity cost of operating flights between these four destinations outweighs the gains from simply focusing on traffic from Bangkok down to each. It must also be remembered that Bangkok features as a major international gateway, and Nok Air has positioned itself as an LCC for travel to these southern tourism destinations.

Table 3.6 Nok Air connectivity network with distances (complete coverage)

	Bangkok	Udon Thani	Loei	Chiang Mai	Mae Hong Son	Phuket	Trang	Nakhon Si Thammarat	Hat Yai	
Bangkok	0	280	254	351	425	428	445	381	480	337.2
Udon Thani	280	0	70	269	343	708	728	652	760	423.3
Loei	254	70	0	339	413	682	699	626	734	424.1
Chiang Mai	351	269	339	0	74	779	796	723	831	462.4
Mae Hong Son	425	343	413	74	0	833	870	797	905	517.8
Phuket	428	708	682	779	833	0	873	800	908	667.9
Trang	445	728	699	796	870	873	0	817	925	683.7
Nakhon Si Thammarat	372	652	626	723	797	800	817	0	852	627.6
Hat Yai	480	760	734	831	905	908	925	852	0	710.6
	337.2	423.3	424.1	462.4	517.8	667.9	683.7	627.6	710.6	4854.6

Questions for consideration

1. Why might Nok Air have considerably less connectivity among its southern destinations?

2. How might a new entrant to the aviation market in Asia treat this information on the overall connectivity of jetstar Asia's network? How might this information be useful in deciding which routes to fly?

SELF-REVIEW QUESTIONS

1. Define connectivity and accessibility and explain how both are critical to understanding the relationship between transport and tourism.

2. In your own words, and with one example from your own experience, describe the distance-decay effect.

3. Describe the following equation in your own words and point out its importance for understanding flows between two nodes:

$$I_{ij} = \frac{P_i P_j}{t_{ij}^e}$$

ESSAY QUESTIONS

1. Outline some of the problems of the distance-decay model. When might such a model not be applicable in tourism?

2. Why is it important to even attempt to model demand from a spatial perspective? What can the tourism sector, and more importantly the numerous industries that comprise it, learn from spatial analyses?

3. Outline why mobilities and movement need to consider issues of transport.

KEY SOURCES

Chapman, K. (1979) *People, Pattern and Process: An Introduction to Human Geography*. London: Edward Arnold.

Chapman's text is a classic geography text published at a time when spatial methods in the discipline were popular. It covers network connectivity and accessibility, spatial distribution of activities, and movement and mobility.

Haynes, K.E. and Fotheringham, A.S. (1984) *Gravity and Spatial Interaction Models*. Beverly Hills, CA: Sage.

Following on from the Chapman text, this text examines further gravity models and interactions across geographic concepts of space.

Taaffe, E.J., Gauthier, H.L. and O'Kelly, M.E. (1996) *Geography of Transportation* (2nd edn). Upper Saddle River, NJ: Prentice Hall.

Taaffe *et al.*'s text is one of the key texts in the geography of transport. This text is particularly useful for understanding distribution of spatial elements and concepts, network form and structure as well as applied examples at regional and local levels.

CHAPTER 4:

GROUND TRANSPORT

LEARNING OBJECTIVES

After reading this chapter, you should be able to

1. Describe and interpret the importance of ground transport within the tourism system.

2. Explain the variances in scale and flows of ground transport.

3. Evaluate the differences and similarities in terms of international scope of various forms of ground transport in relation to tourism.

4. Differentiate, if possible, between recreation-based and tourism-based types of ground transport.

5. Understand recent advances in environmentally friendly forms of ground transport.

INTRODUCTION

In this chapter, several types of ground transport and their links with tourism (and the provision of tourist experiences) are discussed. First, the use of coaches (largely associated with packaged tours) is discussed. Rail travel is briefly profiled and linked to various forms of tourism (e.g. ecotourism, heritage tourism) that utilise this mode of transport to both add to the experience and as an effort towards maintaining some degree of sustainability. Following this, 'personal' types of transport (e.g. self-powered or motorised) in the context of tourism (and recreation) are discussed.

Because ground transport, and particularly personal automobile use, is so pervasive in many developed countries, it is difficult to render a clear assessment of tourism and ground transport relationships. Some of the standard definitions of 'a tourist' in the context of pleasure travel or holidaymaking are inherently limiting when discussing ground transport. Instead, the focus in this chapter will examine the use of ground transport that is inclusive of recreationalists, excursionists, commuters as well as tourists. This is partly to illustrate the blurry distinction between transport and tourism when discussing ground transport, but also to focus on those planning and development measures that, while ostensibly focusing on one particular type of user, may inevitably have an impact on other users.

GROUND TRANSPORT AND TOURISM

The various sizes and spatial scales in which ground transport operations operate play a significant role in the extent to which tourism and some recreational activities are facilitated. As a result, and for the purposes of simplifying this relationship, two broad types of transport under the ground mode that involve tourism can be identified:

1. Personal transport, including not only motor vehicles (which would generally be associated with tourist activities as means of transport) but also off-road, motorised 'quad' (four-wheel) vehicles and pedal-powered bicycles (which would be more properly associated with recreational pursuits and, thus, act as a focus of the experience itself).

2. 'Supplied' transport, including transportation relating to (a) package tours that, for example, make use of coaches or buses and (b) rail travel, including both intra-regional, inter-regional and local transport provision.

As complex as the various types of ground transport are, so too are the means by which they are measured. In urban environments, for example, transport is recognised as a key ingredient for economic development (e.g. Banister, 1995), leading to consideration of the relative cost of congestion, emissions and commuting time on overall productivity. Ground transport development can also be used for the purposes of urban revitalisation (Priemus & Konings, 2001). In the context of tourism, Page and Hall (2003) note that it also plays an important role in hosting visitors: not only is it often the most dominant means by which tourists enter a destination, it also represents a significant proportion of the mode of transport used within a destination. For example, of all the UK tourists travelling to Scotland, over two-thirds arrive by car (Tourism Scotland, 2003). Of course, the spatial proximity of destinations in the context of one's usual residence plays an important role in determining whether ground transport is used. Thus, as the majority of Canada's population lives along the international border with the United States, it is perhaps not surprising to learn that, in 2000, 97% of same-day and 56% of overnight travel by Canadians to the United States was by car (Transport Canada, 2001).

Perhaps unlike other modes of transport (such as air travel and marine transport), ground transport encompasses numerous types. In fact, one could argue that the provision of ground transport types is almost as complex and multi-faceted as the users of these systems. The provision of various types of ground transport is often governed by non-tourism related policies and planning measures. Thus, while the development of certain ground transport types may not be explicitly for the purposes of enhancing tourism, this does not mean that the overall provision of tourist-related activities and use is not possible. This at once raises the difficult question as to what actually can be classified as tourist-related transport provision. In Chapter 2, the blurry distinction between modes of transport and their association with tourism was shown to be problematic. This blurry distinction is perhaps most complicated when discussing various types of ground transport, primarily because it is difficult to separate out tourist use from non-tourist use of ground transport. There are several reasons for this. First, many recreational activities utilise the same mode of transport used for patterned, daily use (i.e. transport to and from work). As a result, the family car may be used as a means of transport for economic reasons (i.e. to and from one's place of employment), but may be used on weekends for family getaways or for afternoon trips to a local attraction such as the beach or a theme park. Trains and buses, for example, can be used not only by commuters, but also by tourists and local recreationalists. In one sense, the multi-purpose usage factor of various types of ground transport best highlights the problem of ascribing one particular type of ground transport almost exclusively to tourist use only. Most types of ground transport, therefore, service multiple functions both within and outside of the leisure/tourism/recreation realm. In many respects, this explains their pervasiveness in developed countries, and increasingly in some developing countries.

It is perhaps best to think of a continuum of association and use. At one end are those types of ground transport used almost exclusively by tourists. A good example is escorted coach tours, which are normally patronised by international tourists, although domestic coach package services can be found in some countries. The other end of the continuum is represented by those types of transport that are used by both tourists and non-tourists (e.g. taxis or other forms of public transportation). A second feature of ground transport that highlights the blurry distinctions involving tourism is the fact that travellers may elect to use, when available, multiple modes of ground transport. In the case of London, business travellers may well utilise a taxi service, the tube, and a city bus all in the same trip. The decisions made with respect to which types of transport are available for use can be based on a number of criteria, not the least of which may include the distance to be travelled, the cost of a particular type of transport (i.e. taxi versus the tube), and the particular time-budget available to the traveller.

A vacationing family to Disneyworld in California may utilise small shuttles around the park itself, or elect to use the park-specific train. For those visiting from nearby states such as Arizona or Oregon, they may elect to drive to the theme park along an interstate highway, while those visiting from further away may initially fly and then utilise ground transport for local mobility. Similarly, a traveller from Europe visiting family in Cairns,

Australia may find that renting a car is more convenient for getting around. The purpose of these examples is to demonstrate that characterising all types of ground transport as either tourist-focused or otherwise can be misleading.

A third and final consideration is the extent to which the motivations of travellers come into play when discussing ground transport. For example, the tourism literature is voluminous with theories of tourist motivation, but many of these are associated with making decisions in terms of which destination to visit. As shown in Chapter 2, the decision-set for modes of transport carries with it numerous elements, but it is the manifestation of that choice that is of interest. For instance, because most modes of ground transport are, in effect, multi-purpose, measuring the extent to which tourists utilise particular types of transport can be problematic. One example is the provision rail transport from Glasgow to Edinburgh in Scotland (Figure 4.1). As the trip is roughly one hour in duration, it is used by residents in both cities to commute, yet it is also used by tourists visiting Scotland as an inexpensive and reasonably convenient way to travel from one city to the other. With respect to managing a service such as this, what becomes critical is a functional assessment of the ridership. Planners tasked with managing types of ground transport need to incorporate tourism-related utilisation into long-range initiatives and operational

Figure 4.1 Multi-use ground transport: The Queen Street Station, Glasgow, is used by both tourists and commuters

decisions (such as seasonality, promotions, etc.). In August 2006 the recently appointed Director of the Glasgow–Edinburgh collaboration project announced that one of the project's goals over the next two years is 24-hour train access between the two cities. The Director noted: 'People coming from the US, from the continent, who are here as tourists, would just assume that, like other metropolitan centres, they would be able to get between the two much later at night. It is essential for tourism alone, as well as other travellers' (Johnston, 2006). By recognising the demand by tourists and other travellers for the nature of ground transport, governments and private companies will be in a better position to assess the incidence of use of specific modes of transport and thus tailor future growth appropriately.

Table 4.1 provides a ten-year overview of the various means of ground transport within the United Kingdom.

Table 4.1 Mode of passenger transport, UK (1992–2003) (in billions of passenger kilometres)

	Buses/ coaches	Cars, vans, taxies	Motorcycles	Pedal cycles	Rail
1992	43	583	5	5	38
1993	44	584	4	4	37
1993	44	607	4	4	37
1994	44	614	4	4	35
1995	43	618	4	4	37
1996	43	625	4	4	39
1997	44	632	4	4	42
1998	45	635	4	4	44
1999	46	641	5	4	46
2000	47	639	5	4	47
2001	47	654	5	4	47
2002	47	677	5	4	48
2003	47	678	6	5	49
2004	48	678	6	4	50
2005	48	678*	6	4	52

Source: Department for Transport (2006)

Note: * Preliminary results.

The important point from Table 4.1 is not necessarily the sheer number of passenger kilometres, but the variability and varying levels of growth in the use of ground transport within the United Kingdom. For example, the use of buses and coaches experienced steady growth over the ten-year period from 1992 to 2002, yet the use of pedal cycles for transport remained relatively flat. In the case of cycling, as will be discussed later, efforts are being made to encourage the use of bicycles for not only regular personal transport but also for recreational purposes. Table 4.2 expands on this and shows historic trend data on the average distance travelled for use of specific types of ground transport.

Table 4.2 Average distance travelled by selected modes of travel, UK (in miles per person per year)

	1975/ 1976	*1985/ 1986*	*1991/ 1993*	*1996/ 1998*	*2002*
Bicycle	51	44	39	38	33
Car only – driver	1849	2271	2993	3319	3356
Car only – passenger	1350	1525	1951	1973	2000
Motorcycle/moped	47	51	38	30	33
Surface rail	289	292	311	290	373

Source: Adapted from Department for Transport (2004b)

When examining Table 4.2 and Table 4.1 together, some interesting trends emerge. First, while the number of passenger trips by bicycle has remained steady during a ten-year period from 1992 to 2002, the average distance travelled has fallen. Similarly, while the number of rail trips has increased, the average distance travelled by rail has fluctuated. In the case of cycle tourism the increasing congestion of vehicular traffic on the UK roadway system may play a role in how far people journey.

In 2000 the UK government announced significant funding increases for transport infrastructure (BBC, 2000). This included:

- Rail – £60 bn:
    - investment to allow 50% more passengers, with faster and safer journeys;
    - £7 bn rail modernisation fund;
    - 6000 new carriages and trains.

- Roads – £59 bn:
    - £30 bn to eliminate backlog in road maintenance;

- 80 major trunk road schemes to improve safety and traffic flow at junctions;

- 100 new bypasses on trunk and local roads and 130 other full-scale local road improvement schemes;

- lower-noise surfaces for 60% of trunk roads;

- a 10% increase in bus passenger journeys and an extension of bus priority schemes;

- widen 360 miles of motorway and trunk road network.

- Local transport schemes – £26 bn:
  - build 100 new bypasses;
  - £20 bn for London, excluding roads – potentially including a cross-London rail link;
  - up to 25 light railway schemes in urban areas;
  - more cycle and walking routes and more 20 mph areas.

In November 2005 the Confederation of British Industry argued that £60 billion was needed to improve Britain's transport infrastructure in order to keep up with demand (Bloomberg News, 2005). Hidden within the data in Tables 4.1 and 4.2, however, and perhaps most importantly, are the reasons for the types of travel listed. Table 4.3 provides a breakdown of the trip purpose for UK travellers. What is remarkable is the growth in distance travelled for holidays, day trips and shopping, reflecting the provision of interconnectedness and accessibility of amenities relating to these activities. While not all of these relate directly to ground transport, it is important to position these figures in the context of government policy for transport infrastructure. Policy decisions need to be made with respect to the use and demand for infrastructure renewal and expansion, hence the reason that most national governments undertake comprehensive surveys and statistical modelling in order to identify probable causes of action.

Table 4.3 Distance per person per year by trip purpose, UK (in miles)

	1985/1986	*1991/1993*	*1996/1998*	*2002*
Commuting	1086	1207	1341	1294
Business	544	678	681	683
Education	161	183	190	195
Shopping	611	778	860	857
Visiting friends	1165	1357	1400	1364
Holiday: base	338	490	474	513
Day trip	307	373	343	368

Source: Adapted from Department for Transport (2004b)

COACH TRANSPORT

Coach transport is often associated with pleasure travel that encompasses numerous destinations (often in circuit routes, involving destinations or places that are usually geographically proximate to one another) over a specific period of time. As such, coach transport is unique in that it involves 1) multiple nodes or destinations; 2) prescribed flows and networks; and 3) highly structured itineraries. As such, one of the more common manifestations of coach travel is large buses shuttling tourists across a wide geographic region with stops at popular attractions and destinations (Figure 4.2). The length of these stops ranges from a few hours to several days.

Coach transport and its relationship to tourism can be significant in some areas and less so in others. In the Royal Borough (UK) (home to Windsor Castle and Legoland Windsor, two attractions that appear in the UK's overall top 20 attraction list) coach tourism is worth some £22 million per year (http://www.windsor.gov.uk/education/educat_index. htm; accessed 6 December 2005). By comparison, coach tours only account for 12% of arrivals to the city of York (www.cityofyork.com/econfact/keystats.htm; accessed 6 December 2005) and 3% of arrivals to Scotland (Tourism Scotland, 2003). The economic

Figure 4.2 Guided tours and coach transport: Tourists disembarking a coach at Te Anau, New Zealand

benefits of coach transport can be even less substantial for some smaller destinations or attractions that exist within a wider national network. These destinations, therefore, may experience regular visits from coaches bringing substantial numbers of tourists, but feature only as a brief stop within a larger itinerary. Without these stops, in some cases, visitor traffic at these destinations could, at best, be reliant upon pass-through traffic using personal modes of transport (i.e. passenger cars). In those destinations where the overall attraction and general infrastructure base is small, coach transport can often represent the most significant manner by which tourism is manifested. This is especially the case in rural destinations where other modes do not offer service, and especially smaller places along a prescribed route from one major destination to another. In the south island of New Zealand, for example, the small town of Gore sits almost half-way between Dunedin and Invercargill (Figure 4.3).

Coach tours make regular stops at Gore, allowing passengers the opportunity to use the toilets and purchase snacks. Despite the high numbers of coach visits, the overall eco-

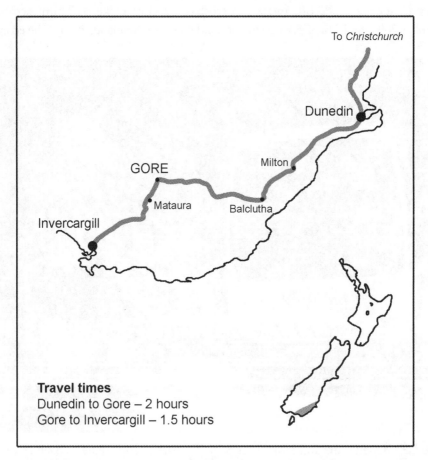

Figure 4.3 Location of Gore relative to Dunedin and Invercargill, New Zealand

nomic impact from these types of tourists is rather small, largely because most stops last little more than 30 minutes. As such, in those locations where coach tours make frequent 'rest stops' or 'tea breaks', it is not uncommon for local policy-makers to study different solutions designed to encourage visitors to stay longer. Yet this type of initiative does not solely rely on persuading coach tourists to stay longer. It requires fervent planning and decision-making processes to be in place and an understanding of how coach tour companies plan itineraries.

RAIL TRANSPORT

Rail transport played a significant role in the economic growth of many countries during the 19th century and, to some extent, the early 20th century (Turton & Black, 1998). In Romania, for example, the development of rail systems prior to the First World War resulted in new patterns of mobility between cities (Turnock, 2001). Rail travel in the United States was developed in a rather haphazard manner, with different gauges of track (i.e. different spacing between rails) being utilised and inefficient network planning that resulted in confusing (and laborious) itineraries for many travellers. As Cocks (2001) notes, 19th-century rail transport in the United States linked major urban areas and played a key role in their growth. For passengers using these early services, they had a choice between rather cramped and rudimentary carriage cars or more upmarket luxury coaches will full service. Using Pullman cars (specialised luxury travel rail coaches) for the luxury offerings, Cocks' (2001) historical research recalls how African-Americans were employed as porters, thus demonstrating how rail travel reflected a particular slice of social history of the time.

When examining how tourism can be linked with rail transport, two main sub-modes can be broadly identified:

1. Inter-destination rail transport (i.e. rail transport that links regions, cities, destinations and even attractions; this can be international or cross-border as well).

2. Intra-destination rail transport (i.e. rail transport that is designed to move passengers within a destination or attraction).

Of course, characterising rail transport this way may be problematic as there is often little to distinguish between these two from an operations/management perspective. For example, some organisations offer both intra-destination and regional rail transport. One example is the Loch Lomond area of Scotland, which lies just outside of the city of Glasgow. The area is frequented by tourists for the purposes of hiking and other outdoor recreational pursuits, and is accessible by car (40 minutes), by bus (40–45 minutes) or by train via services provided by ScotRail. The train service to the area involves a stop at Balloch after starting from downtown Glasgow. The service makes numerous stops in residential neighbourhoods and is primarily used, therefore, by residents to both enter and exit the city during the week. Because of the accessibility it affords, however, it can also be used by tourists as an alternative type of transport to the Loch area.

Many countries are turning to high-speed rail networks (using MAGLEV trains that can run at 500 km/h) to facilitate passenger and cargo movement. At present, Japan, France, Germany and Spain offer the service, although Korea recently launched the KTX service from Seoul to the port city of Busan (BBC, 2004a). A US$20 million study by the California High Speed Rail Authority recently suggested that 220 mph trains could be shuttling some 68 million passengers between San Francisco and Los Angeles by 2020 at a cost of US$37 billion (San Mateo County Times, 2004). Japan's high-speed rail network (Figure 4.4) is designed almost exclusively for the transport of passengers. While the network covers a significant portion of the country, it is designed primarily to link urban areas. Caution must be exercised, however, in determining how much of an impact high-speed rail can have. Unless the overall network is constructed such that rural or periurban areas are factored into the infrastructure, and thus connectivity, of the network, linking urban areas may serve to marginalise smaller regional centres. Gutiérrez *et al.* (1996: 238) suggested that the wider European high-speed network certainly helps link peripheral regions to urban centres, but potentially at the expense of the immediate urban 'hinterlands':

> Until the appearance of the high-speed train, ground transport systems shaped the space in a relatively continuous way: those places geographically located in the more peripheral regions were the least accessible. The high-speed train is changing this situation and is creating a space that is becoming more and more discontinuous (Plassard, 1992) in which the spatial distribution of accessibility depends less and less on the geographical location of the nodes and more and more on the type of infrastructures they are linked up to. Stations on the high-speed lines are at hundreds of kilometres distance from each other, thus creating 'islands' of greater accessibility and, in fact, a space that is becoming more and more discontinuous...

Hensher (1997) explored the demand for a high-speed rail network between Sydney and Canberra, a route that is dominated by the personal vehicle modes of transport, and noted that fierce competition between airlines with cheap airfares may affect profitability of the rail network (although see Chapter 8 for a discussion on Virgin Rail's efforts in the UK). Hensher's study found that varying levels of elasticity are evident in determining the demand for high-speed rail, yet the important lesson is the actual measurement of induced demand for such a product. As governments are more and more interested in funding profitable and efficient services, many will ask whether funding high-speed rail networks in this, an age of affordable air travel, is viable and, perhaps most importantly, politically popular. González-Savignat (2004) utilised experimental design techniques in demand modelling between

Figure 4.4 Moving passengers across extensive distances: High-speed train network in Japan

Source: By D.A. Follett, reproduced with permission

Barcelona and Madrid to answer this, and found that high-speed rail could impact on air travel markets as long as travel time is competitive between the two modes.

Demand for rail services must be measured along numerous variables, including convenience (i.e. factors relating to accessibility and connectivity; see Chapter 3) and price (see also Gutiérrez, 2001). Crockett and Hounsell's (2005) study of public rail transport in South Hampshire (UK) found that convenience related to access, station facilities, service frequencies and schedules, and interchange with other services. Bieger and Laesser (2001) found that factors such as safety, travel time, punctuality, flexibility and comfort were key when their 1999 survey of Swiss consumers considered journeys over 100 km. The advantage of rail transport as identified in Biefer and Laesser's study seems to rest primarily on lower costs, frequency and punctuality. This may be the case in some areas or countries, but with increasing competition from airlines in, for examples, Europe or the United States, the advantages of rail travel with respect to cost and frequency may be eroding. The issue, then, is one of viability. In New Zealand, the lack of demand (perhaps a result of poor marketing) for routes between regional urban centres such as Dunedin and Christchurch meant eventual closure. As the Washington Times reported in 2004, Russian President Putin is seeking to invest substantial amounts of capital in the railway system, which is likely one of the largest in the world. Part of this is to disentangle the process of strong regionalism that has been at the centre of Russian politics for decades, but it could also be used as a means to move people for recreation or leisure purposes (Washington Times, 2004b)

Rail excursion/tour services can be packaged either by wholesalers or directly by the rail company itself. Both work closely with local service providers (primarily accommodations and, in some instances, restaurants and other attractions) to build all-inclusive rail holidays, some of which are escorted. One wholesaler, Maupintour, offers what it calls 'premier all-inclusive worldwide escorted packages' (Table 4.4), which represent a unique tour package in that the mode of transport is both the attraction as well as the key means of mobility. Canada's VIA Rail has long offered package trips across the country that more or less focus on the experiences on the train, but more recently has adopted a comprehensive package service that includes options for festivals, events, meetings and conventions, and adventure activities. In many ways, rail companies are seeking to build on the successes of air transport firms in providing packages and 'value-added' in order to drive business. Now that air transport options can often be just as cheap, if not cheaper, as rail transport, this represents a threat to ground transport provision.

RAIL TRANSPORT AND HERITAGE TOURISM

While the sub-modes of rail transport discussed above go some way to characterising the rail transport/tourism interface, there is another aspect of this interface that deserves mention. In recent years there has been a surge of interest in historical rail travel, where the means of transport also functions as the attraction itself (see Chapter 2). The growth in historic rail travel has, in a large part, been associated with the recognition of the importance of heritage tourism experiences (see Dann, 1994; Dickinson *et al.*, 2004). To this end, Halsall (2001: 152) suggests that '[p]reserved railways and tramways provide opportunities to view heritage transport in museum situations, and also to participate in the re-creations, accurate or otherwise'. Further, '[g]rowing numbers of people enjoy seeing and travelling on preserved historic transport modes, and are fascinated with steam propulsion, railways, and the images and emotions they stimulate...' (Halsall, 2001: 152).

To some extent, such interest is localised within the United States, where there is an abundance of little-used or completely abandoned track, but there has been growth in this area in the UK as well. Several examples from around the world highlight the growing interest in heritage rail travel:

• As an example of the integration of transport with heritage tourism development, in Pennsylvania excursions are utilised in tangent with 'battlefield tourism' as excursion trains wind through part of the Gettysburg battlefield site (Washington Times, 2004a).

- In Dunedin, New Zealand, the Taieri Gorge Railway (www.taieri.co.nz), jointly owed by the Dunedin City Council and the Otago Excursion Train Trust, shuttles day excursionists on The Taieri Gorge Limited along a 60 km route formerly operated by New Zealand Railways in the late 19th century. The railway is a popular choice for tourists as it introduces elements of the history of both the railway itself as well as its historic use in opening up Crown lands for development.

Table 4.4 Selected rail-based Maupintour journeys as at 2 September 2004

Journey	Key features
Europe Railway Journeys (8-day independent holiday)	• Private car airport transfers • 3 nights at the Thistle Marble Arch Hotel in London • Full-day London sightseeing • Transfer to Waterloo Station for first class Chunnel train to Paris • Private car transfer to Hotel Regina • 3 nights at the Hotel Regina in Paris
Rocky Mountaineer Escape (7-day escorted rail tour)	• 2 nights in Vancouver at The Sutton Place Hotel • 1 night in Kamloops at rail-assigned hotel • 2 nights at The Fairmont Banff Springs • 1 night at The Fairmont Chateau Lake Louise • Carriage ride through Stanley Park • 2-day rail journey aboard the *Rocky Mountaineer* • Visit Grouse Mountain Refuge for Endangered Wildlife • See Banff National Park • Sulphur Mountain gondola ride • Visit Columbia Icefield Parkway and take a Snocoach ride across Athabasca Glacier
Trans-Canada by Train (12-day escorted tour)	• 1 night at The Fairmont Royal York in Toronto • 2 nights aboard VIA Rail's *The Canadian* between Toronto and Jasper • 2 nights at The Fairmont Jasper Park Lodge • 1 night at The Fairmont Chateau Lake Louise • 2 nights at The Fairmont Banff Springs • 1 night in Kamloops at a rail-assigned hotel • 2 nights in Vancouver at The Sutton Place Hotel • 11 breakfasts, 7 lunches, 10 dinners

Source: www.maupintour.com (accessed 2 September 2004)

- Fostoria, Ohio was the first electric rail traffic control system to be demonstrated any-where in the world and was also the first place where centralised railroad traffic was established. Plans to revitalise a unique historical tower by the Fostoria Train Tourism Group follow from the recognition that accommodation accounts for only 11% of all visitor expenditures for Fostoria, thus leaving a significant amount (some US$8.5 million) that tourists spend in attractions, food and entertainment (Advertiser-Tribune, 2004)

These examples further demonstrate how a mode of transport can become closely inte-grated with the tourism experience, especially in an historical context. Mowforth and Munt (1998) captured this with their proclamation that tourism is now more concerned with travelling, trucking and trekking, as opposed to sun, sand and sea. This is an impor-tant consideration because it points to changing motivations for leisure and recreational activities, and in this particular case highlights how historic modes of transport get caught up in swaying tourist preferences.

The provision of historical train travel is often fraught with difficulties. First, decisions need to be made on how the project will be capitalised. Significant funding is often required to refit historic locomotives and refurnish passenger cars. In the case of the latter, this is often joined with enhanced safety protocols in order to meet modern safety regulations. In light of this significant investment, the other reality is simply whether enough demand can be realised for this type of experience.

PERSONAL TRANSPORT

Personal transport permeates numerous forms of human mobility, ranging from daily trips to supermarkets to lengthy family vacations, and features types of ground transport such as the personal automobile, motorcycle or bicycle (the latter of which is of course non-motorised). The range of personal transport available demonstrates the range of mobility that people have available to them. In other words, with motorised personal transport being made available at more affordable prices, the geographic scale of accessi-bility has grown. This section examines two particular personal transport types: the personal (or often 'family') automobile and the bicycle.

To a large extent, personal modes of transport can best be characterised and differenti-ated on the basis of provision or use:

1. Modes of transport provided by (or under legal ownership of) the individual; in rela-tion to tourism, this can mean the utilisation of personal automobiles for day or overnight trips at varying lengths.

2. Modes of transport that are provided to the individual at a set cost (i.e. through a rental or lease agreement); tourists may opt to rent a vehicle for transport to and from a destination or within a single, or between two or more, destinations.

Specific modes of personal transport used for leisure and recreation can fall into either category. Tourists in New Zealand may opt to rent 'Maui vans' and explore the country according to their own schedule, which is usually more flexible than packaged tours. These mobile caravans can be almost entirely self-contained for basic services (i.e. food preparation, showers and toilet), but this may also encourage the exploration of environmentally sensitive areas. Alternatively, individuals, couples and families may use their own personal vehicle for weekend or short-break holidays. The Dorset area of the United Kingdom, particularly Bournemouth and Weymouth, features as a popular vacation destination in the summer season. As such, congestion is not uncommon on the main motorway that stretches to London. Because of the popularity of utilising personal automobiles for the purposes of tourism, various road-side services are increasingly becoming available throughout the road networks of many countries. Throughout the UK, Canada and the United States, for example, 'services' (also known as 'rest stops' in North America) areas are common, where travellers are able to stop, use the toilet, stretch their legs, and purchase food and beverage items. In Canada, as in the United Kingdom, such rest areas are normally populated by retail or food and beverage chains. In the UK, RoadChef offers fast food, while in Canada Wendy's (fast food) and Tim Horton's (coffee and donuts) dominate most rest stops along the Trans-Canada highway (Figure 4.5).

Figure 4.5 Rest stop along the Trans-Canada Highway: Services for personal automobiles

PERSONAL AUTOMOBILES

Without question, the use of personal automobile for the purposes of tourism reflects, to a large extent, the proliferation of automobile use in developed countries. The growth of the use of personal automobiles in EU countries surpassed most other forms of transport between 1995 and 2002 (Table 4.5).

Table 4.5 Transport passengers per mode 1995–2002, EU (billions of passenger kilometres)

	Passenger cars	*Bus and coach*	*Railway*	*Tram and metro*
1995	3702.70	462.234	318.617	50.778
1996	3773.58	466.878	325.481	51.623
1997	3843.51	467.439	327.125	52.209
1998	3931.65	474.107	329.361	53.145
1999	4009.02	475.880	338.729	54.613
2000	4074.20	480.053	346.354	56.340
2001	4117.52	482.551	348.091	57.269
2002	4202.59	485.779	346.172	57.317
Percent change (1995–2002)	14%	5%	9%	13%

Source: European Union Road Federation (2005)

In the United Kingdom, the number of trips per person per year by mode and purpose varies considerably, but a pattern emerges that favours the use of personal automobiles for most forms of mobility. Table 4.6 demonstrates the pervasiveness of the automobile in personal mobility in the United Kingdom, with an average of 21 holidays or day trips per year being taken either as a driver or passenger.

In the United States, personal vehicles are used for more than 80% of long-distance domestic trips and over 90% of commuting trips to work, and the 2001 US National Household Travel Survey revealed that 90% of pleasure travel utilised personal automobiles (BTS, 2005c). As far back as 1996, 60% of American households already owned two vehicles, according to one survey (see Edmondson & Du, 1996). In the UK, that figure, as of 2002, was estimated to be only 22% (UK National Statistics, 2005), although two-thirds of all UK households own at least one car (European Commission, 2001). In 1998, over three-quarters (79%) of all passenger traffic in the EU is carried by some form of road transport.

Personal vehicles are used in many countries to access second home cottage areas (Weaver, 2005; see also Box 3.1). Cottage or cabins in North America, generally in wilder-

Table 4.6 Trips per person per year by mode and trip purpose, UK 1999/2001

	Bicycle	Car/van (as driver)	Car/van (as passenger)	Motorcycle	Other Private Vehicle
Commuting	6	89	18	2	1
Business	*	25	3	*	*
Education	1	3	18	*	3
Escort education	*	20	6	*	*
Shopping	2	81	49	*	*
Other escort	*	48	23	*	*
Other personal business	1	44	24	*	1
Visiting friends at home	2	50	43	*	*
Visiting friends elsewhere	*	12	14	*	*
Sport/ entertainment	1	25	22	*	1
Holidays/day trips	2	9	12	*	*
Other, including just walk	*	1	*	*	*
All purposes	15	407	232	2	6

Note: * No data available

Source: http://www.transtat.dft.gov.uk/tables/tsgb02/1/section1.htm#1.1

ness areas that are proximate to smaller communities that provide basic services such as food and petrol, are normally accessed via automobile as well. Dijst *et al.* (2005) note that personal automobiles are the dominant type of transport used to access second homes in the Netherlands and Germany, and Müller (2002a, 2002b) demonstrated empirically that proximity (and thus accessibility) played a key role in determining attractiveness of certain regions as weekend leisure zones in Sweden.

The use of personal transport for recreational purposes is also worthy of mention. Recent surges in sales of SUVs and light trucks might suggest that leisure activities and vehicle ownership have finally merged to the point where specific types of vehicles are being purchased with the intent of using them for specific recreational purposes, yet as Niemeier *et al.* (2001) point out, many owners utilise SUVs in much the same way as they would use regular passenger vehicles. See also Box 4.1 where Porsche Match production with tourism.

CYCLE TRANSPORT

Lumdson and Tolley (2001) make reference to UK General Household Survey research that suggests that Britons use a bicycle more for the purposes of leisure as opposed to a primary means of transport. What is interesting, however, is that this appears to be changing. The same survey suggests that more Britons are engaging with recreational cycling (Lumsdon & Tolley, 2001). What this means is that, as a mode of transport, the bicycle plays a key role in some tourism and recreational behaviour. Lumsdon *et al.* (2004) examined the NSCR, a 6000 km cycle trail that circles the North Sea, which was established in 1998 and officially launched in the year 2001. The project was funded by the European Union and local governments in Norway, Sweden, Denmark, Germany, the Netherlands, England and Scotland. The focus of the network is almost exclusively to generate multi-country touring:

> The tourism experience focuses on the North Sea, and an exploration of the rich maritime culture shared by the countries which border it, a theme which is recurrent throughout its entire length. These images are reinforced by a specific brand identity, which positions the route as a crossnational tourism offering where the use of transport is an integral part of the tourist experience. (Lumsdon *et al.* 2004: 13)

Using an intercept survey at key points along the route, the authors were able to profile the trip purpose of cyclists. Table 4.7 summarises the results of their survey. As shown, a significant number use the cycling route for recreational purposes, particularly for short trips.

Table 4.7 Purpose of journey (cycle trips), NSCR

	*Percentage**
Specific purpose (work, shopping)	8
Short, circular recreational trip (less than 3 hours total)	25
Short, out-and-back recreational trip	23
Day ride	10
Cycling short break (two or three days)	16
Other (including linear and circular touring holidays)	17

Source: Adapted from Lumsdon *et al.* (2004)

Note: * Numbers do not sum due to rounding.

BOX 4.1 Porsche AG weekend getaway: Matching production with tourism

In recent years, many auto manufacturers have sought to bridge the gap between the production process and their end customers. For example, BMW offers tours of its Spartanburg plant in the United States for potential Z4 owners, offering a glimpse of production lines and detailed presentations on product features. These are designed to enhance the experience of vehicle ownership and provide potential (and existing) customers with a value-added benefits and build relationships. It is, in effect, an attempt to build emotional ties to a brand (see Coles, 2004b).

Without question, Porsche automobiles straddle the border between luxury and performance and thus target a particularly exclusive corner of the automobile market. For many, the cost of outright ownership may be prohibitive, but one option available is to rent a Porsche for a weekend getaway. Porsche AG has taken this one step further by offering potential customers the chance to spend time with a product. Among other touring trips designed to acquaint customers with its product, Porsche offers a packaged, weekend trip from Ludwigsburg to Weyberhöfe in Germany, a package that includes a two-day test drive in a Porsche Boxster S (two persons per vehicle), one night's accommodation at the Schlosshotel Weyberhöfe Vital Resort, lunch on both days, a formal evening dinner, a complete tour pack with route map, and a Porsche Travel Club tour guide in German and English (other languages also available on request).

This example represents an interesting, and intentional, blending of lifestyle with tourism. In one sense, it is designed to offer potential buyers of a Porsche to have an extended test of the product before purchase. At the same time, however, it offers individuals who are not all likely to purchase a Porsche to experience an 'exclusive' vacation. It is one example where a particular mode of transport is used as the primary focus of a holiday or vacation.

Questions for consideration

1. Are there any other modes or types of transport that attempt to blend lifestyle with tourism?

2. How might this case study be used as an example of relationship marketing?

Management and marketing implications are apparent from a study such as this. As cycling often takes individuals or groups into areas where travel by motorised transport is otherwise less appealing, there is room to consider what the impact of a cycle route such as this may have on semi-rural areas (i.e. those areas in the near vicinity to larger urban areas). Lumsdon *et al.* (2004) suggest that marketing of the cycle route in urban and regional areas may bring added expenditures to smaller, fringe areas not normally included in day-trip itineraries using motorised transport:

> In overall terms, day visitation by the near market remains the most important seg-ment in terms of volume. The internal regional market is key to the success of this type of route. There is, therefore, a need to promote cycling on the route as a group activity, for example, to parties of family/friends living in residential areas on or near the NSCR. (Lumdson *et al.* 2004: 21)

In the UK, the 1996 National Cycling Strategy was launched with the intent of promot-ing the benefits of cycling. Immediately prior to this, Sustrans (a charity group) was awarded funding from the UK Millennium Commission (using lottery fund) to establish an NCN (Cope *et al.*, 2003). The network currently supports 8000 miles (or 14,400 km) of both cycling and walking routes, although this is projected to grow to approximately 18,000 km by 2005. According to Sustrans (www.sustrans.org.uk), almost one-third of the network is limited strictly to cycling and walking, and thus closed to vehicular traffic. The other two-thirds of the network shares smaller road networks that are open to vehicular traffic, as noted by Cope *et al.* (2003: 6):

> The NCN is a composite of over a thousand local sections ranging in length from a few metres to several kilometres. Each one is designed to be of benefit to local people, with on average two-thirds on minor or traffic-calmed roads and one-third on traffic-free paths. The sections are designed, built and signed by Sustrans and its part-ner bodies. There are over 400 active partners involved in the NCN project, including local authorities, countryside and utility bodies, landowners, central government, amenity groups and community groups.

The network was established both as a means to allow people the freedom to chose an alternative, and even sustainable, manner of transport to and from work, but also to encourage long cycle holidays and family rides that are shorter in duration. Many private companies affiliated with Sustrans offer full services for the traveller, including baggage care, accommodation booking along the route, and advice and tips on attractions. Some tours are even guided, thus bringing a wider package tour element to this mode of transport.

In discussing cycle tourism, an examination of the links between it and specific forms of tourism can be highlighted. For example, outdoor recreational activities and adventure tourism have become quite popular for many tourists. As such, examples such as bungy jumping and hiking highlight, once again, different motivations for undertaking such activities. With cycle tourism, some destinations utilise these outdoor activities for the purposes of reinvigorating tourism in the off-season. For example, in Whistler, British Columbia, off-road cycling (or 'mountain biking') is heavily promoted in the summer season as a recreational activity. In this particular case, tourists need not bring their own bike as many shops that rent skis during the peak winter season also rent mountain bikes. What this demonstrates is the further 'blurry distinction' (introduced in Chapter 1) of the mode of transport also being the central or core activity, but it also shows how a particular mode of transport, in this case a personal mode of transport, can be used to 'counter' seasonal affects of tourist flows. Box 4.2 provides another example of how cycle tourism is used to promote a destination, but at the same time offers tourists a simple mode of transport designed to enhance the experience of a destination.

Box 4.2 Enhancing the experience and vehicle for promotion: The Central Otago Rail Trail, New Zealand (by Sam Hepburn)

Many rural regions have successfully resorted to rural tourism as a way to preserve the rural community lifestyle from the demise of unemployment, urbanisation and development (see, for examples, Butler *et al.*, 1998, Sharpley, 2004). In recent years the transformation of abandoned railway lines into recreational rail trails has become an increasingly popular mechanism for attracting tourist expenditures and mending rural economies. The Otago Central Rail Trail (http://www.centralotagorailtrail.co.nz/) in the Central Otago region of New Zealand is an excellent illustration of a contemporary success story.

Once a vital link between Dunedin and New Zealand's major goldmines, the Otago Central Branch railway line generated a steady flow of commerce and activity to a number of towns and communities throughout Central Otago. However, as time progressed so did the evolution of modern transport, ultimately resulting in the slow demise of this 150 km stretch of railway. By the year 1990 the railway was permanently closed (Central Otago Rail Trail, 2004). The New Zealand Department of Conservation realised the potential of the redundant railway and subsequently acquired the line in 1993 with the vision of developing a unique recreational facility for walkers, cyclists and horse riders (Department of Conservation, 2004). After funding was granted by the Otago Central Rail Trail Trust, and extensive preparation and upgrading of the deteriorating railway was completed, New Zealand's first rail trail was opened in the year 2000 (Central Otago Rail Trail, 2004).

The landscape that the Otago Central Rail Trail traverses cannot be seen from the highways, thus providing a chance for users to enjoy the distinctive Central Otago environment from an exclusive perspective. The rail trail, which is steeped in a sense of history and remoteness, has played a significant role in preserving an important part of Otago's heritage (Otago Rail Trail, 2004). The rail trail now attracts thousands of visitors to the region every year, and although the trail is open to the public free of charge, it acts as a catalyst for economic development for those destinations and attractions spread along the trail (Cycling Advocates Network, 2004).

In conjunction with the Department of Conservation, Ross (1996) explored the types of people using the Otago Central Rail Trail and the activities the trail is used for. A total convenience sample of 47 New Zealand respondents was taken, with the majority being in large groups. Ross (1996) reported that the average age of those using the trail was 49 years, with 34% of the users indicating that they were between 60–69 years of age. The gender of the participants was skewed in favour of females, whom made up 62% of those interviewed.

According to Ross (1996), the vast majority (89%) of those surveyed revealed that they resided in the greater Otago region, with a total of 70% of these people indicating that they lived in Dunedin. The accommodation facilities most frequently used was private home followed by motel. A total of 40% of respondents indicated that they were not staying. Of those that stayed overnight, a high proportion specified that they were staying in either nearby towns such as Middlemarch or Hyde, which at the time of the survey were largely the only available accommodation options (Ross 1996). This has changed somewhat in recent years although anecdotal information provided by local residents and publicans near the rail trail suggests that accommodation provision is still a concern. Although the rail trail has benefited from significant exposure within New Zealand, and thus attracts a large number of residents, its international appeal is still rather limited, partially because of the lack of funding for the marketing of the trail, but also because the proportion of international visitors to the Central Otago region is small compared to domestic visitors.

Questions for consideration

1. How and why might one argue that all levels of government (i.e. local, regional, national) should be involved in the development and maintenance of attraction such as a rail trail?

> **2.** How might the concepts discussed in Chapter 2 benefit the development of a rail trail, particularly with respect to the strategic placement of ancillary attractions, accommodation and other features?

SUSTAINABLE GROUND TRANSPORT

The overall sustainability of ground transport necessarily means consideration not only for the actual environmental impacts, but also the degree to which transport can factor into the overall sustainable nature of tourism development. This includes the level of integration of transport within the tourism system and efficient (economically, as well as environmentally) movement of tourists. Much has been written in the transport literature on the benefits of adopting an overall sustainable approach to transport in general (e.g. Holden & Høyer, 2005). The economic benefits are paramount as costs can be reduced and efficiencies of network operations (whether goods-based or passenger-based) are established.

The degree of mobility offered by ground transport is staggering and relates to the concern over the sustainability of ground transport. In the EU, for example:

- the number of cars in the EU trebled over the 30 year period between 1970 and 2000, from 62.5 million to almost 175 million, with some 3 million cars per year growth;
- 10 hectares of land are covered by new roading each day;
- more than half of the oil consumed in the EU is from personal vehicles, and over two-thirds of the demand for oil is attributed to road transport in general (including transfer of goods);
- 84% of transport-related CO_2 emissions in Europe can be accounted for by general road transport. (Adapted from European Commission, 2001).

It is not entirely clear whether one particular type of ground transport is more sustainable than another type because the environment (and operating procedures, including the network served) varies considerably and may render direct comparison difficult. The use of private automobiles by tourists and recreationalists (either their own or rented) certainly can cause road congestion along main motorways and arterial routes in some countries. Alternatives include the provision of (light) rail in some urban areas or rail carriage in general to link regions and the promotion of other public services such as buses (Walton, 2003). Rail may be an option, but it shares many of the same physical environmental impacts with road travel (e.g. the use of non-renewable fuels and needs an infrastructure and network that is not unlike existing road networks), and varies depending on the level of integration that it has within national economies (see, for example, Plakhotnik *et al.*, 2005).

In Great Britain, Shaw and Farrington (2003: 108) note that rail travel is perhaps 'the most space and energy efficient way of moving large volumes of people and freight':

1. Rail may be up to four times more efficient per passenger kilometre than other types of ground transport, thus it has the ability to move more people at once as opposed to having one individual per personal vehicle on a roadway.

2. Increase in rail utilisation may mean less road traffic congestion, thus a more integrated means of transport between urban areas.

3. Rail is ostensibly safer than personally operated vehicles.

There has been some movement towards improving the environmental sustainability of rail transport. In October 2006, it was reported that Virgin Trains in the UK may have won concessions from the Treasury for the use of biofuels on the company's cross-country Penzance to Aberdeen franchise (Milmo, 2006). Between 1990 and 2005, VIA Rail Canada has cut greenhouse gas emissions by 11% and cut fuel consumption per passenger by almost 30% (Webwire, 2005). The World Bank (1996: 62) notes that a 'fully-loaded train requires only one-third of the energy per passenger-kilometre of a fully loaded car and little more than one-tenth as much as a fully loaded airplane...'. Electrically operated rail is, of course, an option, but this needs to be developed from sources that are renewable (e.g. wind, solar). As noted by the World Bank (1996), above anything else, the suitability of rail travel for passengers (including tourists) needs to be met from a satisfaction point of view before any ecologically friendly benefits are realised.

CHAPTER SUMMARY

The manifestation of coach transport in relation to networks, flows and the resultant user base is somewhat more straightforward. In this chapter, examples of coach transport being used on long excursions by tourists in particular destinations have been provided. In this sense, coach transport features as a particular mode of transport that is usually associated with intra-destination travel by pleasure travellers. At the same time, however, coach transport can be used to transport passengers from origin to destination, but as will be discussed later, this is proving to be less popular now that cheaper domestic airfares are being offered in many countries. In contrast to coach and rail transport, cycle transport is generally restricted to intra-destination or intra-attraction transport, largely because there are limits to which humans are able to supply transport under their own power (excluding, of course, professional athletes). To some extent, and as discussed above, cycle transport is somewhat more in line with recreational and leisure, although the example of the Otago Central Rail Trail shows that some tourist utilise bicycles both as a mode of transport and a recreational activity.

Although the general discussion in this chapter reflects the use of ground transport by tourists, the blurry distinctions introduced in Chapter 2 should be kept in mind with respect to the nature of transport provision not directly targeted at tourists. As such, urban and regional transport network planning also affects the scope of network and flows of tourists.

Box 4.3 Case study – The National Railroad Passenger Corporation (Amtrak)

In some form or another, various levels of government in the United States have been linked to the ownership and management of rail transport operations since the 19th century. The extent of this relationship has varied (i.e. outright owner-ship of operations and infrastructure or policy directives aimed at direct regulation) and depends largely on the level of government in question. In gen-eral, however, the Federal branch of government takes more of an interest in ensuring the overall viability of rail as a major mode of transport that is inher-ently tied to the wider manufacturing sector and is thus a significant means by which goods and raw materials are transported.

It is important to note, however, that the intense interest in rail operations by the United States is not unusual. As recently as June 2004, a secret report sur-faced that suggested the British government is considering renationalising the rail industry in the face of increasing burdens (by some estimates, £3 billion a year) on the taxpayers to subsidise the industry (The Express, 2004). Page (1993) sug-gests that when government has a direct ownership role in rail transport, the resulting organisation charged with its management is characterised by:

- a close relationship with central government due to state ownership and the provision of operating subsidies;
- periodic changes in the political will of national governments to subsidise rail;
- constraints on investment planning and a forward-looking approach to development;
- a lack of management freedom prior to the 1980s;
- an inability to prevent loss of market share in rail travel in the 1970s and early 1980s. (Adapted from Page 1999: 84)

Although Page's example is somewhat specific to the European Union, a number of his listed characteristics can be applied to the role the US Federal Government has played in the corporate lifespan of the National Railroad Passenger Corporation, better known as Amtrak. Amtrak has been a formidable force in linking many places in the country. From a passenger operations perspective, it is an excellent example of the role of government in the provision of rail transport for the general population. It also serves to demonstrate the turbulent nature of rail transport provision, especially with regard to shifts in management structure and focus, combined with increasing competition from other modes of transport.

Amtrak services more than 24 million passengers a year in 46 states, utilising some 22,000 miles of track, which are mostly owned by independent freight railroad companies (Hoover's Company Profiles, 2004). Established in 1971 and funded by the US Federal government (a relationship that continues to this day), the 1990s were a turbulent time for the company, as it faced numerous external threats (Hoover's Company Profiles, 2004), including:

- flooding in the US Midwest;
- significant competition from airlines able to offer cheaper airfares; and
- ongoing safety concerns following severe, high-profile accidents.

Revamped strategic plans, including one that would result in the complete privatisation of the company in 2002, have generally been unsuccessful due to managerial 'bloat' along with declining revenues and, by extension, shrinking cash reserves. A bailout of US$300 million in 2002, following years of increased costs but shrinking revenues, helped, but it was the subsequent managerial reforms that allowed the railroad to cut costs and increase efficiencies. As of 2004, the company says it still needs significant amounts of funding to stay alive, but this is something the government is increasingly becoming wary of.

In July 2004, the US House Appropriations Committee approved an $89.9 billion fiscal 2005 spending package for the Transportation and Treasury departments, which included $900 million for Amtrak, a $300 million cut from 2004 funding, which was considerably less than the $1.8 billion Amtrak officials have sought. Ernest Istook Jr, a Republican congressman from Oklahoma and chairman of the House Appropriations transportation and treasury subcommittee, pointed to the problems with ongoing Federal government support for Amtrak:

> The administration believes, and I agree, that realistic Amtrak reform language must be enacted before we start putting more money into [the service], that includes acceptance of [financial] responsibility at the state and local level, which they have not been willing to have… Unfortunately, we have too many places in the country that say 'we want Amtrak, but we don't want to pay for it, we want Uncle Sam to be the only one to pay for it. (The Bond Buyer, 2004)

The company is facing heavy capital expenditures in the near future, not the least of which include replacing older carriage cars and locomotives and substantial infrastructure repairs and maintenance. To date, the government has injected some US$24 billion into the company (National Journal, 2004) and this has caused many to suggest several options to relieve the Government of the burden of running a massive transport operation such as Amtrak (see also www.saveamtrak.org):

- full privatisation, and at the same time encouraging competition;
- split the company into several companies (one of which would focus almost exclusively on passenger provision, while another would look after maintenance);
- transfer the operations of the company to individual states, although many states believe that the provision of interstate rail transport falls under Federal jurisdiction. (Adapted from National Journal 2004)

The future of Amtrak is very much in doubt, but it is unlikely any US Federal government will allow the company to completely cease operations. If anything, and not unlike many governments around the world, public transport is often heavily subsidised. In the US, the massive US interstate highway system and the airline industry receive substantial subsidies for operations, largely because the cornerstone of major economies is the ability to transport both people and goods.

Questions for consideration

1. Why do some people feel that a national rail network should be managed by the national government?

2. If a scenario of full privatisation is realised, how would competition be encouraged in such an environment? What would be the impacts for Amtrak in terms of operations?

SELF-REVIEW QUESTIONS

1. What are main types and key characteristics of ground transport in relation to tourism?

2. Why are some types of ground transport more likely to feature bi-directional flows than others?

3. Why can it be said that rail transport covers both national and local needs with respect to tourism-related transport?

ESSAY QUESTIONS

1. Why is it difficult to distinguish between tourist and non-tourist use of ground modes of transport?

2. Discuss the relationship between coach transport and touring holidays.

3. Give an argument outlining the importance of passenger rail transport in an age where relatively cheap air travel is available in most developed countries.

KEY SOURCES

Banister, D. (2002) *Transport Planning* (2nd edn). London: Spon Press.

Banister's book (although not really a textbook per se) is an excellent one generally outlining transport planning issues, including policy issues which inform plans. This book has a slight UK bias (despite the presence of a dedicated chapter on international experiences), but is nonetheless an excellent source.

Essex, S. (1994) Tourism. In R. Gibb (ed.) *The Channel Tunnel: A Geographical Perspective* (pp. 79–100). Chichester: John Wiley & Sons.

The entire Gibb volume provides an excellent overview of the development of the Tunnel itself. Essex's contribution is useful starting point, and should be read in conjunction with news items from early 2006 highlighting the financial problems the Tunnel is having. Readers should also consult news sites (e.g. Google News) for items on the Eurostar, particularly in relation to passenger loadings, markets and alliances with destinations/attractions such as EuroDisney.

Turton, B. and Black, W.R. (1998) Inter-urban transport. In B. Hoyle and R. Knowles (eds) *Modern Transport Geography* (2nd edn). Chichester: John Wiley and Sons.

The Hoyle and Knowles text is standard reading in many transport geography courses worldwide. In addition to Turton and Black's overview of inter-urban transport linkages, Page's chapter on tourism and transport is also worth a read as it encapsulates some of the salient issues.

CHAPTER 5:

MARINE TRANSPORT

LEARNING OBJECTIVES

After reading this chapter, you should be able to

1. Assess and evaluate the connections between globalisation and transnationalism and the global cruise industry.

2. Identify and describe various types of marine transport in the context of recreation and tourism.

3. Explain how certain marine tourism products are packaged and marketed.

4. Describe and critically assess models of market segments associated with cruise tourism.

5. Identify and evaluate the environmental impacts of marine transport.

INTRODUCTION

Much like the varying types of ground transport and their relationship to tourism, exploring the integration between marine transport and tourism also reveals significant variations and complexities. Marine transport, as a whole, can generally split between the transport of goods and the transport of people, which is, on the surface, not entirely dissimilar to a wider segmentation of the uses of ground and air transport (and, quite often, goods and people share transport in many situations). It also demonstrates many of the same blurry distinctions that were discussed in Chapter 2. Some island archipelagos rely on marine transport for the transfer of goods and people from island to island. Similarly, tourists often utilise various types (and forms) of marine transport for either functional reasons or directly integrated within a wider tourist experience.

The management of marine transport is difficult due the vast array of transport types that can be utilised for the purposes of recreation and/or tourism. In many cases, such as cruise tourism, international regulations may exist but are regularly seen to be lacking in the ability to be enforced because of lack of monitoring or policing at the international level. Much of this is related to environmental issues, but regulations relating to enterprise operations (business practices) are not overseen, generally, by international bodies to the extent where level playing fields exist (see Lester & Weeden, 2004, for example). Modern tourism is represented in marine environments in several ways, each of which is discussed in this chapter:

1. Cruise tourism, where worldwide growth in the past several decades has been substantial; growth post-11 September has also been significant, with the Cruise Line Industry Association reporting some 10.5 million cruise vacationers in 2005 (up 40% from 2001) (PE.com, 2005).

2. Functional marine transport mechanisms, such as ferries and other forms of propelled vehicles that move tourists (and non-tourists) from locality to locality.

3. Personal water transport, including powered and non-powered watercraft.

CRUISE TOURISM: RESORTS AT SEA

Mobility by sea is at the heart of human expansion. Early masted ships carried migrants and goods destined for newly colonised places. Prior to technological and economic realities that realised efficient, fast and (comparatively) cheap air travel, journeys were undertaken using ocean liners (in the tradition of *Titanic*). Such extended periods of holiday are comparatively fewer today, where it is not unusual for itineraries to begin and end at the same point, and often last little more than several weeks (Douglas & Douglas, 2001). 'Cruising', in other words, has become a commodity, and linkages to the predictability and routinisation of the experience have been made (Weaver, 2005). Douglas and Douglas (2001) outline a truncated history of cruise tourism:

- the initial offering of round-trip tickets from England to the Mediterranean onboard P&O ships in 1844;
- the first cruising yacht in operation in 1881 by the Oceanic Yachting Company in response to a British Medical Journal report suggesting that sea cruises provided significant health benefits;
- the first custom-built cruiseship, luxurious in ornamentation, launched in 1900 by the German company Hamburg–Amerika;

- the popularity of 'booze cruises' during the Prohibition Era in the United States in the 1930s;
- the conversion of migrant ships to cruise vessels to accommodate demand for cruising post-World War Two;
- the retrenchment of many line cruise companies during the 1950s and 1960s when air travel become more affordable, although this spawned increased popularity of package 'fly and sail' experiences.

As of early 2005, global cruise tourism shows few signs of stagnation, although annual and seasonal fluctuations continue to exist. The small Dutch island of Bonaire in the Caribbean reported an annual 6% growth in arrivals for the first six months of 2004 (Bonaire Department of Economic and Labour Affairs, 2004). The member companies of the CLIA (2005) carried 2,604,544 worldwide guests in the second quarter of 2004, representing an 11% increase over the same period the year previous.

Cruise tourism has, in recent years, also expanded internationally. The Indian cruise market has recently expanded to the point where the national government has opened up the sector to foreign direct investment (Express Travel and Tourism, 2005). In July 2006, Costa Cruises (part of Carnival) is scheduled to begin five-day cruises for Chinese travellers departing from Shanghai (Katz, 2006). Star Cruises, a Malaysian-based cruise line, has a number of large cruise ships visiting larger ports of call throughout Southeast Asia. Star Cruises ships are generally targeted at local Asia markets, but they also attract Australians, Britons and Americans (Gunderson, 2005).

The provision of cruise experiences is often measured in terms of the available berths, which is one way of describing total accommodation available on a ship. Berths can generally refer to individual bunks, and thus give a reasonably accurate indication of the total number of people that a ship is able to hold. As of 1 January 2004 there were 339 active ocean vessels plying water routes around the world, representing slightly less than 300,000 berths (Ebersold, 2004). The sizes of these vessels varies considerably, which can be somewhat explained on the basis of target market, ownership criteria and integration with other tourism attractions, destination and other forms of transport. Roughly one-third of ships have a total berth capacity of less than 500, and approximately 11% have berths of less than 100 (Ebersold, 2004). Larger ocean-going vessels often carry upwards of 2500–3500 people, whereas shorter, more coastal, cruise ships will carry far fewer and have nautical ranges which are considerably smaller (even though, by design, they may be able to traverse ocean waterways). Ships with fewer berths may also target a high-net income market segment, and thus offer more luxurious on-board amenities. As of 2003, the size of the global cruise market is said to be roughly 11 million to 12 million passengers (Ebersold, 2004: 2). Table 5.1 outlines the growth of worldwide cruise tourism over the course of the 1990s.

Table 5.1 Worldwide cruise demand (millions)

	North America	Europe	Rest of world	Total
1989	3.29	0.53	0.20	4.02
1993	4.48	0.88	0.25	5.61
1997	5.05	1.36	0.46	6.87
2000	6.88	1.95	0.78	9.61

Source: Adapted from WTO 2003a, based on data from CLIA (US and Canada), Passenger Shipping Association (Europe), GP Wild Ltd International (Rest of world)

Cruise demand is greatest in North America, and this is one of the more obvious reasons why regions such as the Caribbean have witnessed substantial cruise tourism growth over the past few decades (Wood, 2004a) (Table 5.1). Cruise companies have even diversified the products and ports/destinations served when targeting the North American market, with some companies (e.g. Cruise West – www.cruisewest.com) offering cruises up the west coast of the continent to Alaska and several tour companies in the United States marketing package tours that feature cruises to Antarctica. Although the cruise tourism product can be said to have been globalised as a consequence of substantial international demand for international cruise experiences (Wood, 2004a, 2004b), the North American market is currently served by three major companies (Carnival Corporation/Carnival PLC, Royal Caribbean Cruises, and Star Cruises), which account for over 80% of the cruise market capacity. These companies are simply keeping up with demand, and this demand has been fuelled by the desire of many North Americans (primarily United States residents, which account for 95% of all cruise passengers from North America) to holiday close to home rather than travel long distance (Travel and Tourism Analyst, 2004).

CRUISE MARKETS AND PRODUCTS

Cruise tourism has witnessed remarkable growth over the past few decades. The Caribbean, long popular with the North American market, carries the highest berth capacity (well over half) (Table 5.1), although there is significant growth in the European/Mediterranean area, Alaska, the trans-Panama canal route and Mexico (Dwyer & Forsyth, 1998; see also Ebersold, 2004). The largest source markets for cruise tourism are the United States and the United Kingdom; according to the UNWTO (WTO, 2003a) the United States alone makes up approximately three-quarters of the worldwide cruise market, although the main source markets in the United States are areas that are proximal to main base ports for many cruise companies (e.g. the Pacific, where Los Angeles and San Francisco serve as base ports, and the South Atlantic, as many cruise companies serving the Caribbean base ships in Florida). The city of New York, by its sheer size, also supplies a significant number of cruise tourists (WTO, 2003a). Post 11 September 2001, the popularity of cruise travel among United States residents waned slightly, largely due to fears of terrorism, although there are signs that this is improving (Conroy, 2004). As well, some peripheral countries that rely on US cruise tourists (such as New Zealand) have been worried that global unrest will result in US cruise tourists staying closer to home (New Zealand Herald, 2003).

BOX 5.1 Cruising with the cargo: A new segment in cruise tourism? (by Gregory S. Szarycz)

At present, the global cruise industry has approximately 40 new cruise ships on order worth around a total of $500 billion (Peisley, 2000: 12–13). The projection for the number of global annual cruisers is 16 million by 2009. This represents a 60% growth from 2000, an indication of a major boom in this sector of the tourism industry. In order to accommodate this growth the industry is building larger and fancier ships; Cunard recently celebrated the inauguration of the *Queen Mary 2*, the world's largest, most anticipated, and most expensive (at approximately US$800 million) ship ever constructed.

However, for a select group of discriminating travellers, opulent meals or movies on the deck with popcorn and a cocktail are not significant draw cards. Many former 'mass cruisers', dissatisfied with the artificiality and rigidity of these floating palaces, are turning to different 'kinds' of cruising experiences. One such alternative to the 'typical' cruise experience is travel aboard cargo ships. These 'freighter cruises' are essentially scheduled container vessels that have made available space for a limited number of passengers.

Freighters are one of the lifelines of the world economy; they acquire, transport and deliver a wide range of cargo across a global network of shipping routes, but of the 29,000 large ocean-going ships in the world, only about 1% carry both passengers and cargo. Freighters operate on tight schedules and, fundamentally, are not designed to accommodate tourist interest or demand. They are guided by where the cargo, not the passenger, needs to be. In terms of entertainment, there is not much social life, particularly on the freighters that take up to a maximum of 12 passengers. Nevertheless, although standardised, pre-packaged cruise holidays are still the norm, agencies booking freighter cruises are now appearing throughout the world, offering a variety of itineraries to match consumer interests.

The freighter market is a special niche serving those travellers with the time and temperament to sail long itineraries – anywhere from a few weeks to several months – and who don't mind doing without the amenities on a modern cruise ship. Passengers tend to be retired and well-travelled, but owing to trip length and usual lack of an on-board physician (on ships carrying fewer than 12 passengers), many vessels set an upper limit on passengers' ages – typically around 80 though sometimes as young as 75 – and passengers are required to present a doctor's note saying they are fit for this kind of travel.

So what motivates a tourist to choose a freighter over a luxury cruise ship, and where does this interest come from? Maybe it is bragging rights they are after – the trip nobody they know has taken. Or maybe cities at sea just remind them of cities at home, which is what they are looking to escape from. Passenger-carrying container vessels and their various itineraries appear to be in a favourable position from an overall consumer demand standpoint for many reasons. Preliminary investigations (Szarycz) into consumers' perceptions of freighter travel seems to indicate that travellers are actively seeking out freighters to experience 'real' ocean travel, providing novelty, adventure, challenge and insight about local peoples and their cultures in 'authentic' versus "over-touristified' (see Tucker, 2002) ports of call. Freighters also provide an ideal environment within which passengers can 'escape' from mass cruise tourism and the 'over-touristification' and commercialisation of the cruise experience.

Although it is very difficult to accurately define the size or scope of this market, all indicators show that it is expanding in size and becoming increasingly popular. In the United Kingdom, Germany and USA, an ever-increasing range of shipping companies are accepting paying passengers on board their ships. With rapidly growing numbers, this subculture may eventually represent a financially viable market in its own right.

Questions for consideration

1. How might an intermediary (such as a outbound tour operator) market these kinds of experiences?

2. It is noted above that the market segments targeted with these types of cruises are a 'subculture'. To what extent do you think other forms of cruise travel might be considered a subculture?

The UNWTO (WTO, 2003a) reports that the most rapid growth of cruise tourism was during the 1990s, when demand for international cruise trips grew at a cumulative rate of 7.9%, compared to 4.3% for overall world demand for international travel. Miller and Grazer (2002) note that the North American market alone has experienced an annual growth rate of over 8% since 1980. The CLIA, an organisation that represents companies that cover nearly all of the United States residents' marketable (i.e. influenced through advertising and promotion) cruise itineraries, claim that cruise tourism will be worth some US$85 billion in 2007 (Douglas & Douglas, 2004). The popularity of cruising certainly explains these increases, but the actual size of ships, which has increased as well, can also explain this growth. As Wood (2000) notes, economies of scale (i.e. larger ships = greater economies of scale) have translated into substantial profits for many cruise companies:

> These huge vessels have spawned a new category 'Post-Panama' ships because they are too big to go through the Panama Canal. Upping the ante further, Cunard has announced its project Queen Mary, to build the 'grandest and largest liner ever built' (Goodman, 1998). Further, in the wings although some doubt whether they will ever be realised are long-standing plans for the 250,000 ton America World City, which with three skyscraper-type towers on an aircraft carrier-style hull, would carry 6,200 passengers. (Wood, 2000: 349)

Traveltrade.com monitors the construction of new ships and reported several 'newbuilds' in January 2005 that feature massive size and berths (Table 5.2). Many of these newbuilds are of such a size that it raises issues of port suitability. Many are simply so big that manoeuvrability becomes difficult in ports that may have previously played host to numerous smaller ships. With bigger ships, however, cruise companies can reduce the cost per berth and, at the same time, maximise revenue per berth, but for some destinations larger ships may mean fewer ports of call. Some ports therefore have few options available to them when it comes to ensuring cruise lines continue to call. One is to ensure that harbour re-developments are adequate. Barbados, for example, has in recent years upgraded its harbour area, including the dredging of the inner basin to accommodate larger passenger and cargo ships (see www.barbadosport.com). However, with renewed dredging comes the problem of what to do with the material that is brought to the surface. In some instances, land reclamation can be undertaken, as in Barbados, or may simply be stored, as in the case of Tampa Bay, Florida. Even the regular maintenance of ship channels, often requiring dredging, is a challenge and a significant cost, both financially and in terms of adequate disposal of dredge materials.

Table 5.2 Cruise line newbuilds

Cruise line	Ship name	Tonnage	Berths	Launch date
NCL America	Pride of America	72,000	2000	June 2005
Carnival	Carnival Liberty	110,000	2974	July 2005
Princess	Crown Princess	113,000	3100	May 2006
Costa	Costa Concordia	112,000	3000	June 2006
Royal Caribbean	Freedom of the Sea	158,000	3600	April 2006

Source: From www.traveltrade.com (accessed 12 January 2005)

THE CRUISE TOURISM 'PRODUCT': EVOLUTION

Weaver (2005) applies the notion of 'McDonaldization' (where the core principles of production can be found in the elements of efficiency, calculability, predictability, control and the "irrationality of rationality') to the cruise experience. Interestingly, Weaver (2005) argues that the cruise experience exhibits characteristics that are more reminiscent of post-Fordist principles of customisation of experiences. Put another way, where Coles (2004a) notes the post-Fordist tendencies of resort production/consumption relationship in the Caribbean, Weaver (2005) points out that cruise companies recognise the shifts in their target markets worldwide:

> Indeed, some cruiseship companies own ships that are built to serve certain national markets. One owned by P&O Cruises, Oriana, can accommodate over 1,800 tourists and caters mostly to the British market. On board, the breakfast buffet features a number of popular British "delicacies": kippers, smoked haddock, and baked beans. That afternoon tea is served on board reinforces the ship's British extraction (Scull, 1996). The dinner menu features quite essentially British fare such as steak and kidney pie and roast beef. (Weaver, 2005: 359)

This represents, in many ways, the maturation of the cruise industry such that the common view of homogenous ships offering banal on-board experiences to similar groups of passengers may perhaps be outdated. Notwithstanding the question of whether cruise tourism is sold as a product or an experience (see Chapter 8 for a more detailed discussion of the complexity of the transport 'product'), in the past decade, cruise companies have certainly diversified the list and range of components that comprise the package or experience on offer, thus resulting in luxury-branded cruise options as well as budget cruises with fewer amenities. Wood (2004b) notes that three broad markets can be identified based on these variations of amenities: mass, niche and luxury. Ward (2005) notes that the current 'mass market product' that characterises modern cruise tourism can be divided into three categories (cf. Box 5.2):

1. The 'budget' segment: characterised by small companies (which may be operationally smaller but may also be tied to other distribution chains such as agencies or wholesalers) and the use of older ships.

2. The 'contemporary' segment: manifested in most major cruise lines (e.g. Carnival, Royal Caribbean, Disney, Star Cruises) and, as such, target 'the hedonistic orientation of the aging baby boomers'. Ships in use are generally larger and allow for significant economies of scale, and are thus akin to being described as 'floating destinations' or 'floating resorts' (Teye & Leclerc, 1998: 155) because of the wide variety of amenities, attractions and services on board.

3. The 'premium' segment: not unlike the contemporary segment in terms of operations (e.g. ships and perhaps price), but more or less targeted toward upscale markets (and perhaps older demographics) and may feature specific destinations in line with the demands of these markets [see Box 5.2].

Classifying cruise line companies and the products on offer is difficult because of the wide range of products, destinations (ports of call), market access and size of ships. The Berlitz Guide (www.berlitzpublishing.com; or internet search: Berlitz Guide), however, offers a classification system that is recognised as an industry standard. In examining variables such as the ship itself, the accommodation provided, the cuisine, the service and the programme (itinerary plus on-board activities), Berlitz assigns a star rating system to cruises, and not necessarily to actual ships or companies. The star system is not unlike standard accommodation ranking systems such as those used in North America and Europe: 5 star plus represents the highest rating, with 1 star the lowest. Related to this is Berlitz's lifestyle classification, which encompasses the following groups:

1. Standard: generally in the lower proportion of the price scale for cruises.

2. Premium: the middle band of the price scale, featuring better amenities such as accommodation and restaurants.

3. Luxury: the upper echelon of the price scale where amenities are considered to be excellent.

Niche markets

Beyond the mass market cruise profile is a more niche orientation in some of the product offerings. Wood (2004b) also outlines several of these, including multiple-mast ships (under sail), specialty destinations (e.g. Antarctica, the Arctic [see Marsh & Staple, 1995]), and other adventure-related products. Cartwright and Baird (1999) also identify 'freighter cruises' (sometimes referred to as merchant marine cruises; see Box 5.1), which use cargo vessels outfitted with a small number of passenger cabins, and 'river cruises' such as those along the Nile, the Danube and the Rhine, which generally do not feature accommodation on board.

The niche market segment, comprised of substantially more and often luxurious product components, is comparatively smaller and may perhaps not benefit from substantial economies of scale, but can be profitable nonetheless. The luxury cruise market is broadly characterised by higher prices as well as premium on-board products and services. To some extent, there is a blurring of this market with the 'premium' segment in Wood's (2004b) characterisation as many 'mass market' cruise ships may offer superior accommodations or additional hospitality at an extra cost. As a result, some of the larger cruise operations feature tiered service levels available on single ships. Each tier is targeted towards a distinct market that carries differing demand elasticities of price and service levels. Star Cruises in

BOX 5.2 *The World*: Luxury cruise ship or mobile residence?

One example of the luxury cruise segment is *The World*, introduced by ResidenSea in 2002 (Sometimes referred to as *The World of ResidenSea*). More of a permanent 'apartment complex' (Wood, 2004b: 136) than a cruise ship, it is marketed as floating resort complex where both permanent and temporary accommodation is available, thus calling into question whether the permanent residents on board are indeed 'tourists' (see Chapter 1). Services include, for example, a medical centre, concierge service, an art gallery and theatre, eight options for dining, a night club, a tennis court and conference rooms (see www.aboardtheworld.com). Apartment options, for purchase, include studios (beginning at US$725,000), one-bedroom units (starting at US$1,247,500) to three-bedroom units (starting at US$4,170,000). The target market to which *The World* is directed is clearly upscale, but the experience on offer is one that the company asserts is unlike other forms of cruise tourism. From the website, the company suggests: 'Because *The World* is our home – one of our holiday homes – it is a far cry from what people might imagine. It should be clear that *The World* differs from a cruise ship. Our travelers prefer an alternative to ordinary cruising or yachting.'

Questions for Consideration

1. To what extent could it be argued that *The World* is not, in fact, a cruise ship, but merely a permanent residence and/or second home?

2. Given the itinerary for *The World* (available on the website noted above), what similarities and differences are apparent when comparing to the itinerary/networks for major cruise lines such as, for example, Carnival?

India, for example, launched a specialised luxury cruise offering from Mumbai to Goa in 2005, designed specifically to target the domestic Indian market by offering distinct Indian features on board, including restaurants and brief stops to historic landmarks. Interestingly, and not generally in keeping with the typical profile of luxury cruise products, Star Cruises is marketing its product to multiple market segments, including middle-class and upper-class passengers as well as families (Hindustan Times, 2005).

A more recent 'step' along the evolutionary line of cruise provision is the close mirroring of some cruise operations to no frills air travel. easyCruise, launched in March 2005, operates a 'low-frills' service, with itineraries that are flexible and on-board facilities generally limited to one or two bars, a café, a shop and a gym. Compare this with some mass cruise experiences, which may feature multiple instances of each type of amenity. The target market for this kind of operation is, according to the company, 'independently minded' people in their 20s, 30s and 40s. Virgin also announced in October 2005 that it also plans to enter the cruise market.

TRANSNATIONAL CORPORATE REALITIES OF GLOBAL CRUISE LINES

The effect of globalisation on the status of multinational corporations such as those involved in cruise tourism has been noted (e.g. Wood, 2000, 2004a, 2004b). Wood (2004b) argues that cruise ships are themselves 'deterritorialised' on the basis of their transient routes and itineraries, but also with respect to ownership and substantive control of operations. The on-board experience is deterritorialised because, according to Wood (2004b: 140),

> [t]he mass-market cruise ship experience is deliberately manufactured to be detached from the region in which it cruises... Port information sessions are almost always exclusively about what things to buy and which shops to spend your money in. Entertainment is 'Las Vegas' style. Food and drink on board do not reflect the cruise region. By and large, generally with only minor exceptions, the regional culture where cruising takes places is almost invisible in terms of what is experienced onboard the mass market ships.

Further, ships are deterritorialised because of the multinational labour force on board. For most ships, the division of labour not only follows traditional descriptions (officers, staff, crew), but also the ethnicity within each is remarkable, as Wood (2000: 353) notes:

> On most vessels there is a clear ethnic cast to this hierarchy: Norwegian or Italian officers, Western European and North American staff (mainly the cruise director's, hotel management, entertainment, and business staff), and Asian, Caribbean, and Eastern European crew. While Eastern Europeans (with a significant proportion coming from areas of ethnic conflict) appear to occupy a somewhat ambiguous role in this hierarchy, there tends to be a quite sharp line between the first two groups and the third. Among 'the most common mistakes made', a cruiseship employment guide informs its intended audience of North Americans and Europeans, is 'applying for a job that is traditionally held by a Filipino or by someone from other "third world" countries' (Landon 1997: 48).

That cruise companies operate in multiple markets and service multiple international destinations is appropriate given that their companies themselves, and the organisational behaviour, are often situated in multiple jurisdictions (e.g. country of ship's registration, sales base, labour force). Unlike airlines, there is often little impetus to have a 'flag cruise company', and thus many companies have little formal association with any one single country. Not surprisingly, cruise companies are often incorporated in countries where corporate tax is minimal, if present at all (Wood, 2004b), or have relatively lenient employment laws and regulations (Carnival, for example, is registered in Panama, and Royal Caribbean is registered in Liberia, although both companies have offices in the United States). Unlike the airline industry, under the open-registry, or flag-of-convenience, structure of the global cruise industry, ships registered in countries where employment, health and safety laws are either absent or unchecked can hire crews at low wages (below minimum wage in the country of their home office [Wood, 2000]) and fewer benefits. As Wood (2004b) points out: 'The flag of convenience regime is at the heart of the cruise industry's economic competitiveness.' Cruise companies argue that such regimes allow for the negotiation of wage and compensation packages that are truly global, thus reflecting the global nature of the operation or enterprise, yet some, including Wood (2000: 351), recognise this as the manifestation of the very policies of globalisation as espoused by institutions such as the International Monetary Fund and the World Bank.

It is probably more accurate to speak of transnational corporate structures as there is considerable amount of horizontal international movement between large and small corporate enterprises (Table 5.3). Smaller companies and fleets are regularly purchased by larger companies, assets are sold or traded and networks, at which one time may have been concentrated within certain geographic area, are expanded. The nature of cruise company operations (largely in international waters) often means that national laws do not apply, and international regulations and governance are scarce. This becomes particularly significant in the context of waste dumping, which is discussed later in the chapter.

Table 5.3 Main companies/cruise lines (as at January 2002)

Group/cruise company	Number of ships	Gross tonnage	Berths
Carnival	46	2,323,110	61,597
RCC	23	1,841,244	47,184
Star Cruises Group	19	905,254	25,210
P&O Princess	18	1,083,062	27,420

Source: Adapted from WTO (2003a)

Interestingly, worldwide demand is being satisfied by a small number of companies. As Ebersold (2004: 1) notes, almost two-thirds of the global berth capacity is in the hands of ten companies. According to the WTO (2003a), four corporate groups of companies (multiple brands under a single corporate umbrella) hold nearly 80% of the world's supply of cruise berths: Carnival, Royal Caribbean Cruises, P&O Princess and Star Cruises. With market access and penetration being quite important, one assumption is that cruise companies are generally in competition with other companies. While this is true, most are, more accurately, in competition with ground tourist resorts as cruise ships are structured and operated to be not dissimilar to all-inclusive resorts: many contain casinos, shopping malls, libraries, meeting rooms, discos and nightclubs, jogging tracks, saunas, gyms, movies theatres, bars, a variety of restaurants (some with 24-hour service), beauty salons and a host of other activities designed to entertain guests. These larger ships, often classified as VLCV, then, are not only a means of transport *to* a destination, but are in many ways *the* destination itself. As well, all-inclusive cruise tourism and all-inclusive resort (or enclave) tourism feature heavy integration among all aspects of the production cycle. For example, not unlike enclave tourism, cruise companies will offer packages that feature integration and dependencies with other commercial operations, including airlines, hotels and various activities at certain ports of call. Such integration is also a feature of ground tourist resort products (Cartwright & Baird, 1999; Laws, 1997).

PROFILING CRUISE TOURISTS

While the issue of demand for various modes of transport was assessed in Chapter 2, there are several aspects that need to be considered when exploring the level of demand for cruise experiences. As cruise tourism is one of the fastest growing forms of leisure-based travel (WTO, 2003a), accurate forecasts are important for a variety of reasons:

1. Knowledge of future demand enables corporations to make more profitable decisions to be made, such as routing, port planning, alliances, ship development (and decommissioning) and corporate structure.

2. Coastal and island destinations that court cruise tourism are better able to plan for future (re)development of ports and their associated activities.

3. Horizontal corporate entities that are allied with cruise corporations will be able to act strategically in terms of the benefits and potential pitfalls of entering into alliances with these companies.

The most logical indicator of demand for cruise vacations would be the proportion of cruise vacations taken against the overall demand for leisure trips within a certain market. However, there are a number of problems with this:

1. Demand for leisure travel within a source market can be varied, and thus include international, national and local travel, thus complicating which market is of prime importance for assessing demand; with cruise lines offering multiple products, demand will need to be forecasted for each of these separate segments (i.e. budget travellers, luxury travellers, educational cruises, etc.).

2. Because cruise lines operate out of a base port (i.e. Miami for most major Caribbean lines), the cost of transport to and from Miami for those markets outside of the city itself must be built into the demand model. For instance, the relative cost of transport to and

from Miami for tourists on the northern part of the Eastern Seaboard (e.g. the Carolinas) must be factored into the demand for cruise vacations from this segment (especially given that many airlines that enter into arrangements to supply cruise companies with passengers through alliances often themselves embark upon marketing in the same market segments that encourage long-haul or short-haul travel that rivals cruise vacations).

Relating to demand, the motivation for cruise travel is complex, and thus not unlike the motivation for leisure or pleasure travel in general. Given that cruise tourism is a rapidly expanding segment of international tourism, considerable effort has been devoted to profiling potential cruise tourists, and indeed learning about those who already partake in this form of holiday or vacation. Indeed, Wood (2000: 349) notes that 'destinational cruising', where the ports of call often play a key role in potential customers' decisions, demonstrates how transport can assist with the promotion of destinations. This represents somewhat of a paradox: while cruise companies are certainly interested in generating as much on-board sales as possible, they are inextricably bound to offer itineraries that entice people to take a cruise. Not surprisingly, then, many cruise companies are quite interested in learning more about their markets (see, for example, Table 5.4).

Table 5.4 Selected demographic characteristics of Caribbean cruise tourists

		Percentage
Country of residence	United States	82.3
	Canada	8.5
	United Kingdom	5.5
	Others	3.7
Age	Under 25 years	17.6
	26–35 years	13.1
	36–45 years	18.7
	46–55 years	21.2
	56–65 years	16.7
	66 years and older	12.8
Annual income	Less than $25,000	6.7
(US$)	$25,000–$40,000	16
	$40,000–$60,000	17.3
	$60,000–$75,000	13.9
	$75,000–$100,000	22.6
	Over $100,000	23.4
Marital status	Single	17.2
	Married	76.3
	Widow(er)	3.2
	Divorced	3.3

Source: Adapted from WTO (2003a) based on data from Florida–Caribbean Cruise Association (2001)

Note: Not all figures total 100% due to rounding.

Market segmentation (often in relation to understanding motivation) in cruise tourism has been undertaken in numerous studies both by private associations and companies as well as scholars with a general interest in marketing and tourism. The CLIA (1995–2001), for example, identifies six segments of the United States and Canada cruise market:

- 'Restless baby boomers' – eager for novel experiences; represents approximately 33% of the total cruise market, with nearly 1 in 6 being first-time cruisers (WTO, 2003a).
- 'Baby boomer enthusiasts' – familiar with the potential cruise experience and use it as a break from routine; represents approximately 20% of the total cruise market with nearly half being first-timers.
- 'Lovers of luxury' – demand for high-cost, high-quality cruise experiences and opt for more specialised cruise operators; as expected, they only make up 14% of the total cruise market.
- 'Demanding buyers' – search for the optimal price-quality ratios, so discounts and specials are taken advantage of; represent 16% of the total cruise market; high degree of repeat customers (80%).
- 'Explorers' – high degree of travel experience, and thus perhaps more interested in the destinations along the itinerary; represent 11% of the market.
- 'Boat enthusiasts' – important repeat market with strong familiarity of cruise experience; nearly 90% are repeat buyers, but only represents approximately 6% of the total cruise market.

The WTO (2003a) has identified several niche markets for cruise tourism. The younger segments of the baby boomer population in several Western countries, for example, may actively search for modestly priced vacation experiences that can include the family. Thus, many cruise companies are encouraging children to partake in cruises. Not only does this capture the market for families, but it also introduces younger generations to the cruise experience, and thus companies are hopeful that they can eventually capitalise on this market as repeat customers. In Europe, one of the largest growing niche markets for cruising is the senior citizens market, primarily because of the strong purchasing power they hold. Another market that may hold considerable potential is one that focuses specifically on the conference and incentive market. MICE tourism is a rapidly growing segment worldwide, with many destinations actively involved in the promotion of their facilities and logistical/organisational arrangements to large corporations or organisations. Many newer ocean vessels feature the necessary conveniences and attributes, such as large halls with state-of-the-art audio/visual facilities, dining options and on-board accommodation.

Another niche market segment is one which is closely allied with adventure cruises. Tauck Tours (based in the United States), for example, was one of the first to offer regular chartered exploration cruises to the Antarctic. Such cruises are often hosted by experts and scientists. This segment is small, but it often commands a high price point for the services on offer (ranging from 'hard adventure' to 'soft adventure', Ryan & Trauer, 2003) because of the unique environments that are included on the itineraries. A related form is the utilisation of specialist ships for cruising. Not unlike heritage railways and their renewed popularity (see Chapter 4), sailing ships, designed to replicate historic ships, offer experiences not unrelated to those programmed under the wider characterisation of heritage tourism (e.g. museums). Further some specialist ships (e.g. Windjammer cruises) allow for passenger participation in the operation of the vessel itself.

Qu and Ping (1999: 241, Table 2), in their study of 330 residents of Hong Kong, found that key motivators for cruising included 'escape from normal life', 'social gathering', and 'beautiful environment and scenery'. Duman and Mattila (2005) examined several affective factors associated with the perceived value of cruise tourism to passengers. Using a mail panel in the United States to identify previous cruise passengers, their sample of 1500 residents found that hedonistic elements of cruising generating the strongest perceptions of value. This reinforces the perception held by some that cruise ships are often seen as resorts at sea, and thus demonstrates why cruise tourism is often in direct competition with enclave tourism.

Cartwright and Baird (1999: 94–101) surveyed 100 'cruisers' and found that the typical cruise profile consisted of seven elements that demonstrated the propensity to undertake a cruise:

- *Partygoer*: takes a cruise for the purposes of 'activities and nightlife' on board a ship and be 'less concerned with the higher density often associated with high-activity, intensive-destination oriented vessels' (Cartwright & Baird, 1999: 97).
- *Relaxer*: generally less concerned with ports of call, and thus may elect to spend minimal amounts of time in port partaking in activities or shopping; more content with lounging; this group is 'likely to be less comfortable with high-density vessels where private space may be at a premium' (Cartwright & Baird, 1999: 97).
- *Enthusiast*: high repeat visitation as this group typifies the type of cruiser who may be 'addicted' to cruising. Cartwright and Baird (1999) found that UK cruise travellers were more likely to stick with one company brand, which underlines the importance that cruise companies put on encouraging repeat visitation. Cruise enthusiasts are likely to be 'very knowledgeable about the industry and the cruise companies. The companies use this enthusiasm through their various loyalty schemes to ensure that the cruisers feel a part of the family…' (Cartwright & Baird, 1999: 97).
- *Stroller*: this group encompasses individuals for whom formalities (such as dressing for dinner on board) are important. It is likely, however, that few individuals would travel for the purpose of dressing up or experiencing formalities, but Cartwright and Baird (1999) point out that the characteristics of this group can be found in other groups. To

a large extent, this group represents a means by which passengers escape routine (see also Krippendorf, 1987).

- *Seeker*: this group represents individuals for whom cruises are perhaps not ideal. A seeker is one who is interested in learning about the destination visited, including cultures and history. To a large extent, 'mass' cruise ship itineraries do not typically allow for large amounts of time in ports of call, largely because it is in their interest to have passengers spend money on board rather than in duty-free areas in ports of call. Seekers, then, are likely disproportionately small compared to the other types listed above.

- *Explorer*: the explorer category is not dissimilar to the seeker category in that numerically they are small. However, there is evidence to suggest that many companies are catering for this particular grouping, who are interested in visiting places 'few have seen before' (Cartwright & Baird, 1999: 100). It is this grouping that is targeted with such cruises as those featuring Antarctica, for example.

- *Dipper*: this group represents the vast majorities of cruisers, and is characterised by Cartwright and Baird (1999: 100) as follows:

    A little bit of culture, a small taste of a different lifestyle and then back to the welcoming cultural bubble of the ship. The dipper is truly the 'been there, seen that, experience this and bought the T shirt' person. The dipper will be, in the main, satisfied with an explanatory leaflet, a briefing from the port lecturer and a tour of the highlights.

EVALUATING THE ECONOMIC IMPACT OF CRUISE TOURISM

One of the most pertinent questions concerning cruise tourism is the overall economic impact of these activities. Dwyer and Forsyth (1998: 394) pose several questions that speak to these concerns:

    To what extent does foreign ownership of cruiseships limit the economic impacts of this form of tourism? Will these impacts be concentrated in stopover port areas or distributed nationally? What are the potential net benefits (as compared with economic impacts) of cruise tourism regionally and nationally?

Cruise passengers will normally spend on a variety of elements associated with their cruise experience, including air transport to and from the cruise base (where the ship is either stationed or where the cruise begins), food on board the ship itself (assuming that main meals are not covered in the overall cost of the cruise itself), and in port when a ship makes a temporary stop (Dwyer & Forsyth, 1998). Added to this is the money spent by the tourists before they leave for their holiday. For example, expenditures may be lodged on items needed for the trip itself, including clothing, toiletries, safety items (including medication) and assorted accessories.

Passenger expenditures exist within a delicate balance with the marine-related charges and fees levied against cruise companies. As a result, some will elect to utilise ports where such charges are comparatively small, thus reducing their operating costs. By extension, some ports may actively chase cruise business under the assumption that ships bring in tourists eager to spend money on souvenirs or other items. Finally, crew expenditures can include staff for ship maintenance, marketing of cruise offers and products in key market areas, and various taxes levied. In short, there are a variety of elements that need to be fully considered when examining the economic impact of cruise tourism, however many of these are difficult to calculate so a complete picture of that impact is often difficult, if not impossible, to ascertain. In many ways, this is not at all dissimilar to the challenges in measuring the economic impacts of tourism in general (e.g. Wagner, 1997)

What is perhaps most critical is the extent to which individual ports of call benefit from cruise tourism. Not all destinations benefit directly from cruise tourism and the expenditures of cruise tourists. Many cruise ships put into port for a limited time, often only a few hours (Dwyer & Forsythe, 1998). Thus, the window of opportunity for cruise tourists to generate a significantly beneficial economic impact on the destination is quite small, unlike those tourists who stay at least one night (often more) where expenditures are less concentrated and may be more significant. Further, the vast majority of cruise expenditures take place within duty-free areas or areas specifically controlled by local port authorities (Douglas & Douglas, 2004). Thus, the sale of goods (and prior distribution) may be highly regulated, and may only benefit certain retailers. Some port calls may include, however, short excursions outside of the port area, but because of the limited timeframe their economic impacts may be limited.

Overall, it is still not entirely clear whether cruise tourism is economically beneficial to many destinations. In the Bahamas, much like many other island states in the Caribbean, the number of cruise tourists as a proportion of all tourists is significant, but an important study by Wilkinson (1999) illuminated the very real financial impact of cruise tourists when compared to stayover tourists (i.e. those who stay at least one night). Wilkinson (1999) utilised real dollar expenditure data (calculated using the relative purchasing power of overall cruise tourist expenditures) and found that, between 1980 and 1996, visitor expenditures of cruise tourists declined almost 44%. Wilkinson (1999: 269) concluded that 'cruise visitors have little potential economic impact (in either current or real terms) on the average Bahamian compared to stayover visitors, even though the ratio between stayover and cruise real expenditures dropped from 17.5: 1 in 1980 to 13.2: 1 in 1996'. This is critical given that Wilkinson demonstrates that the average expenditure per person in port has risen. Wilkinson's study raises a important issue, and not just in the context of the Caribbean. Policy-makers and government officials need to carefully consider the real economic gain from cruise tourism not on its own, but rather in the context of whether ground FITs would be more lucrative for local business and the overall economy.

FUNCTIONAL MARINE TRANSPORT

The above section primarily considered those types of marine transport that acted, more or less, as both attractions and the mode of transport. However, there are examples of marine transport in which their function is more practical as opposed to operating for the purpose of providing entertainment or leisure value. Douglas and Douglas (2001: 331) refer to this kind of usage as 'line voyage', where the primary purpose is to travel from point A to point B.

Numerous examples of line voyages exist, although the geographic scope is somewhat more limited than it was in, for example, the 19th century when they were used for the transport of migrants from Europe to many areas around the world. Between the north and south islands of New Zealand several companies offer ferry services. One such company, the Interislander, offers a vehicle service where passengers board the vessel in their vehicle at Picton in the south island and disembark in Wellington in the North Island. This complements many touring holiday itineraries as it lessens the need to make separate arrangements for both islands. Other ferry services operate within the Mediterranean and provide a means by which regular labour mobility is facilitated between separate islands or an island and some mainland countries such as Italy and Greece. In all these cases, functional marine transport exists within a blurry distinction of transport-for-tourism and transport-and-tourism (Chapter 2), as both tourists and non-tourists can make use of these services.

In many areas of the world, watercraft are used as a functional means by which other tourism experiences are provided. Whale watching is an excellent example of this, although one of the key considerations is the impact on whales by the presence of marine vessels (e.g. Beach & Weinrich, 1989; Blane & Jaakson, 1995). Worldwide, concerns have been raised over the rapid over-crowding of near-shore waterways, which have significant impacts on whales and other aquatic species (Figure 5.1). Although visitor numbers to whale watch operations have been increasing, the concern is that increased interaction between boats and whales can have negative consequences for whale breeding and general welfare. In Boston harbour, the use of bigger and faster boats has certainly allowed more tourists, in greater comfort, to view whales, but accidents in which some whales are injured by these watercraft occur. Whale welfare concerns are not new. Japanese tour operators were concerned back in 1996 about growing numbers of tourists in boats and the impact this was having on local whale populations. In fact, some Japanese fishermen have turned to whale watching as an alternative economic activity in the summer, when available fishing stocks were traditionally low. In Iceland, it is hoped that promoting whale watching to domestic and foreign tourists may go some way to shifting attitudes towards the hunting of various species (Sykes, 1997) given international concerns over the Icelandic whaling industry. In September 1998, a boat carrying tourists from a whale watching expedition off the

Figure 5.1 Crowding at the attraction? A touring vessel in Milford Sound, New Zealand approaches a group of sea lions resting on shore

coast of Massachusetts struck a large Minke whale; tourists watched in horror as the body came to the surface in the wake of the boat (Bayles, 1998). The increase in popularity of whale watching means that consideration must be given to more appropriate forms of transport operations associated with the activity, including:

- the distance from boat to whales, with attention devoted to monitoring of chasing;
- the length of the trip/excursion (i.e. the length of time the mammals are exposed to watercraft);
- the most appropriate number of watercraft that can operate in an area (in effect, the carrying capacity);
- the type of watercraft used, and whether it is considered to be the most appropriate (i.e. number of engines, sound proofing);
- the behaviour of the boat in terms of direction, in order to minimise potential collisions and hardship caused to whales.

One answer to the problem of vessel proximity and presence is to completely remove the vessel from the experience. 'Wings over Whales' is a New Zealand-based company that offers flights for passengers to acquire a unique view of whales in their natural habitat. As their website (www.whales.co.nz) proclaims, passengers can observe several whales species, as well as other natural and man-made local features.

FERRY SERVICES

Also to be considered under the cruise product, but somewhat outside the traditional segmentation of cruise experiences based on amenities, is the role of hybrid cruise provision within multi-modal itineraries. Ferry systems are common in several parts of the world and offer transport for either practical reasons (i.e. to remote areas not easily accessible on land or air) or for other reasons (i.e. in order to allow people to sightsee whilst in transit on a ferry). Intra-destination ferries, such as the Brisbane CityCat service, serve as alternative means of mobility within a destination. In other cases, ferry services are critical for maintaining accessibility to more remote destinations. Along the north coast of British Columbia, Canada, the March 2006 sinking (in which two people lost their lives) of the *Queen of the North* (one of two ferries operated by BC Ferries to remote destinations along the coast) has meant that access to areas such as Prince Rupert has been cut in half (CBC, 2006).

Høyer (2000) notes that, in Norway, many Norwegians cross the Skagerak by ferry for leisure purposes. Preliminary data from Statistics Norway (2005) confirms the massive growth in marine transport for 2004. Car ferries can be used by FITs on self-directed holidays. One example is the Irish ferries systems offering ferry services to and from Britain under the slogan 'The Low Fares Ferry Company!' (http://www.irishferries.com/). In these examples, there is a marked departure from what is traditionally viewed as cruise tourism, but each feature amenities and distribution mechanisms that are not entirely dissimilar to more 'traditional' forms of cruising.

Challenges facing the ferry industry

Dunlop (2002) notes several challenges facing the ferry industry, particularly in Europe:

1. Safety issues will continue to be paramount after the Estonia disaster in the Baltic Sea in September 1994; since then regulations have been introduced governing stability and loadings, the cost of compliance to which, according to Dunlop (2002: 115), can be millions of pounds.

2. The creation of fixed land links can render some ferry operations financially unviable. After the Channel Tunnel opened, prices for ferry services across the Channel fell but some operations were closed permanently (Dunlop, 2002).

3. The price of fuel is also a consideration for the financial viability of ferry services, but because, as a mode of transport, these would compete with other modes where fuel costs are also increasing, this is perhaps less of a challenge.

4. The biggest challenge, according to Dunlop (2002: 115–116), is loss of duty-free sales with the introduction of common markets across Europe:

Perhaps even more bizarre is the position of a ship that trades from the UK into two separate Continental countries. P&O's Pride of Bilbao trades to both Spain and France. In its shop one area displays items that can only be sold when the ship is on the Spanish route and another area displays goods that can only be sold when the ship is on the French route. No wonder the passengers are baffled.

Another issue that has arisen from ferry services is the environmental impact. In parts of Finland (especially Tallin Bay; see Andersson & Eklund, 1999) and New Zealand (the Foveaux Strait), concerns over wave damage from ferry crossings have been raised. In the New Zealand case, environmentalists claim that wave action is destroying sensitive underwater environments and contributing to erosion onshore. Other concerns with ferry operations include the dredging of harbour areas at ferry ports, pollution from ferries themselves (and thus not unlike cruise ships), and the safety of smaller vessels operating near larger, high-speed ferries. In Hawai'i, ferry services compete with inter-island air connections. Hawai'i Superferry Inc. (www.Hawaiisuperferry.com) recently announced high-speed ferry services between Kaua'i, O'ahu, and Mau'i beginning in 2007, with full daily services to the Big Island (Hawai'i) in 2009. According to the website:

> For half the price of flying, families can travel interisland with the convenience of their cars. School groups, hula halau, canoe clubs, and cowboys can travel to events with their vans, equipment, canoes, and horses in their trailers. Neighbor island farmers can deliver fresh produce and cattle to market daily. Hawai'i Superferry will soon be linking islands and connecting families ... economically, comfortably and with care for our environment.

In the early stages, however, the service has not been without its problems. In April 2006, state senators expressed concern over the lack of public input into the endeavour and have threatened to withhold public funds earmarked for harbour redevelopments (Star Bulletin, 2006). The battle for inter-island accessibility, however, is likely just beginning. Mesa Air Group Incorporated, responsible for the highly successful regional carrier Mesa Air in the continental United States, announced in March 2006 that it would launch a subsidiary airline (called 'go!') that would fly between Honolulu and airports on Kaua'i, Mau'i, and the Big Island for introductory fares of US$39 one way (Song, 2006). Compare this with the published fares (as of April 2006) on the Hawai'i Superferry website that show advanced, web-based fares of US$42 (US$52 during peak periods such as weekends) and it is clear that traditional 'fare wars' in Hawai'i have extended beyond single modes of transport.

PERSONAL WATER TRANSPORT

Outside of the formal aspect of tourism, it is important to identify where transport may fit in with leisure and recreation experiences as there may in fact be instances of cross-over (where a wide variety of recreationalists and tourists may participate in similar activities). Several examples of personal water transport are used in the context of leisure environments where factors such as safety, environmental sustainability and the quality of the experience are paramount:

1. Non-powered personal watercraft: Canoes are prime examples of non-powered personal watercraft currently used in a variety of leisure environments. In North America, recreational canoeing is common in seasonal cottage areas (see Box 3.1), where families may spend several weekends and perhaps one or two weeks a year on annual holiday leave. The British Canoe Union (www.bcu.org.uk), for example, indicates on its website: 'In an activity such as canoeing, safety and keeping people safe is all about risk assessment and minimising the risks involved at all levels of participation.' Canoeing, as with other forms of non-powered personal water craft, may cause comparatively fewer ecological impacts, but codes of conduct may also be in place in some areas that limit the potential for damage. For example, the presence of canoes can cause significant harm to nesting waterfowl. Due to the popularity of canoeing in some areas where canoes and their occupants must share the waterways with other recreational and non-recreational users, efforts are usually made to make people aware of the implications of actions.

2. Powered personal watercraft: This type is generally associated with motorised (or propelled) watercraft that can be used individually or to tow a rider behind using some other recreation device (such as an inner-tube, boogie board or waterskis). These types of watercraft can be dangerous to the operator due to the often high speeds involved, although many federal and local jurisdictions around the world have introduced mandatory operator certificates that are provided after short courses. In both Missouri and Pennsylvania, for example, a boating licence is not required, but operators of personal watercraft are required to obtain Boating Safety Education Certificate, which applies to personal watercraft (see www.boat-ed.com). Some concern over the impact of powered personal watercraft on waterfowl has been raised, and Burger and Leonard (2000) found that, because of their speed and agility, these craft can actually be more harmful than larger boats.

Regardless of the mode of power, codes of conduct, some of which are enforceable through legal means, can be found in many areas where outdoor water recreation activities are popular. For example, the New South Wales (Australia) personal watercraft code of conduct (http://www.maritime.nsw.gov.au/pwc.html) covers a significant array of issues, including operational elements (such as restricted zones), safety considerations and noise considerations. What is not mentioned, however, is whether certain areas, such as wildlife refuges, are de facto 'off limits' to operators, but local jurisdictions may require these areas to be clearly marked.

Personal watercraft can also be used as the mode of transport along specific trails. For example, in the Chesapeake Bay area in the United States (http://www.baygateways.net/watertrailtools.cfm), marine trails have been developed where recreational users are allowed to utilise personal watercraft for transport purposes. However, allowing these recreational pursuits means installing proper regulation and policy structures such that environmental damage, as well as social problems due to noise and perhaps trespassing, are managed. One of the problems this brings is the jurisdiction under which these trails may fall. In some cases, water trails may be contained within local or national parks, and thus their management may be comparatively easy, yet when these trails cross numerous jurisdictions, the management of recreation activity, and indeed the kinds of transport utilised (which may be a mix of terrestrial and marine transport options), trail-wide regulations become difficult (and the corollary to this is air transport emissions, which is explained in Chapter 9).

SUSTAINABLE MARINE TRANSPORT

Many of the environmental concerns over marine transport centre around the damage done to sensitive aquatic environments as a result of the presence of watercraft of varying sizes. Waste management on cruise ships is a concern for both cruise companies and the local jurisdiction in which they operate. They are a concern for the company because the handling of waste needs to be done in a manner such that the cost does not significantly impact upon the overall profitability of the operation (in other words, barring any social responsibility felt by the firm, if the cost of dumping waste is economically beneficial, it is likely to happen). Of course, some, if not most, companies wish to preserve the very environments in which they operate and thus act as responsible corporate citizens. On-board waste management is a concern for destinations, of course, because of the damage caused by pollution emanating in varying forms from cruise ships. The wider question, however, is one of monitoring. Johnson (2002) notes that the International Convention for the Prevention of Pollution from Ships (the 'MARPOL Protocol') has six annexes, of which four relate directly to ship waste, suggesting that is an important problem. Ocean Conservancy, an environmental NGO, has characterised ship pollution as 'unsolicited contributions', and can include:

1. Oil pollution resulting from groundings, collisions or the pumping of bilge water. Bilge water is a 'mixture of salt water and leaks from water cooling circuits, fuel, lubricating oil, drainage, from sedimentation tanks and particles of dirt and soot' (WTO, 2003a: 167). Cruise companies are generally required (or at the very least encouraged) to filter out impurities in bilge water before discharge.

2. Sewage, which is generally diluted with smaller amounts of water compared to sewage on land (largely because of limitations of space on a ship), and is thus more concentrated;

can carry bacteria and diseases which can affect aquatic wildlife and enter the local food chain.

3. Grey water, or waste water not from sewage sources (sinks, drains, showers, etc.), which can include 'fecal coliform, food waste, oil and grease, detergents, shampoos, cleaners, pesticides, heavy metals, and, on some vessels, medical and dental wastes' (Ocean Conservancy, 2002: 15).

4. Hazardous wastes can come from on-board photo processing chemicals, paints, solvents, printer cartridges, etc. Ocean Conservancy notes (2002: 16) that 'a typical cruise ship with 3,000 passengers and crew generates approximately 15 gallons of photo processing chemicals, one and a half gallons of PERC and other chemicals, and one and a half gallons of paint waste per day'.

5. Ballast water, which is often 'taken on' by some ships in order to balance the overall weight distribution of the vessel. Where this can become problematic is when a ship takes on water for this purpose in one area and releases it in another area, thus potentially introducing non-native aquatic species. As Ocean Conservancy (2002: 3) outlined, the legal implications of waste management for cruise operations are perhaps not as stringent as needed:

> Although cruise ships generate a tremendous amount of waste from the thousands of people on board, they are not subject to the same wastewater regulations that govern municipalities of comparable size. Under the Clean Water Act, cities must treat their wastes, limit the amount of pollution they discharge, and monitor and report on discharges from sewage treatment facilities. Yet cruise ships are not required to obtain Clean Water Act discharge permits, nor to monitor or report on their discharges. Gray water from on-board laundries, galleys, baths, and showers is essentially unregulated. And even where regulations are in place, enforcement is lax.

6. Solid waste: for 3000 passengers and crew, approximately 50 tons of solid waste is generated over a week-long cruise (Ocean Conservancy, 2002).

7. Air pollution is not normally thought of when discussing the environmental impact of cruise ships, but the reality is that diesel engines used to propel a large ship generate significant amounts of sulphur dioxide, nitrogen oxide and particulates.

8. Physical damage to aquatic ecosystems results from all of the above combined. This can include damage from anchors and the direct removal of habitats from dredging (in order to provide sub-surface 'space' for ship manoeuvres). In the Caribbean, for example, anchor damage has been known to be responsible for significant damage to coral reefs (Allen, 1992). In Tasmania, Ellis *et al.* (2005) developed a model that considered a range of variables (including boat length, draft, tonnage and propeller size) to predict the impact of turbulence from a ship on benthic marine habitats.

Local jurisdictions are increasingly putting tighter controls over cruise activities. In 2001, for examples, the state of Alaska became the first US state to impose pollution controls on cruise ships (Schulkin, 2002). Despite the tighter controls, infractions still occur. The Surfrider Foundation, a not-for-profit foundation, lists several instances of fines and infractions from the past decade:

- From 1993 to 1998, cruise ships were involved in 87 confirmed cases of illegal discharges of oil, garbage, and hazardous wastes into United States waters, and have paid more than $30 million in fines. Some of these cases involved multiple incidents of illegal dumping that numbered in the hundreds over the six-year period.
- In 2001, Royal Caribbean admitted in court it had installed special piping to bypass pollution control devices and pleaded guilty to dumping toxic chemicals. Royal Caribbean was levied fines and penalties totalling $33.5 million to settle dumping complaints that occurred between 1994 and 1998.
- In April 2002, Carnival Corporation pleaded guilty to falsifying records to cover up pollution by six ships over several years. They were assessed an $18 million fine and were placed on probation.
- In July 2002, Norwegian Cruise Lines paid a $1 million fine and agreed to pay $500,000 to environmental organizations in Florida for falsifying Coast Guard records regarding discharge of oily waste and hazardous waste into the ocean.
- In September 2002, a fired Carnival Cruise Lines executive filed a 'whistle-blower' lawsuit, alleging a host of environmental violations, including toxic chemical dumping.
- In October 2002, Carnival Corp. disclosed that officers from one of its ships had been subpoenaed to testify before a federal grand jury in Alaska regarding a 40,000-gallon wastewater release.
- In April 2003, a lawsuit filed by Bluewater Network, Environmental Law Foundation, San Diego Baykeeper and Surfrider Foundation against Carnival Corp., Princess Cruise Lines, Royal Caribbean, Holland America and others for illegal discharge of ballast water into shoreline waters was settled after the cruise lines agreed to pay $75,000 to research alternative ballast water management technologies. Carnival Corp. admitted to breaking the law. (http://www.surfrider.org/a-z/cruise.asp; accessed 18 October 2005)

Of course, the size and distribution of cruise activity needs to be considered. In Alaska, despite increasing concern from environmental groups, the proportion of pollution from cruise ships versus other ocean-going vessels and port-bound tugs is small (Crenson,

2001). In the Caribbean, where cruise tourism is more popular, concerns about environmental stability continue to grow as the number of ships and passengers either grows or remains constant. Ocean Conservancy (2002: 4) has outlined a series of recommendations for the United States cruise industry, although some of their recommendations could be applied in other jurisdictions:

- *Reducing and regulating cruise ship discharges to improve water quality*: Cruise ship discharges should be regulated under U.S. environmental laws just like similar sources of pollution. Consequently, Congress and the Environmental Protection Agency (EPA) should repeal the exemption of gray water discharges under the Clean Water Act and ban the discharge of untreated sewage from cruise ships in U.S. waters. In addition, treated sewage and gray water should be discharged only while the vessel is underway at a minimum of six knots speed and 12 miles from shore. No cruise ship discharges should be permitted within marine protected areas or other sensitive and important ocean habitats such as marine sanctuaries, refuges, or parks. Finally, the EPA should establish water quality standards and allow states to establish no-discharge zones to protect special ocean sites.
- *Improving monitoring and inspection*: Cruise ship wastes should be comprehensively monitored, sampled, and reported. Congress should increase U.S. Coast Guard funding for more aerial surveillance and surprise inspections, and the EPA's expertise should be used to ensure proper monitoring and testing of discharges and pollution control equipment. The data gathered should be made available to the public so that citizens can make informed choices about cruise ship operations in their communities.
- *Strengthening enforcement mechanisms*: The U.S. Justice and State Departments should take measures to ensure that cruise ships flying foreign flags – as all cruise ships in U.S. waters currently do – are not permitted by their governments to violate U.S. environmental laws. Penalties and fines for violations should be increased to effectively deter scofflaws. Moreover, passengers, crew, and the public should be encouraged to report violations through educational materials and rewards.
- *Improving air quality controls*: The EPA should issue regulations to reduce emissions from cruise ship smokestacks in U.S. waters, and cruise ships should be encouraged to use local electrical grids when in port to reduce emissions. The EPA and the Coast Guard should also work with states to develop air-sampling programs. To reduce air emissions from ships worldwide, the United States should ratify Annex vi of the MARPOL Convention [which addresses air pollution].
- *Developing education and training programs*: Cruise line companies should educate their passengers and crews on complying with U.S. and international anti-pollution laws, and develop 'green' training and education programs for onshore operators and guides. Portside waste reception facilities should be assessed, and where inadequate,

145

they should be improved to accommodate the large amount of trash generated by cruise ships.

- *Improving research and development:* All new cruise ships should be designed with the latest pollution control equipment to eliminate waste discharges into the marine environment. The cruise line industry should continue to research and develop state-of-the-art waste processing technologies and design and implement sampling programs to demonstrate that discharges are not harming the marine environment.

There are regulatory bodies that oversee activities such as cruise waste management systems. The IMO (a subset agency of the United Nations) has established two committees: the MEP and the Maritime Safety groups. Both oversee the regulatory environment within which cruise (and freight) traffic operate. As well, the ICCL promotes environmental awareness among its members. In recent years, the ICCL has worked towards establishing a series of waste management practices for ships. Revision 2 to Standard E–01–01 outlines many of these practices (ICCL, 2003):

1. Photo Processing, Including X-Ray Development Fluid Waste: Member lines have agreed to minimize the discharge of silver into the marine environment through the use of best available technology that will reduce the silver content of the waste stream below levels specified by prevailing regulations.

2. Dry-Cleaning Waste Fluids and Contaminated Materials: Member lines have agreed to prevent the discharge of chlorinated dry-cleaning fluids, sludge, contaminated filter materials and other dry-cleaning waste byproducts into the environment.

3. Print Shop Waste Fluids: Member lines have agreed to prevent the discharge of hazardous wastes from printing materials (inks) and cleaning chemicals into the environment.

4. Photo Copying and Laser Printer Cartridges: Member lines have agreed to initiate procedures so as to maximize the return of photo copying and laser printer cartridges for recycling. In any event, these cartridges will be landed ashore.

5. Unused and Outdated Pharmaceuticals: Member lines have agreed to ensure that unused and/or outdated pharmaceuticals are effectively and safely disposed of in accordance with legal and environmental requirements.

6. Fluorescent and Mercury Vapor Lamp Bulbs: Member lines have agreed to prevent the release of mercury into the environment from spent fluorescent and mercury vapor lamps by assuring proper recycling or by using other acceptable means of disposal.

7. Batteries: Member lines have agreed to prevent the discharge of spent batteries into the marine environment.

8. Bilge and Oily Water Residues: Member lines have agreed to meet or exceed the international requirements for removing oil from bilge and wastewater prior to discharge.

9. Glass, Cardboard, Aluminum and Steel Cans: Member lines have agreed to eliminate, to the maximum extent possible, the disposal of MARPOL Annex V wastes into the marine environment. This will be achieved through improved reuse and recycling opportunities. They have further agreed that no waste will be discharged into the marine environment unless it has been properly processed and can be discharged in accordance with MARPOL and other prevailing requirements.

10. Incinerator Ash: Member lines have agreed to reduce the production of incinerator ash by minimizing the generation of waste and maximising recycling opportunities.

11. Graywater: Member lines have agreed that graywater will be discharged only while the ship is underway and proceeding at a speed of not less than 6 knots; that graywater will not be discharged in port and will not be discharged within 4 nautical miles from shore or such other distance as agreed to with authorities having jurisdiction or provided for by local law except in an emergency, or where geographically limited. Member lines have further agreed that the discharge of graywater will comply with all applicable laws and regulations.

12. Blackwater: ICCL members have agreed that all blackwater will be processed through a Marine Sanitation Device (MSD), certified in accordance with U.S. or international regulations, prior to discharge. Discharge will take place only when the ship is more than 4 miles from shore and when the ship is traveling at a speed of not less than 6 knots.

Part of the problem with the management of cruise ship pollution or waste is the short-term economic forecasting undertaken by many destinations. Following on from some of the economic considerations of cruise tourism discussed earlier, it is not difficult to see that some destinations, such as those in the Caribbean, desire cruise tourism in some form for the foreign exchange that it brings. It also can be one of the more significant forms of tourism in some destinations, as access by other means (particular air) is difficult and expensive. Cruise tourism, then, and in some situations may represent the 'easiest' and economically logical form of tourism. Although Johnson (2002) calls for greater integration of policy and management between cruise companies and destinations in order to preserve local environments and thus escape setting goals that merely aim for short-term economic gain, this is often difficult in fragile economies where the planning for the long term is simply not feasible because of rapid shifts in existing economies and unforeseen externalities. The issue, then, is the price of sustainability. NGOs such as Ocean Conservancy undoubtedly have excellent suggestions for the management of waste and pollution from cruise ships, but part of the problem is the degree to which cruise companies serve destinations where alternative forms of tourism may be difficult to establish for a variety of reasons (e.g. initial development costs, foreign direct investment incentives, taxation and incentives, labour, market saturation due to nearby competition).

CHAPTER SUMMARY

Cruise tourism may be considered to be prevalent in a tourism context, but personal watercraft numbers have been growing (particularly in North America) and thus represent an important consideration of transport in recreation and leisure. Segmenting markets for cruise tourism is complex but the more common means include demand estimation, availability of transport alternatives or based on sub-type (such as 'budget' or 'luxury'). Assessing demand is equally complex and may depend on proximity to suitable source markets. Where these source markets exist outside of main ports, the factoring of additional transport (often manifested in cruise packages) needs to be considered. The economic impact of cruise tourism raises issues of adjusted or real dollar measurements of cruise activities in ports of call, and begs the question as to the real value of cruise tourism where other types may (or may not) achieve stronger social benefits. The environmental impact of cruise tourism is contentious, with many examples to found of both infractions and efforts to improve overall environmental responsibility.

Box 5.3 Case study – Corporate consolidation in the global cruise industry (with Adam Weaver)

There are several unique aspects of the global cruise industry that allow it to sit alongside other economic sectors (e.g. manufacturing and finance), thus demonstrating the extent to which global cruising for pleasure can be exceptionally profitable (Weaver & Duval, in press). The first is the massive amount of corporate consolidation that has taken place within the cruise industry since the 1980s, resulting in the creation of two substantial cruise companies: Carnival Corporation and Royal Caribbean Cruises Ltd. Together, they account for over 80% of total market share within the global cruise industry (Marti, 2005). Carnival Corporation, which owns Carnival Cruise Lines, acquired Holland American Line and P&O Princess Cruises in 1989 and 2003 respectively. In 2000, Carnival Corporation became the sole owner of Costa Cruises, an Italian-themed cruise line. Royal Caribbean Cruises Ltd, the company that owns Royal Caribbean International, acquired Celebrity Cruises in 1997. Efforts made by Royal Caribbean Cruises Ltd to merge with Princess in 2001 were thwarted by Carnival Corporation's successful takeover bid. The primary impetus behind these acquisitions and consolidation activities is the desire on the part of larger corporations to diversify. For example, each separate cruise line owned by Carnival Corporation intentionally serves a different market, and most acquisitions made by Carnival Corporation in the past decade have enabled it to serve specific markets previously untapped. The reasons large cruise conglomerates go through the process of acquiring other firms/companies is because acquisitions and mergers allow corporations to obtain less tangible assets, namely companies with reputable brand names. From a business standpoint, it is also seen to be very efficient in that it is often less expensive for a corporation to acquire an existing business than to start an entirely new product or service line itself.

Interestingly, in spite of the consolidation that has occurred within the global cruise industry, brand identities and ship names have often been preserved. For example, when Carnival Corporation acquired Holland America Line, ships within the Holland America fleet were not renamed. Initially, there were fears that Holland America would be 'Carnivalized' after the acquisition (Slater & Basch, 1989) because Carnival Cruise Lines markets its vessels as 'fun ships' and caters to consumers who enjoy a more festive environment on board. Holland America, however, actively markets to an older, more up-market clientele (Sarna & Hannafin, 2003). The reason Carnival has kept the Holland brand separate is primarily to retain existing market segments that are valuable and responsive to the brand itself.

As mentioned, the acquisition of different brands by a corporation may also be part of a strategy to serve different national markets. Some brands, for example, may only have value in certain national markets. A good example is when Carnival Corporation acquiring AIDA Cruises after the former consolidated with Princess Cruises in 2003. AIDA Cruises serves the German market, and the brand has been in existence since 1993. Carnival Corporation apparently plans to preserve the brand's current identity (Blenkey, 2005) in order to allow AIDA to preserve its market presence.

Questions for consideration

1. What variables might one company take into consideration when deciding whether to retain the brand name of a product/service produced by a company that it recently acquired?

2. To what extent could it be argued that such 'mega-mergers' are somewhat anti-competitive?

SELF-REVIEW QUESTIONS

1. What are the main forms of waste from cruise ships?

2. What are some of the more common cruise market segments?

3. Which areas have seen the biggest growth in cruise tourism?

ESSAY QUESTIONS

1. Write an essay that discusses why marine tourism, in general, may or may not be more likely to experience elasticity in demand among certain demographic segments.

2. Why could the argument be made that the environmental impacts of cruise tourism are more concentrated that land-based forms of tourism?

KEY SOURCES

Cartwright, R. and Baird, C. (1999) *The Development and Growth of the Cruise Industry*. Oxford: Butterworth-Heinemann.

This text is, in many ways, somewhat simplistic for university-level education, but it nonetheless offers a comprehensive overview of cruise ship operations, including pricing structures, markets and marketing, and itinerary construction. One criticism is the lack of acknowledgement of wider political or governmental influences.

Wood, R. (2000) Caribbean cruise tourism: Globalization at sea. *Annals of Tourism Research* 27 (2): 345–370.

Wood's article has acquired a certain status within the academic community in that it isolates regulatory and governmental considerations in the global cruise industry. Wood outlines how cruise tourism is effectively 'deterritorialised' in that it is manifested in global routes yet is based in countries where laws and governance are less than rigourous. A noteworthy angle of Wood's article is the nature of cruise employment and how this represents a true picture of globalisation of the industry.

Cruise Line International Association – www.cruising.org and International Council of Cruise Lines – www.iccl.org

CLIA and ICCL are large trade organisations representing the global cruise tourism industry. Both sites feature reports and recent facts/figures relating to the industry.

CHAPTER 6:

AIR TRANSPORT

LEARNING OBJECTIVES

After reading this chapter, you should be able to

1. Recognise the importance of air transport in global tourist flows.

2. Critically assess the extent to which air transport has contributed to tourism development both globally and in specific localities.

3. Appreciate the network structure of airline operations and how these relate to tourist flows.

4. Understand the nature of global airline alliances and the resulting marketing and operational considerations for tourism.

5. Identify and evaluate sustainable forms of air transport.

INTRODUCTION

Graham (1995) notes that the air transport industry has two discerning, but broad, characteristics: 1) returns are marginal on what is an extremely capital-intensive business; and 2) it is one of the most regulated transport industries, although this is slowly changing with the introduction of more liberalised air transport regimes worldwide. The economic uncertainty that follows global air transport may not be entirely unique (some rail operations, as discussed in Chapter 4, have experienced similar hardships), but it is followed closely in the international media because of the potential consequences for local and national economies if air services are culled due to negative financial performance. Immediately following the 11 September 2001 attacks in the United States, the United

States Federal government introduced the Air Transportation Safety and System Stabilization Act which provided some US$5billion in direct compensation and almost US$10billion in loan guarantees to assist carriers with the forecast slump in the air travel market (Guzhva & Pagiavlas, 2004). Unfortunately, the losses were too great. Almost four years later, almost to the day, Delta Air Lines and Northwest Airlines, two of the largest carriers in the United States, declared bankruptcy. According to both, one of the main reasons was the high cost of air turbine fuel (or jet fuel) prices, but some news reports also pointed to the fact that it is becoming increasingly difficult for many 'legacy carriers' (i.e. those airlines that have been operating for quite some time and operate under a model that is not necessarily 'low cost') to compete against LCCs or, perhaps more accurately, LCLF carriers. As a sign of turbulent times, Delta and Northwest were the third and fourth airlines to declare bankruptcy in the United States following the 11 September 2001 hijackings/terrorist attacks (the other two being United Airlines and US Airways), the immediate aftermath of which challenged airlines' operating profitability.

Without question, the operating environment for airlines in the early part of this century is contentious and fraught with uncontrollable externalities that are seemingly insurmountable. Examples that illustrate this are the rising cost of airline fuel and increased competition from other airlines (primarily LCLF carriers, as discussed in this chapter) with somewhat different business models. Curiously, some airlines have weathered recent world events without much trouble, and this is largely attributable to their existing financial health and ability to react to changes in the marketplace.

The chapter begins with an assessment of the role of aviation in tourism development. Following this, network structures (which should be read in conjunction with Chapter 3) are discussed. The nature of the LCC/LCLF 'revolution' is discussed next, followed by some consideration of the increase in mergers and alliances being formed worldwide. As is demonstrated, the issue of alliances is one in which the global map of mobility is demonstrated. The next section considers a very important aspect of global aviation operations, namely the nature of regulation and governance concerning, quite literally, who can fly where, how often, when and for what reason. These agreements effectively govern global mobility and thus play an enormous role in the ability of destinations to register themselves on potential tourists' perceptual 'radar screens'.

AIR TRANSPORT AND TOURIST FLOWS

Air transport is of interest to tourism development and tourist flows because it is a fundamental cog in the global tourism interaction sphere. The state of the global aviation sector is therefore important to understand as it can often single-handedly shape tourist flows where air access is part of the dominant network provision for accessibility and connectivity. This chapter, then, provides an overview of air services worldwide, with an emphasis on the nature of these services as they relate to tourism.

For the most part, two strands of services have emerged over the past few decades: FSAs (or network carriers, as they are sometimes called) and the value-based service option (often referred to as the LCC), which may often target the same markets as served by network carriers. There are other service offerings, however, including regional carriers and charter airlines, each of which can have different impacts on tourism flows (see, for example, Bieger & Wittmer, 2006) (Table 6.1). Charter airlines have been called the original LCCs (Buck & Lei, 2004) for they often opt for 'seat only' sales in conjunction with vertically-integrated alliances with tour operators. Papatheodorou (2002) notes that vertical integration is advantageous because the cost of marketing and promotion, commissions and commercial risk is generally borne by the tour operator, although in Europe as of late some charter operations have been replaced by LCC/LCLF carriers (Bieger & Wittmer, 2006).

Table 6.1 Air carrier typology and associated impact on travel flows

	Network carriers	*Regional airlines*	*LCC/LCLF airlines*	*Charter carriers*
Target market	Extensive coverage and attempt to increase market share; impact of formal alliances often critical in maintaining market share	Geographically smaller than network carriers, but often serves to feed larger networks; generally service geographically niche markets	Specific niche markets, perhaps where demand is price elastic	Tours and mass tourism flows to holiday resorts (e.g. UK to Mallorca).
Critical success factors	Market share	Share of niche market	Strong traffic growth that is almost self-generating	Relationships with tour operators; seasonal loadings
Network type	Generally hub-and-spoke	Several smaller hubs possible; some point-to-point routes	Generally point-to-point, but some smaller hubs can emerge	Generally point-to-point

Source: Adapted from Bieger and Wittmer (2006)

It is important to point out that the introduction of VBAs or LCCs (the terms are often used interchangeably, although they admittedly have slightly different meanings) has been significant in that it can explain much of the increased accessibility of some destinations worldwide (e.g. Forsyth, 2006). As outlined in Chapter 3, accessibility is one of the key factors in tourism development. As many destinations seek to target international markets, air transport has largely replaced other forms of long-distance transport (cruises or boats) and this has, in turn, literally opened up many destinations to multiple markets worldwide.

THE ROLE OF AVIATION IN TOURISM DEVELOPMENT

It is appropriate to begin with the potential role that aviation plays in tourism development. Several key points can raised at the outset:

1. The technological developments related to aircraft engineering, particularly with respect to efficiencies realised in power plant manufacturing and overall fuel efficiencies, but also in the area of improved aerodynamics, have resulted in overall reductions in operating costs. For the most part, this technological 'revolution' in air travel took place with the introduction of the jet engine technology, allowing aircraft to travel faster. This was a significant factor in explaining the increasing accessibility of some destinations to key markets. One particular example is the Caribbean, a region that has traditionally relied upon North American markets for the bulk of its tourist supply. With the introduction of jet-powered aircraft, many destinations in the Caribbean (particularly Hispaniola, Jamaica and Puerto Rico, but others as well) became more accessible to key urban markets in North America as a result of the reduced travel time. This was one factor that contributed to the expansion of Caribbean tourism in the 1950s and 1960s. Another example of the efficiencies brought about with the introduction of jet-engine technology was the ability to fly non-stop between two nodes in a network (with a good example being the introduction of non-stop services, for example, across the Pacific Ocean between Australia/New Zealand and the United States).

2. The ability of an airline (whether charter or RPT) to fly from one place to another is often carefully regulated by international bilateral agreements in air services (commonly called air service agreements). Although this is examined in more detail below, it is worthwhile to point out at this point that the political relationships between two or more countries (multilateral or plurilateral (Holloway, 2003), if more than two) can often dictate the flow of tourists between places, regardless of the physical ability of an airline to service a destination. In other words, an airport at a destination may be fully capable of servicing one or more international flights from an operations perspective, and a viable market may exist to justify a certain frequency of service, but the appropriate permissions and certifications at the government level must be in place before services are initiated.

This is an important consideration in that it is not always lack of demand that determines whether a carrier opts to run services on a particular network.

3. The expansion of air networks has meant the introduction of tourism to many non-Western countries. For example, Turton and Mutambirwa (1996) note that the government of Zimbabwe has needed to provide regular subsidies to Air Zimbabwe in order to ensure a regular flow (or as regular as possible) of tourists to the country (and certainly within the country). Not surprisingly, global economic conditions play an important role in the provision of flights to and from some destinations. Domestic and international travel in China has recently surged, largely a consequence of the country's strengthening economy. In 2004, some 28.5 million residents of China (PRC) travelled overseas (PATA, 2005). Data from 2003 indicate that the most popular destination was Hong Kong SAR, and, to meet this demand, China Airlines announced in October 2005 that it would be increasing flights frequency between Taipei and Hong Kong. Inbound international arrivals to China (PRC) have increased an average of almost 7% between 2000 and 2004 (PATA, 2005). In 2004, China (PRC) received almost 13 million international visitors by air, with the majority coming from Chinese Taipei, Japan, Korea and Hong Kong SAR (PATA, 2005) (Table 6.2), suggesting a largely regional market in terms of draw.

Table 6.2 Arrivals to China (PRC) by air, 2003 and 2004

	2003	*2004*	*Percent change*
The Americas	645,022	1,118,206	58
Europe	869,589	1,523,913	45.4
South Asia	71,058	122,733	30.6
Northeast Asia	5,265,475	7,653,418	16.5
Southeast Asia	872,127	1,575,039	54.6
Other Asia	157,795	293,600	54.2
Australia/New Zealand	151,013	246,418	52.6
South and Central Pacific	1176	1720	−0.1
Other areas	47,240	82,197	66.6
Total	8,080,495	12,617,244	19

Source: PATA (2005)

4. Air service expansion has also meant considerable expansion of domestic services in both developed and developing countries. In Mozambique, domestic passenger numbers have increased considerably in recent years, prompting the introduction of a new private carrier, Air Corridor, serving Nampula to Maputo with regular low-fare service and nearly 100% loadings (Stevens, 2005). In New Zealand, Air New Zealand's introduction of its Express Class service, ostensibly a three-tier fare structure with the lowest fares being below NZ$100 has resulted in increases in domestic travel (New Zealand Herald, 2005).

NETWORK STRUCTURE: HUB-AND-SPOKE VERSUS POINT-TO-POINT/GRID NETWORKS

Airlines must make critical decisions as to where and how often they fly into and out of certain destinations. Much of this is based on demand, of course, but the cost of service also enters the equation. In other words, demand may be strong to serve particular origin–destination pairing (hereafter 'O–D pair'), but it may not be strong enough to warrant continuous service, taking into consideration seasonal fluctuations, economic fluctuations in both origin and destination, and the overall viability of tourism development at the destination. For example, an airline's market research may indicate that demand exists to service a particular O–D pair, but only at certain times of the week. Thus, in order to launch this service, the operations or scheduling department must decide which aircraft to use and, most importantly, what the aircraft will do when not serving that particular O–D pair. In other words, it may be entirely unprofitable for an airline, already sufficiently geared to meet its existing network, to purchase or lease new equipment to service a new O–D pair even if adequate demand exists. If, however, the new aircraft could be utilised efficiently in other routes during off-peak demand cycles on the new O–D pair, then it may be profitable. This goes some way to explaining decisions to reduce or eliminate services on some O–D pairs.

To some extent, the smart management of networks can help realise efficiencies of operation. As discussed in Chapter 3, networks are equated to potential accessibility and connectivity, which is a significant hurdle to ensuring that tourist flows are maintained and economically sustainable. As Bowen (2000: 27) notes: '…changes in the accessibility of a place within international airline networks can either enhance or detract from its ability to attract tourists'. What was not explored fully in Chapter 3, however, is the nature of networks and how they are involved in providing accessibility. There are three primary types of networks in passenger air transport:

1. Hub-and-spoke networks: this type of network involves a central hub into which traffic from outlying spokes feed the hub. Hub-and-spoke networks can exist in domestic networks (the United States also being an example as many airlines utilise hub operations at major US cities) as well as international networks (for example, Singapore Changi and London Heathrow are considered major international hubs) (Figure 6.1a). Since deregulation in the late 1970s in the United States, the use of hub-and-spoke networks has been increasingly popular (although Taneja (2003) points out that they were developed in the early 1950s).

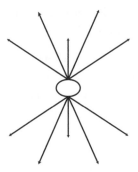

a) Hub-and-spoke network design

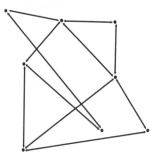

b) Grid network design

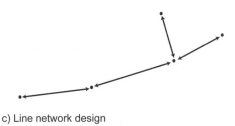

c) Line network design

Figure 6.1 Network options – air transport

Source: Adapted from Hanlon (1999)

2. Grid networks: generally a feature of domestic travel in some countries, particularly in larger countries such as the United States, this type of network has the advantage, according to Hanlon (1999: 84) such that 'they make it easier to achieve high rates of utilisation, of both aircraft and crews. Flights can be scheduled to operate on a number of different routes without backtracking, which helps to minimise the time for which aircraft are idle on the ground and which also means that crew stopovers and slippage can be minimised'

(Figure 6.1b). Fixed costs associated with this type of network are generally higher because of the cost of operating more links (Pels *et al.*, 2000).

3. Line networks: Hanlon (1999: 83) notes that a line network is realised if an aircraft 'sets out from its base airport and makes a number of intermediate stops en route through to its ultimate destination. The intermediate stops are made either to refuel or to pick up traffic' (Figure 6.1c).

THE HUB-AND-SPOKE CONCEPT

Air carriers, in recent years, have adopted hub-and-spoke networks for a variety of reasons. First, the nature of a hub-and-spoke network design (Figure 6.1a) is such that it allows for numerous permutations of O–D pairs served (Hanlon, 1999; see also Borenstein, 1989, Dresner & Windle, 1995, O'Kelly, 1998). In other words, when compared to a line network (Figure 6.1c), there is a substantial increase in the availability of O–D pairs that can be serviced (Hanlon, 1999). Nodes can be connected to only one hub (a single assignment model) or multiple hubs (a multiple assignment model) (Bryan & O'Kelly, 1999). As Taneja (2003: 39) notes, a hub-and-spoke system 'enabled an airline to increase frequency to thin markets by connecting the thin markets (the spokes) to a central airport (the hub)'. Hub-and-spoke systems became a popular way for airlines to manage their networks because the presence of a central hub meant the centring of operations. Passengers from the spokes would ultimately be routed at the hub, thus providing an increase in the potential market for the other spokes. The system meant airlines had the ability to service numerous markets, and is one of the reasons smaller aircraft (such as regional jets) are used on spokes and larger aircraft are used to move passengers from hubs to other hubs and internationally (Savage & Scott, 2004). American and United Airlines, for example, serve well over 100 smaller cities/markets using this system (Taneja, 2003). Some airports/cities now serve as critical hubs for specific airlines: American Airlines at Raleigh Durham and Nashville, Lufthansa at Frankfurt and Munich, and Singapore Airlines at Singapore. Dominance tends to be limited to the smaller regional or 'mid-size' urban centres. Thus, major international airports at London, Los Angeles or New York are not operated by airlines as market-capturing or network-enhancing hubs, rather, the network incorporates these as spokes in the wider network.

Second, the ability to market to wider segments must be considered when considering the power of a hub. Dresner and Windle (1995: 202), for example, found that hubbing by North American carriers resulted in increased passenger numbers across the Atlantic Ocean to Europe: 'A carrier with a strong hub at the home end of a trans-Atlantic route can expect to carry significantly more traffic than a carrier with a weak hub, all other things being equal.' Moreover, they (1995: 202) suggest: 'Combining regional and longer distance international traffic at a single hub airport would appear to be one policy that would facilitate hubbing and generate traffic.' Third, there may be significant cost efficiencies in routing traffic through a centralised hub airport in that the network can see profitable returns on operations if it is maintained and market demand is met. Aircraft utilisation efficiencies can also be realised in that inbound aircraft to a hub can be redeployed to other spokes. In the case of a line network, this is generally not as easily achieved, and aircraft can simply be relegated to flying back and forth between an O–D pair. This can result in considerable downtime on the ground between flights, which is an unprofitable exercise in that the aircraft, while not being used, is not earning revenue (Button, 2002).

In short, hubbing as a concept means that airlines have the opportunity to increase the utilisation of expensive aircraft. The reality, however, is not as simple as increased aircraft utilisation through hubbing. Doganis (2002) suggests that the economics of hubbing rests with the unit costs of flying short sector lengths between spokes and the hub: the higher frequency of short flights can be costlier because of additional aeronautical charges at airports through increased aircraft rotations (take-offs and landings) and the increased costs in handling passengers throughout the network must be considered.

What does hubbing mean for the passenger? In the positive, it means that the number of potential destinations served is exponentially greater. The hub concept can also reduce overall costs to passengers when compared to linear or point-to-point networks (Button, 2002). As discussed earlier, accessibility is critical in tourism development, and the ability of a carrier to integrate a destination with a series of other destinations utilising hubbing can be powerful in terms of marketing and the ability to secure positive market share. What it could also mean, however, and potentially as a negative consequence, is the dominance of one carrier at a particular hub. As Doganis (2002: 257) notes:

> Once airlines have established dominance at a hub through control of a disproportionate share of the flights offered and traffic uplifted, it is very difficult for another airline to set up a rival hub at the same airport, because it is unlikely to get enough runway slots to offer a similar range of destinations. In the United States the hub operator will also control most of the terminal gates. If the new entrant chooses to compete on just a few direct routes from the hub airport, it will face a competitive disadvantage *vis-à-vis* the hub airline in terms of ensuring adequate feed for its own services. Thence the notion of the 'fortress hub'.

The fortress hub can result in pricing that may not entirely favour the passenger at the hub airport, largely because there is little incentive to compete on price when the threat of competition is small (Borenstein, 1989). Although not a fortress hub in the strict sense of the definition, US Airways utilised Pittsburgh International Airport as a national hub for many years, but announced in late 2004 that it would no longer be doing so. The result was the introduction of new services from LCLF carriers and new services from existing network carriers, and fares for flights originating in Pittsburgh decreased dramatically (St Paul Pioneer Press, 2005).

Another problem highlighted by Doganis (2002) as well as Rietveld and Brons (2001) is that an efficient hubbing system must rely on precise frequencies of operations to and from the hub. If one flight is late there is a chance that other sectors and routes emanating from the hub can be negatively affected. In some cases, numerous flights may need to be held at the hub because an incoming flight from a spoke, carrying passengers en route to various other spokes via the hub, is delayed (Doganis, 2002). The power of hubs in creating increased accessibility (from a geographical perspective) across a particular network may, however, need to be carefully considered in the overall convenience of forcing passengers to transit through often congested airports. As Button (2002: 182) indicates, managing congestion is an exercise in economics:

> The issue is not about whether there should be no congestion at all, it would simply be wasteful of airline resources and airport capacity if congestion was zero, but rather of what is an optimal level of congestion. In economic terms, an optimal level of congestion is derived at that point where the benefits of an additional user just balance with the costs that the user imposes on the system (including those on other users). In the case of many transportation infrastructures, such as airports, the pricing system is not used to allocate resources but rather is used as a revenue-collecting device. As a result, it is argued, there will be excessive use of the infrastructure because users do not allow for the effects of their actions on other users.

Button goes on to note that the optimal scheduling of hub operations is generally a balance between the costs borne by the airline and by the passengers. The result is that scheduling through hubs is often done in 'banks'. These banks may result in periods of increased congestion at hub airports, but according to Button (2002: 183), they also mean lower fares and an increase in choices for services/routes as passengers can change aircraft at the hub and be on their way relatively quickly, or what Button (2002: 183) refers to as the 'convenient concentration of connecting services'.

Not all airlines operate a hub-and-spoke network. Southwest Airlines, for example, largely operates as a point-to-point network operator, and does so quite profitably. One question remains: When is an airline to adopt one network type over another? O'Kelly (1998: 176) helps answer this:

> Point-to-point systems have an advantage in short-haul market pairs with a dense level of demand, and work well provided the carrier does not try to offer service between all pairs of places. Hub-and-spoke systems in contrast seem to be the ideal solution (from the carrier's viewpoint) whenever the flows can be channeled through convenient switching points (such as Chicago or Atlanta).

In more recent years, the plethora of what have come to be known as 'sixth-freedom carriers' (e.g. Emirates, Etihad, Singapore Airlines) has meant substantial international point-to-point services being developed with less reliance on the hub. These carriers offer competitive fares over traditional legacy carriers with strong international networks, and their interest is primarily in shuttling passengers from major city to major city.

THE LCC/LCLF 'REVOLUTION'?

If recent worldwide press (but predominantly in Europe) is any indication, LCCs or LCLF carriers (which are sometimes even called simply 'no frills') seem to have limitless expansion possibilities. Francis *et al.* (2004: 508) explain this as follows:

> With the spread of deregulation of airline markets worldwide, low-cost carriers have begun to emerge in countries as diverse as Canada, Brazil, South Africa, Slovakia, Australia and New Zealand. There is little doubt that the growth in low-cost travel is the result of an innovative business model that has successfully reduced air fares to a point where they are often cheaper than surface transport alternatives (Doganis, 2001; Caves & Gosling, 1999; Barrett, 2000). Competitive price advantage is a key factor behind the low-cost carriers continuing to experience growth post-September 11, in stark contrast to many of the traditional scheduled carriers (Goodrich, 2002).

Traditional FSAs (or what are sometimes called 'network carriers' as they have armed themselves with expansive networks for maximum coverage and, thus, market share) have differentiated themselves on the basis of service and networks. Thus, an FSA may offer various classes of service (e.g. economy class, business class and first class, or any combination of the three) and serve major hub or international airports. In fact, the strategy that generally drove most FSAs was one of convenience and service, and thus many of the 'traditional' FSAs, or network carriers, boast enormous grid or hub-and-spoke networks with high levels of connectivity. A passenger on board an FSA could expect full meals (sometimes two if the sector length was long enough) and often free beverages. Often, alcohol is provided free on international flights. Other amenities provided include pillows, blankets, hygiene kits (including a toothbrush, toothpaste, socks; and these kits are often much more substantial in business and first class), complimentary newspapers and access to personnel at either the origin or destination to help with further travel arrangements.

Contrast this to a 'typical' LCC/LCLF (hereafter LCC) airline: services such as those listed above are reduced, if not completely withdrawn. Some aspect of service may still be emphasised, but this is more generally along the lines of non-aeronautical means of service provision, such as providing access to destinations not normally covered by the FSAs. Price, however, is the driving message in marketing and advertising (hence the designation of the VBA, and forms the basis of the service provision of these types of carriers. Some of the more common attributes by which LCCs are identified include:

1. *Cost control.* This is perhaps the single most effective means by which an LCC will compete with network carriers or FSA (Francis *et al.*, 2004). Profit is realised by cutting costs.

By keeping costs low, LCCs are generally in a position to turn higher profits or compete strongly along routes dominated by one or more FSA. Cost control encompasses a number of measures that can more or less be attributed as a means by which efficiencies are utilised. Costs are kept under control in a number of ways. One of the most common is the outright homogenisation of the fleet. A carrier such as Ryanair or easyJet, for example, may elect to utilise only Boeing 737 aircraft, thus enabling commonalities of crew training and rostering. B737s are extremely common (and thus financially successful for Boeing) and many LCCs utilise older aircraft that may be ex-lease from FSAs, and thus acquired at an attractive price. As well, this may mean that maintenance costs are kept low because only mechanics who can work on these type of aircraft are needed, although it should be pointed out that some LCCs (as well as FSAs) may elect to contract out maintenance procedures in a bid to save costs rather than keep permanent mechanics and service personnel on the payroll. It is, however, not entirely uncommon for an LCC to begin operations with new aircraft. Holloway (2003) notes that US LCC JetBlue initiated services with new Airbus A320 aircraft because of the aircraft's reputation for reliability and low maintenance costs. Another means by which costs are kept low is the almost constant utilisation of the aircraft. As aircraft are expensive and a significant part of the operating cost of an airline, an aircraft that is idle or parked at an airport waiting for the next flight is not earning revenue. As most LCCs tend to lease (or 'wet lease', meaning cockpit crew are contracted as well) aircraft, the cost of the lease applies whether the aircraft is flying or not. Thus, an LCC will endeavour to utilise an aircraft as much as possible such that it is able to maximise revenue generation.

One way to ensure maximum 'up time' in operations is to minimise the turn-around time at airports. It is not unusual, therefore, for LCCs (particularly in Europe) to have incredibly quick turn around times (some as short as 30 minutes) (Doganis, 2001). As Holloway (2003) cautions, however, the utilisation of aircraft may not necessarily help to control or reduce costs, but it does serve to minimise resource waste. Costs within the passenger cabin are also reduced as much as possible on LCCs, with the most common being the absence of free meals and most beverages. Most LCCs adopt what is termed a 'buy on board' policy, where passengers are offered snacks or sandwiches at a cost. Some beverages, such as tea and coffee, may continue to be offered free, but these are generally loss leaders in that the airline hopes that a biscuit or other snack item is purchased (perhaps at a substantial markup) to accompany the free beverage. More 'drastic' measures of cost control can occur. In 2005, for example, Ryanair removed the seat pockets from seats in order to save time and money in cleaning.

2. *Routing.* Like FSAs, LCCs generally do not rely upon any single model of network design. In many regional situations, such as the European Community, many LCCs opt to fly linear networks. Linear networks tend to allow for quick turnaround times (thus not needing to wait for other feeder aircraft from other spokes) and may help to serve a par-

ticular market pair with great frequency (see Box 6.1). Reynolds-Feighan (2001) examined the networks of LCCs in the United States and found that, rather than utilise hubs as a means to transfer passengers to other services, LCCs used instead nodes within their smaller networks as entry and exit points. Overall, Reynolds-Feighan (2001) noted the network density and degree of connectedness is smaller when compared to FSAs.

3. *Use of secondary airports.* The usage cost (e.g. aeronautical charges) at larger airports tends to restrict LCCs to regional airports, often at the expense of convenience for some passengers. LCCs will utilise secondary airports (some of which, such as London Stansted, have traffic growth rates such that they can be considered anything but a 'regional' airport) and thus incur lower aeronautical fees. Ryanair, for example, utilises Frankfurt Hahn airport, located approximately 120 km from Frankfurt itself, and Barcelona Girona, which is approximately 100 km from the city of Barcelona. Travel to and from cities like these to outlying airports is generally borne by the passenger (but provided, or at least organised, by the carrier), which raises the question of the actual savings in transport costs if this additional cost is factored in (Prideaux, 2000a, see also Chapter 2). Hall (2005: 241) notes that a French court decision in 2003 to stop Ryanair flights to Strasbourg (mounted as part of a legal challenge by Brit Air over subsidies paid to Ryanair by the Strasbourg Chamber of Commerce as an incentive to operate and thus generate tourist flows) could have a negative impact on accessibility for Britons owning second homes in France (although the airline did initiate flights to Baden-Baden, some 40 km outside of Strasbourg, following the court action).

4. *Single-class service.* It is not uncommon for an LCC to offer a single class of service (usually economy). Quite often, the seat pitch (which is the distance between a seat back and the seat back in front of it) is smaller than FSAs; it is not uncommon to find seat pitches of 32 inches, and sometimes even 30 inches. However, if one were to apply a cost-benefit analysis of smaller seat pitches, as most LCCs run on shorter routes passengers may be forgiving of less overall comfort if the price paid for the service is favoured. And in relation to seating, some LCCs may elect not to have pre-seating available, thus resulting in a first-on-board-first-selected seating arrangement. An added advantage of operating with a single class of service is that aircraft across the fleet are generally interchangeable, which is another example of efficient resource utilisation.

5. *Distribution channels.* Ryanair was one of the first LCCs to utilise call centres almost exclusively for a significant portion of its ticket sales (although its distribution channel in its early days still relied on travel agents to some extent). This served to minimise the cost of paying out commissions to travel agents on sales volumes, but it also allowed the airline to form direct relationships with its customers (Calder, 2002). In recent years, the internet has generally supplanted call centres for many airlines (not just LCCs), although LCCs generally embrace internet bookings as a means of controlling distribution costs (see also Box 7.1).

6. *No frequent flyer programme.* Most LCCs will neither operate their own, nor subscribe to others', frequent flyer programmes. These can be expensive to maintain and monitor, and thus generally do not figure into the 'low cost' model of the airline. There are, however, some exceptions. JetBlue in the United States, for example, offers TrueBlue points, which are earned on the basis of the length of flight (and when a passenger reaches 100 points on JetBlue, they are offered a free round-trip flight to any destination to which the airline flies):

- SHORT (2 pts): Flights from JFK to Buffalo, NY.
- MEDIUM (4 pts): Flights from Rochester, NY to Ft. Lauderdale, FL or from JFK to New Orleans, LA.
- LONG (6 pts): Flights from JFK to Oakland, CA or Burlington, VT to Long Beach, CA. (www.jetblue.com; accessed 2 November 2005)

In the Pacific region, Virgin Blue, Pacific Blue and Polynesian Blue reward passengers with points towards what they call the 'next generation' of loyalty programmes. In additional to teaming with a credit card provider, the airline's Velocity programme has partnered with several operations that have traditionally avoided linkages with other major loyalty programmes: Emirates, Virgin Airlines and Europcar.

BLURRY DISTINCTIONS BETWEEN LCCS AND FSAS

Despite what may seem to be clear differences between LCCs and FSAs, the question of what defines a 'typical' LCC airline was, at one time, quite easy to answer. Since 2002/2003, however, this delineation has become increasingly blurry. Part of this has to do with the economic realities facing legacy or FSAs following the 11 September 2001 attacks in the United States, but it also represents global economic realities as more and more aviation environments are deregulated and liberalised in terms of access. As a result, what used to differentiate an LCC from an FSA is now less clear. For one, as distinct as some of the features are that distinguish LCCs from FSA or network carriers, there have been recent changes to the industry that have brought about changes in FSA cost management structures that have led to service offerings that have more in common with traditional LCC carriers. In other words, at times it is difficult to identify a true full service carrier today because the nature of the product or service is changing and has become more and more like the typical LCC of the mid- to late 1990s. As Francis *et al.* (2004: 508) point out: 'Some of the major airlines have responded by lowering prices, simplifying their price structure and improving internet sales facilities.' For example, some FSAs have become somewhat truncated and serve only very selective routes where demand is strong. One way this is achieved is the decommissioning of certain routes structures, thus these traditional network carriers have networks that are less connected and thus not as accessible. As well, the price structures of airfares available from traditional carriers have changed to compete with LCCs. Indeed, Vowles (2000) found that the presence of low fare carriers in the United States had a statistically significant impact (in favour of the passenger) on air fares at hub airports.

Another means by which FSAs have sought to reframe their business strategy is to utilise a subsidiary or franchise airline to serve the shorter routes and effectively feed the hubs of the parent carrier (and example being the establishment of GO, a subsidiary of British Airways) (Pender, 1999). This serves two main purposes. First, it allows an existing carrier to compete in a particular market where its existing structure and operations profile may preclude it from doing so. Second, it protects the brand name of the parent enterprise should the LCC subsidiary face future financial problems (Pender, 1999; se also Holloway, 2003).

Cabin service among FSAs is also starting to closely mirror that of LCCs. Many FSAs have adopted a 'buy on board' approach to food items and services, particularly on domestic routes. In some cases, food services have been almost completely eliminated in favour of providing tea/coffee and a biscuit. In an effort to offset the cost of rising fuel, Air Canada announced in October 2005 that, on short-haul flights, the use of pillows and blankets will incur a charge (Airwise News, 2005a).

Table 6.3 outlines some of the similarities (and differences) between LCCs and FSAs. In some cases, it can be argued that there are few conceptual differences between the operating strategies of each, but what may be different is how these strategies are put in to practice or operationalised. For example, LCCs often state that where passengers are connected to another flight operated by the same airline, there is no guarantee that the second flight will be held for them if their initial flight is delayed. A passenger on Ryanair may purchase a ticket from Dublin to Luton and then another ticket from Luton to Milan (this is because Ryanair, like many other LCCs, will not issue a ticket from Dublin to Milan connecting at Luton). If the flight from Dublin to Luton is delayed for any reason, Ryanair may not necessarily guarantee that the flight from Luton to Milan will be held, even if there is more than one passenger on the Dublin–Luton flight destined for Milan. LCCs will do this because it introduces streamlined efficiencies into the system, even though some commentators claim it is at the expense of customer service. The airline will claim that its ability to offer low fares will come at a cost to convenience on some occasions, and thus cannot guarantee such interlining without impacting on fares. Contrast this, to some extent, with a traditional network carrier operating domestic flights as part of a wider network. In some situations, the Luton–Milan flight may be held for a specific length of time in order to capture late inbound Dublin–Luton passengers. Failing to do so may mean the airline, through its own policies, may be responsible for accommodating those late arriving passengers (either overnight or on a later flight).

Table 6.3 Similarities and differences in LCCs and FSAs

Product attributes	*LCC*	*FSA*
Branding	Generally limited to branding exercises associated with promoting low fare levels and resulting benefits to passengers (thus, advertising may feature pricing over other elements)	Generally extends to the brand of the airline as a service provider, but may also feature fare specials and, more recently, any overt fare structures introduced to compete with LCCs
Fare structure	Perhaps less complicated than an FSA, but yield management is still utilised	Moving towards a less complex model of fare structures as presented to the public (e.g. tiered based on permissible changes); yield management utilised
Distribution channels	Preference is online and direct sales	Increasing movement towards online and direct sales in order to reduce commissions paid to third-party channels (e.g. travel agents)
Check-in	Often utilises e-tickets as a cost-saving measure	Domestic operations of some carriers may utilise e-ticketing, but international travel or interlining (e.g. multiple carriers on one itinerary) may require paper tickets/coupons
Airport relationships	Secondary airports prevalent in order to keep aeronautical charges low	Focus still on major airports, although LCC subsidiaries of major carriers may opt to utilise secondary airports
Connections and connectivity	Primarily linear or point-to-point networks, with smaller hubs sometimes present	Hub-and-spoke connectivity models still present, but the main feature here is the ability to interline as a result of using major airports; membership to global alliances (and sometimes smaller alliance agreements) permits expansion of network

Product attributes	LCC	FSA
Class of service	Generally one class, but in some cases business class seating and (limited) amenities provided (e.g. Freedom Air in New Zealand)	Largely based on at least two classes of service (business and economy), but quite often first class service is offered on international or oceanic flights
Inflight service	Buy on board programmes; leather seats on some carriers (e.g. JetBlue in the United States)	Oceanic flights tend to offer full meal service(s), but some domestic services may be limited to complimentary tea/coffee/water with a light snack; improved seating, especially in business class (where the move toward 'lie flat' seating has been rapid in the past few years); buy on board programmes slowly being introduced
Utilisation of aircraft	High	Longer routes may mean different measures of utilisation; some domestic routes feature relatively high utilisation
Turnaround time	Fairly quick and efficient	Can often be just as quick, but may be hindered by scheduling that may not necessarily require quick turnarounds
Frequent flyer programmes	Becoming more common	Very common, and often linked with global programmes such as Star Alliance, OneWorld or Skyteam
Seating features	Often very dense, with 32 inch or 30 inch seat pitches common; some do not have pre-assigned seating	Varies by route: international routes may feature as much as 34 inches of seat pitch, while domestic services may offer less; pre-assigned seating dominates

Product attributes	LCC	FSA
Operational considerations	Expansion of service offerings to include ancillary products/ services such as accommodations and rental cars (and often part of branding initiatives)	Ancillary products/services remain, but also package deals with accommodation, rental cars becoming common

Source: Based on O'Connell and Williams (2005)

Using Table 6.3 as a reference point, the true differences between LCCs and FSAs are perhaps not as great as they were perhaps even five years ago. As discussed above, FSAs are emulating operational procedures (such as simplified fare structures, no frills service) that were once strictly within the operating domain of LCCs. As well, some LCCs have adopted service levels and procedures (e.g. frequent flyer programmes, comfortable seating) that were once limited to FSAs. The result is that even the comparison of LCCs versus FSAs is problematic in 2005, and will perhaps be increasingly blurry throughout the remainder of the decade.

BOX 6.1 Macau SAR and LCCs: Positive correlations between transport provision and tourism expansion (by Simon Rowe)

Macau is an SAR of the PRC with a total population of around 448,500 residents. Macau's tourism is based around its gambling industry which, when combined with general tourism, accounts for 40% of GDP. From July 2004, Macau has become a destination for several LCCs, such as Malaysia-based Air Asia and the Singapore Airlines' LCC subsidiary Tiger Airways. With the vast majority of Macau's tourists coming from Taiwan, China and Southeast Asia (DSEC, 2007), these LCCs play an important role in providing access and, consequently, the future of Macau as a tourism destination.

In July 2004, Air Asia was the first LCC to service Macau, and it now provides 7 scheduled flights per week (14 from December 2005) from Kuala Lumpur and 14 from Bangkok (21 from December 2005). Tiger Airlines began its scheduled daily service from Singapore to Macau in September 2004 and have subsequently added a daily flight from Manila effective from October 2005. Macau's national carrier, Air Macau, also services Bangkok and Manila.

As with many destinations where tourism is a dominant form of export earnings, Macau has been affected by externalities such as 11 September 2001, the bombings in Bali, and SARS. Table 6.4 shows the large reduction in visitor arrivals by air into Macau after 2002, possibly a consequence of the Bali bombings. It is interesting to note that, although arrivals by air were significantly reduced, total visitor arrivals still increased slightly. In 2004 when the two major LCCs began flying to Macau, visitor arrivals by air increased by 24% from 2003.

Macau is an attractive base for LCCs because of relatively smaller aeronautical charges at the main airport and its desirability as a tourism destination. For these reasons, two new Macau-based LCCs are planned to be launched in 2006. MEAS is planning to introduce a new LCC in May 2006. Golden Dragon Airlines is another Macau-based LCC being launched in 2006, with plans to fly to China, Vietnam and Laos. As a result, the ratio of air arrivals to total arrivals can also be expected to increase, as air travel becomes more affordable to Asian consumers. In effect, the continued growth in overall arrivals could spawn increased competition among LCCs servicing the destination.

Table 6.4 Arrivals to Macau SAR

Year	Total visitor arrivals (000's)	Visitor arrivals by air (000's)
2000	9162	834.2
2001	10,279	861.8
2002	11,531	905.4
2003	11,888	654.6
2004	16,673	861.8

Source: Adapted from (http://www.pata.org/patasite/fileadmin/docs/speeches_presentations/050311_Koldowski__PATA_ITB_Final.pdf) and (http://www.dsec.gov.mo/english/indicator/e_tur_indicator_2.html)

A major question here is whether the large increase in LCCs servicing Macau is sustainable. Australia's Virgin Blue was considering establishing a hub in Macau, but decided against it in May 2005. Although they did not provide an official reason for their decision, it is possible that they had concerns that the market was going to become saturated and overly competitive, leading to massive price wars and a subsequent loss of profitability. Price wars would, of course, be good news for the consumers, but also for the tourism industry of Macau, which will benefit greatly from the influx of tourists. If the increasing number of LCCs in Macau can be sustained without pricing each other out of the market, the future of tourism in Macau is very bright indeed.

Questions for consideration

1. What other regions or smaller countries have a strong reliance upon LCCs for the majority of their passenger airlift? Are there similar concerns regarding market saturation in that the total airlift is in the hands of one or two airlines (i.e. a duopoly or monopoly)?

2. To what extent could it be argued that tourism development in Macau is heavily dependent on the LCC model of airline operations?

What do LCCs mean for passengers? The reality is that many FSAs have lost business because of the price advantage, brought on by cost savings, introduced by LCCs. Writing in *The Times*, columnist Cath Urquhart (2005) noted that 'Ryanair has succeeded brilliantly in re-educating travellers as to what a short flight really is: something that is about as comfortable and basic as a bus journey.' This re-education effort has apparently been successful in some regions. In the UK, for example, one in three flights are operated by an LCC (This Is Money, 2005).

MERGERS AND ALLIANCES

Airline alliances have become extraordinarily common in the tight aviation market of the 21st century. Alliances can be global in scope (Table 6.6) or local with a focus on shared services. In many respect, alliances represent 'shortcuts' to otherwise out-of-reach (for aeropolitical reasons discussed below) markets. Indeed, the President and Chief Executive Officer of KLM, Leo van Wijk, once remarked that 'Alliances are ... a reasoned response to an antiquated regulatory system ... [They] permit indirect access to restricted markets' (Staniland, 1998). Tretheway and Oum (1992) have sketched out three types of airline alliances. These are:

- *Type I: a route-by-route alliance.* Such an alliance may involve coordination between two or more airlines on a specific route (i.e. reduction of 'wingtip' flights, where two flights from competing airlines leave from an origin bound for the same destination) or a combination of routes.
- *Type II: commercial alliance.* This type of alliance generally involves the 'coordination of flight schedules and ground handling, joint use of ground facilities, shared frequent flyer programs, code sharing, block seat sales, and joint advertising and promotion'

(Oum & Park, 1997: 138). In some cases, this can involve 'code sharing' where a passenger's ticket may indicate that they are flying with Airline Y when in fact the plane they board is operated by Airline Z. Airline Y, under a code-sharing agreement, is

Table 6.6 Major global alliances: key statistics

	Star Alliance	*OneWorld*	*SkyTeam*
Participating carriers*	Air Canada, Air New Zealand, ANA, Asiana Airlines, Austrian, bmi, LOT Polish Airlines, Lufthansa, Scandinavian Airlines, Singapore Airlines, Spanair, TAPPortugal, Thai Airways International, United Airlines, US Airways, Varig, SWISS, South African Airways	American Airlines, British Airways, Cathay Pacific, Finnair, Iberia, JAL, Malev, Quantas, Royal Jordanian	AeroMexico, Air France, KLM, Alitalia, Continental Airlines, Delta, Korean Air, Northwest Airlines, CSA Czech Airlines, Aeroflot
Global passenger share**	24.7	15.3	22.4
Global operating revenue share**	28.4	17.6	22.4
Jet aircraft in service**	2152	1574	2032

Source: Based on IATA (2004)

Notes: * As of August 2006; ** 2003 data. As part of the 'associate member' scheme (where airlines do not meet full entrance requirements) of SkyTeam, several airlines, including Air Europa of Spain, Panama's Copa Airlines, Kenya Airways, Romanian carrier Tarom, and Portugalia have been recruited; China Southern Airlines expected to join SkyTeam as a full member in late 2007; Turkish Airlines under discussions with Star Alliance in late 2006). MOUs signed between Air China and Shanghai Airlines in 2006.

allowed to sell a block of seats (agreed upon in the terms of the alliance) on a particular flight. This can be a powerful marketing tool for many airlines as such an alliance may serve to instantly expand its network in a virtual sense without the added cost of operations.

- *Type III: alliance of equity.* This type is typified by one airline obtaining a share of holdings in another airline, thus potentially leading to efficiencies of operations (perhaps initially financial, but ostensibly they have the opportunity to realise similar alliance benefits as identified in (I) and (II) above). This should not be confused with the outright share purchase agreements reached where no intention of alliance of operations features in the sale agreement. In other words, the selling of shares to another airline does not always proceed with the intent of forming some form of alliance. Glisson *et al.* (1996: 29) point out that equity alliances can actually be problematic in that they need to acquire regulatory approval in many jurisdictions:

> These equity holdings are looked on by some as a method of cementing the relationship between two airlines for the long run. However, they can be difficult to obtain because of severe restrictions and equity positions held by home governments. Those who oppose equity arrangements do not see the need for the investments, and feel they are nothing more than a waste of management time and an inappropriate use of investors' money. To many, equity alliances are becoming a system of airlines in distress. Others see them as defence postures.

All airline alliances are, in their own right, strategic in that they are usually designed to introduce operational efficiencies and markets to one or more existing carriers' networks (see Box 6.4). Oum and Park (1997) outline several reasons for strategic alliances, regardless of type, which I list here:

1. The expansion of networks can bring with it ease of transfers and integration: For example, passengers can 'interline' between two or more carriers who are members of the same alliance, primarily because of the agreement among them regarding passenger handling and booking (particularly in reference to customer reservation systems).

2. Feeding traffic between partners: An alliance between several carriers can mean an instant (often logarithmic) expansion of a network, although post-alliance business operations may need to be refined such that maximum efficiencies of network service are realised (e.g. decisions regarding which carrier shall remain on certain routes and which may not).

3. Efficiencies of cost: With the expansion of the network, airlines within an alliance can market themselves as being able to service, relatively seamlessly, numerous destinations, such that the cost of any one single airline achieving a network of this size would be prohibitive. Efficiencies of cost can also mean the sharing of maintenance and other personnel across an alliance.

4. Efficiencies of service: Relating to feed traffic discussed above, a good example of efficiency of services within an alliance is the conscious coordination of flight schedules across a network served by an alliance. This is one of the goals of a proposed alliance between several Arab airlines (Saudia, Egypt Air, Gulf Air, Middle East Airlines, Oman Air and Royal Jordanian) to be called Arabesk (www.ameinfo.com, 10 October 2005).

5. Wider choice of itineraries for passengers: Somewhat related, and by virtue of an expanded network and the accrued efficiency of service(s), itinerary choice can often be enhanced between multiple city pairs as passengers may have more than one airline choice.

The most visible alliances are those that are global in their operational outlook. The best known examples are Star Alliance, OneWorld and SkyTeam. Star Alliance captures the largest global passenger share and enjoys the highest operating revenue (Table 6.6). Tourists can often draw significant benefits from these types of alliances. For example, Star Alliance offers several round-the-world fares, the conditions for which are different but can often represent significant value for the passenger. For the Star Alliance, there are four fare levels: 26,000 miles, 29,000 miles, 34,000 miles or 39,000 miles, and in some cases there are no 'high season' surcharges (i.e. when some flights are generally more expensive due to higher demands in peak seasons). On these round-the-world tickets, journeys can be from ten days to one year in duration, and can include as few as 3 or as many as 15 stopovers (depending on the class of service), although there are limitations to the number of stopovers within a particular country or region. The normal conditions of such round-the-world tickets are that the outbound direction must be different from return trip; in other words, departing from Canada a traveller must travel in one direct (eastbound or westbound) and ultimately end up back home. Ticket prices are generally favourable and often fixed, and due to supply and demand for particular routes, quite often cheaper than simple O–D return trips.

Where such global alliances were, at one time, primarily established and agreed to by partner airlines on the basis of enjoying the ability to market their own product as part of a seamless network of airlines around the world (discussed more extensively in Chapter 8), more recently attention has been directed towards the seamless operations across member airlines. Star Alliance, for example, has embarked upon new IT initiatives that would allow for joint electronic ticketing, which would allow passengers to receive an electronic ticket that covers a range of airlines across a substantial network.

REGULATORY ENVIRONMENTS

Worldwide, there are complex sets of arrangements, primarily political, that determine and dictate whether an airline has the ability to fly from one country to another and whether networks are indeed serviceable. Airlines are given permission to fly to destinations through ASAs, which can be considered more or less parallel with economic trade policies between countries. ASAs detail the extent to which air services may be allowed (or not) between two countries (and within countries by foreign airlines) and can include provision for the size of aircraft, the number of seats, and departure and arrival times. ASAs are sometimes referred to as aviation bilateral agreements (when two countries sign an agreement for air service) or multilateral agreements (when more than two countries are signatories), but this is more of procedural classification as bilateral and multilateral agreements themselves are not specific to air services and often include ASAs. Air services are governed ostensibly by what are referred to as 'freedoms of the air', of which there are nine (Figure 6.2).

The first five freedoms were established at a special conference held in Chicago in 1944, where the United States and 54 other countries sought to establish a common practice of regulating air routes. This was difficult given the number of countries involved, so it was resolved that individual nations could grant (or not) reciprocal rights to the use of their air space. Gradually, four more freedoms were added (http://www.icao.int/icao/en/trivia/freedoms_air.htm; see also Doganis, 2001; Holloway, 2003). Using official ICAO descriptions, the nine freedoms are as follows.

First freedom – the right or privilege granted by one state to another state or states to fly across its territory without landing; as such, an airline registered in country A can fly to country B by freely flying over country C.

Second freedom – the right granted by one state to another state or states to land in its territory for non-traffic purposes (e.g. fuel or other technical/mechanic reasons); thus, an airline from country A has the right to land in country C en route to country B, but it cannot pick up or drop off passengers in country C.

Third freedom – the right or privilege granted by one state to another state to put down, in the territory of the first state, traffic coming from the home state of the carrier; that is, the right of an airline registered in country A to carry passengers from country A to country B.

Fourth freedom – the right or privilege granted by one state to another state to take on, in the territory of the first state, traffic destined for the home state of the carrier; that is, the right of an airline registered in country A to bring back passengers from country B (thus, the third and fourth freedoms are normally tied together).

Fifth freedom – the right or privilege granted by one state to another state to put down and to take on, in the territory of the first state, traffic coming from or destined to a third; in Figure 6.2, this means the right of a carrier registered in country A can fly to country C, drop off and pick up passengers, and continue on to country B. This will require an agreement between country A and country B as well as country A and country C. Holloway's (2003: 217) definition is as follows: '[A] privilege allowing an airline to uplift traffic from one foreign state and transport it to another state along a route which originates or terminates in that airline's home state.'

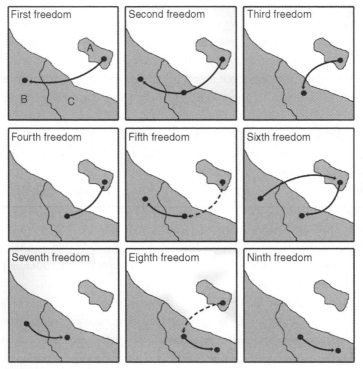

Figure 6.2 Freedoms of the air

Source: Based on Rodrigue (2006)

Sixth freedom – the combination of third and fourth freedoms, 'resulting in the ability of an airline to uplift traffic from a foreign state and transport it to another foreign state via an intermediate stop – probably involving a change of plane and/or flight number – in its home country (e.g. American carrying traffic from London to Lima over its Miami hub)' (Holloway, 2003: 218). Doganis' (2001: 227) definition is as follows: 'The use by an airline of country A of two sets of Third and Fourth Freedom rights to carry traffic between two other countries but using its base at A as a transit point.'

Seventh freedom – the right or privilege granted by one state to another state, of transporting traffic between the territory of the granting state and any third state with no requirement to include on such operation any point in the territory of the recipient state, i.e. the service need not connect to or be an extension of any service to/from the home state of the carrier. In Figure 6.2, this means that an airline registered in country A has seventh freedom rights to carry traffic between countries B and C without stopping in country A. If traffic was required to stop in country A, fifth freedom privileges would be enacted (or required).

Eighth freedom – the right or privilege of transporting cabotage traffic between two points in the territory of the granting state on a service which originates or terminates in the home country of the foreign carrier or (in connection with the so-called seventh freedom of the air) outside the territory of the granting state. 'Consecutive cabotage' is enacted when an airline registered in country A can have service that originates in country A, continues to a first port in country C, picks up and drops of passengers, and continues to another point in country C (Holloway, 2003, who uses the example of a international flight landing first in Hawai'i and then on-flying to Los Angeles). 'Full cabotage' 'is the operation by an airline of services within a single foreign country' (Holloway, 2003: 218), and is referenced by some as the ninth freedom.

Ninth freedom – the right or privilege of transporting cabotage traffic of the granting state on a service performed entirely within the territory of the granting state (also known as 'stand alone' cabotage). This freedom is somewhat rare (as are seventh and eighth freedoms) outside of formal open skies agreements, but it means that an airline registered in country A can operate entirely within, for example, country C without initiating or terminated a service in country A.

The provision of freedoms in air traffic can certainly have influence on competition and, by extension, accessibility and connectivity across networks. For example, the fifth freedom provision to allow airlines to pick up and drop off passengers at intermediate points often triggers discussions regarding protectionist policies of governments towards their own state-designated airlines. One of the most competitive fifth freedom markets in the world is the trans-Tasman route between New Zealand and Australia. Just after 11 September 2001, many fifth freedom carriers (including Thai, LAN, Aerolineas Argentineas, and several others) began flying from New Zealand (usually Auckland) and Australia (usually Sydney, Melbourne or Brisbane). Because these flights originated outside of either Australia or New Zealand, and benefited from the SAM established between Australian and New Zealand, stopping in either destination to drop off or pick up passengers meant access to new markets. Passengers benefited, with traffic across the Tasman growing significantly in the past few years. Overall, it is important to note that the Chicago Convention provided the ground work for the regulation of the airline industry, which was about to enter a period of significant growth. Coupled with this, of course, was the parallel growth in tourist traffic worldwide.

As the nine freedoms of the air demonstrate, governments are responsible for managing their airspace as well as the regulations by which airlines must abide in order to operate (for safety and tax reasons). Many are also responsible (or rather, hold themselves responsible) for managing the business environment in which airlines operate. Many bilateral agreements contain strict clauses governing substantial ownership and effective control regulations pertaining to airline operations between two countries. Airlines designated by a state are thus usually required to be substantially owned by nationals (usually a majority proportion), and effective control may be limited to a certain percentage of non-nationals. These provisions, as Holloway (2003) notes are perhaps the second-most important facet governing international aviation traffic flows and market entry (Holloway, 2003).

Ownership and effective control of a carrier leads to debate over whether they should adopt protectionist mechanisms that effectively protect national 'flag carriers' versus full liberalisation of the operating environment (thus opening up the market to competition). Flag carriers are those that have generally been certified to operate within a country and have thus met often very strict ownership and control provisions. Such provisions state that a certain percentage (often a majority) of ownership of the airline must be in the hands of national companies or citizens. As well, to be a flag carrier it is not uncommon for a percentage of effective control to be registered in the hands of nationals. Thus, it can be argued that by setting these kinds of regulations to which flag carriers must adhere, a certain degree of government protection is realised as they can act to control entry into the market by foreign carriers.

At one end of the protectionist–liberalisation spectrum is full liberalisation, in which governments allow foreign interests to operate or own a majority shareholding in an airline or allow foreign airlines unfettered access to local markets (perhaps even allowing them to operate domestically). In this case, airline operations are only impeded by logistic or infrastructural limitations (i.e. slot space at airports – see Chapter 7). At the other end of the spectrum is a protectionist policy in which a government may elect not to allow foreign carriers access to domestic markets or tightly restrict foreign ownership and control. In some instances, governments may be majority shareholders themselves in a particular airline (for a variety of reasons, but primarily to ensure services are available for the population and, thus, ensuring economic growth in peripheral regions), thus their own interest may dictate the extent to which protectionist policies are enacted. There are problems with a protectionist policy towards air services. As Hanlon (1999: 33) notes:

> Governments traditionally regarded air transport as, in some sense, a public utility. Strictly speaking, it is not. Economists prefer to reserve the term 'public utility' to enterprises that have characteristics of natural monopolies. Natural monopolies exist where the advantages of size are so great that a service can only be provided at least cost if it is supplied by one, and only one firm.

As such, Hanlon (1999) notes that airlines, unlike other service sectors such as telecommunications or electricity grids, have lower fixed-to-variable costs, thus fall outside of the traditional sense of needing full protection from competition.

For the most part, many countries are moving towards a more liberal form of government control of air services. In Europe, bilaterals after 1985 were largely 'open market' types, allowing open-route access and were generally unrestrictive in terms of capacity control (Doganis, 2001). Starting in the early 1990s, India, for example, has radically revised its policies towards competition and access by foreign airlines. Jet Airways was launched in 1993 and now holds a sizeable share of the domestic market.

As a result, Air India, owned fully by the government of India, launched an LCC spin-off, Air India Express, in April 2005. Other airlines, such as Kingfisher Airlines, SpiceJet, Go Airlines and Paramount Aviation also began operations in India in 2005, with two more (IndiGo and Indus Airways) initiated services in early 2006 (Manorama Online, 2005). With a population of over 1 billion, the introduction of competition has stimulated both domestic travel as well inbound and outbound travel (Travel Daily News, 2005).

Another example of the relaxing of protectionist policies is the increase in the number of open skies agreements that have been signed in the past few decades. One example is the SAM between Australia and New Zealand (see also Box 6.2). While not entirely a full open skies agreement, the SAM (established in the late 1990s) is designed to allow New Zealand-based and Australian-based airlines to operate within and between the two countries without barriers or restrictions. Such single market designations are important for two reasons. First, the economic benefits realised can be substantial. In the case of Australia and New Zealand, the SAM is couched in the wider closer economic relations policy that the two countries have adopted. Second, tourism can benefit because airlines, without government restriction, can elect to fly to an airport of their choice (although within capacity restrictions at the airport).

In the past few years, support for further liberalisation across regions and continents has been growing:

1. The European Union, for example, operates ostensibly as a wide, free market area (the EU single air market or more commonly known as the ECAA), designed to allow free movement of goods and peoples across borders. Prior to this, regulatory discretion was focused at the national level. Individual states within the EU continue to ratify bilateral agreements with external states, such as the case with India and the United Kingdom in 2005, which paved the way for increased air services that have been hugely beneficial for passengers in terms of accessibility and fares.

2. The Singaporean government has sought an open skies treaty with Australia for several years. Such an agreement would allow Singapore Airlines fifth freedom access to the United States via Australia. Already, Singapore has the rights to fly to the United States directly, but its fifth freedom traffic is limited to services to New Zealand. Understandably, Qantas, the Australian flag carrier, is concerned that this access could cut into its profitable Pacific routes between Australia and Singapore, particularly as Singapore Airlines is one of the launch customers for the Airbus A380 aircraft (although Singapore Airlines has hinted that it may not utilise the A380 on the Pacific route between Australia and the United States if an open skies agreement is reached, likely to the chagrin of Airbus, which may have hoped to use any operational successes of Singapore Airlines' utilisation in future marketing efforts). Qantas thus continues to urge the Australian government not

to ratify the agreement. Australia may, however, ultimately enter the agreement as part of a wider package of economic negotiations and bilateral agreements in an effort to secure strong relationships with Singapore.

3. Related to the EU's open skies agreement, discussions in November 2005 between the United States and the EU focused on the potential for free market access. New agreements would replace many of the bilateral agreements that the United States holds with many individual countries in Europe, some of which date to the 1970s. In fact, in November 2002 the European Court of Justice ruled that these multiple (and varied) agreements (covering routes, capacity, frequency and even fares in some instances) violated the EU Community Law. Many European nations, particularly Britain, have argued that the United States needs to eventually relax controls on its own domestic market, thus allowing foreign ownership and rights for foreign airlines to operate domestically. On 18 November 2005, a tentative agreement was reached (the first of several to come), allowing European carriers to depart from countries other than their own for the United States (ostensibly, fifth and seventh freedom rights). Some commentators (e.g. Firey, 2003) argue that a full open skies agreement would most likely benefit European carriers by way of mergers and alliances with American carriers as opposed to actually competing with them. However, given that a sizeable number of American carriers operating across the Atlantic are currently under bankruptcy protection, the attractiveness of such mergers with European carriers may be limited.

4. Also in November 2005, the United States and Canada ratified an updated version of their existing open skies agreement where airlines of each country will have complete fifth freedom access. In other words, a United States carrier can, for example, fly from Chicago to Toronto and then onward to London. However, the benefits to passengers may not be immediate as there is already in place numerous trans-Atlantic and Caribbean-bound flights from both countries so it is, at the time of writing, difficult to surmise what new routes may be added as a result of this revision.

Not all countries have embraced the liberalisation of air services. South African Airways, for example, has been attempting to negotiate the right to pick up passengers in Accra (Ghana) bound for the United States. At present, South African Airways flies to New York, Washington and Denver. The Ghanian government, however, is refusing access to the airline and the South African newspaper *Business Report* has suggested that protectionist policies are to blame, with Ghana worried that the airlines will capture, disproportionately, the Ghanian market at the expense of future Ghanian airlines (given that Ghana Airways ceased operations in 2003) (d'Angelo, 2005). A similar example revolves around the actions of the United States government with respect to Unites States designated carriers post-11 September 2001. In September 2005, British Airways PLC Chief Executive Rod Eddington suggested that the United States' protectionist policies of 'propping up' failing US-based airlines is unwise, noting: 'They're operating in protected markets, they're hoovering up public funds and still they can't make a profit' (Forbes, 2005). The implication is that US airlines operating under federal bankruptcy laws continue to operate at a loss across the Atlantic Ocean, thus preventing true competition from Europe-based carriers.

INTERNATIONAL REGULATORY FRAMEWORKS: INDUSTRY ORGANISATIONS

There are other international regulatory frameworks designed to represent common procedures. The most comprehensive of these is a series of articles and regulations established by NGOs such as the ICAO and the IATA. ICAO, a specialised agency of the United Nations, manages several aspects of aviation as agreed to by most world governments. It has recently adopted six strategic objectives for the period 2005–2010:

A. Safety – Enhance global civil aviation safety
B. Security – Enhance global civil aviation security
C. Environmental Protection – Minimize the adverse effect of global civil aviation on the environment
D. Efficiency – Enhance the efficiency of aviation operations
E. Continuity – Maintain the continuity of aviation operations
F. Rule of Law – Strengthen law governing international civil aviation. (www.icao.org)

IATA is a global industry association targeting air carriers by helping to 'ensure that Members' aircraft can operate safely, securely, efficiently and economically under clearly defined and understood rules' (www.iata.org). IATA acts to assist members with customs regulations, governmental regulatory policies and legal causes. It also operates the Simplifying Passenger Travel programme (www.simplifying-travel.org), which is a global programme that is designed to accommodate and promote efficiencies across the entire air travel experience (i.e. not just in the air). IATA's efforts are also designed to be of assistance to consumers/passengers, as stated on their website:

> For consumers, IATA simplifies the travel and shipping process. By helping to control airline costs, IATA contributes to cheaper tickets and shipping costs. Thanks to airline cooperation through IATA, individual passengers can make one telephone call to reserve a ticket, pay in one currency and then use the ticket on several airlines in several countries – or even return it for a cash refund. (www.iata.org)

Understanding the regulatory environment in which airlines operate is important for tourism because it dictates the networks and the performance of these networks. What can be concluded from this section is that tourist flows are highly political in that they are determined ultimately through the negotiations between two or more governments. As a result, it can be argued that the basis through which tourism exists at an international level, at least those tourist flows that rely on international air transport, is essentially an exercise in negotiations of access.

SUSTAINABLE AIR TRANSPORT

Several recent studies have outlined the environmental damage that air travel can cause. In 2000, the United States GAO published a report (GAO, 2000) that outlined the concerns with aircraft emissions and their impact on the atmosphere:

- Jet aircraft are the primary source of human emissions deposited directly into the upper atmosphere. The Intergovernmental Panel on Climate Change and experts noted that some of these emissions have a greater warming effect than they would have if they were released in equal amounts at the surface – by, for example, automobiles.
- Carbon dioxide is relatively well understood and is the main focus of international concern. According to the Intergovernmental Panel on Climate Change, it survives in the atmosphere for about 100 years and contributes to warming the earth. Moreover, as noted, global aviation's carbon dioxide emissions (measured in million metric tons of carbon) are roughly equivalent to the carbon emissions of certain industrialized countries.
- Carbon dioxide emissions combined with other gases and particles emitted by jet aircraft-including water vapor, nitrogen oxide and nitrogen dioxide (collectively termed NOx), and soot and sulfate – could have two to four times as great an effect on the atmosphere as carbon dioxide alone. According to the Intergovernmental Panel on Climate Change the atmospheric effects of these combined emissions will require further scientific study.
- The Intergovernmental Panel on Climate Change recently concluded that the increase in aviation emissions attributable to a growing demand for air travel would not be fully offset by reductions in emissions achieved through technological improvements alone. (GAO, 2000: 5)

The RCEP (UK) issued a report in 2002 that examines the impact of radiative forcing from aircraft emissions in the atmosphere. Radiative forcing is a 'convenient measure of the greenhouse effect of a change in a constituent is provided by the imbalance between solar and thermal radiation at the tropopause when the change in the constituent is suddenly imposed' (RCEP, 2002: Box 3A, 10; see also Egli, 1991). The tropopause is the distinct transition between the troposphere and the stratosphere. Aircraft generally travel between 9 km and 13 km above the Earth's surface, and very close to the tropopause which can vary depending on latitude. The RCEP (2002: 18) notes that '[t]he total radiative forcing due to aviation is probably some 3 times that due to the carbon dioxide emissions alone. This contrasts with factors generally in the range 1–1.5 for most other human activities.' The impacts of aircraft emissions, according to the RCEP's report, can

181

be different in the stratosphere and the troposphere, but the exact implications for the lower bands of the stratosphere are not exactly known. It has been suggested that water vapour from emissions may have some impact on cirrus cloud formations (Seinfeld, 1998), thus suggesting that the height at which aircraft travel could be altered depending on local climatic situations:

> Modern weather forecasting capabilities are increasingly such that the regions of likely supersaturation in the upper troposphere and the height of the tropopause in any region may be usefully predicted some days in advance. When there is more scientific understanding of the various elements involved in the climatic impact due to aviation, it should be possible to route individual aircraft so that, for example, they spend less time in regions where persistent contrails and enhanced cirrus cloud could be formed, or so that they almost always remained in the troposphere where the water vapour effects are negligible. (RCEP, 2002: 19).

In order to reduce the production of airborne particulates and emissions stemming from aircraft, some have advocated the use of an emissions tax, although such a tax was reasoned by Olsthoorn (2001) as having only a marginal impact on carbon dioxide levels. At Zurich airport, emissions charges are levied against aircraft based on the amount of pollutants they are known to emit; thus, larger aircraft are taxed at a higher rate than smaller aircraft. The European Commission is actively investigating the feasibility of introducing a region-wide emissions tax or, most significantly, extending carbon emissions trading to the aviation sector, which was being considered in mid-2004 (Wastnage, 2004; see also CE Delft, 2005) (see Chapter 9 for a discussion on Pigouvian taxes versus absolute emission reductions). The BBC reported in 2005 that aircraft are responsible for 3% of carbon emissions in the EU. Airlines would likely oppose the introduction of a new tax at a time when oil prices, and consequently jet fuel, are impacting upon profits.

The problem, however, is that these kinds of approaches do little to make the air transport industry more sustainable. Taxes or carbon trading may go some way to alleviating the burden of emissions on any one country, but true change may rest in alternatives brought on by technological innovation. One group, the Sustainable Aviation Group, comprised of British Airways, Virgin Atlantic, Airbus UK and the British Airports Authority, suggest that it is in fact technological innovations that are needed to stem emissions, yet some environmental groups (including the 'Green Sky Alliance' in the United Kingdom) and organisations point to the potential for rapid increases in greenhouse gases as a result of aircraft emissions and regardless of improvements made to engine technology (Webster, 2005). Virgin's Richard Branson was reported in November 2005 to be considering a new venture in which alternative aviation fuel can be developed. Using waste material from plants, the 'cellulosic ethanol' will be 100% environmentally friendly, but at the time of writing engine manufacturers such as General Electric and Rolls Royce had yet to speculate the amount of research and development needed to make alternative, sustainable aviation fuel a reality (ABC News [Australia], 2005).

Virgin Atlantic has arguably been at the forefront in environmentally friendly air transport activities. The company's website highlights several examples:

- Virgin Atlantic, the Group's flagship airline, has already developed its own Corporate Responsibility Policy and Strategy for Environmental Sustainability. Virgin Atlantic recognises the importance of minimising the impact of its operations on both the local and global environment through improving fuel efficiency, reducing emissions and noise.
- Virgin Atlantic has one of the youngest (and therefore more fuel-efficient) long haul fleets in the world.
- We have committed to meet the new EU limits on NO2 emissions which come into force on 1st January 2010.
- Over the past 30 years, the average fuel efficiency of an aircraft per passenger kilometre flown has improved by 50 per cent, thanks to improvements in airframe design, engine technology and increased load factors. Fuel represents a colossal proportion of operating costs, so using it efficiently is a commercial as well as environmental priority. Compared with its competitors, VAA's aircraft carry a higher number of passengers per air transport movement, and operate with a higher load factor – which means greater fuel efficiency.
- VAA's entire fleet complies with the ICAO Chapter 3 noise standards. VAA is monitoring its aircraft noise and seeking to set targets for its reduction. (http://www.virgin.com/aware/future.asp; accessed 18 November 2005)

Given the obvious contribution of aircraft to pollution and greenhouse gas production, where does this leave tourism? Ironically, the very mode of transport that carries 'ecotourists' or 'green tourists' may in fact cancel out any gain from their immediate, on-the-ground activities designed to be sustainable, 'eco', or 'green'. Gössling *et al.* (2002: 210) sums this up appropriately:

From a global sustainability and equity perspective, air travel for leisure should be seen critically: a single long-distance journey such as the one investigated in this survey requires an area almost as large as the area available on a per capita basis on global average. This sheds new light on the environmental consequences of long-distance travel, which have rather seldom been considered in the debate on sustainable tourism. Taking these results seriously, air travel should, from an ecological perspective, be actively discouraged.

CHAPTER SUMMARY

This chapter has outlined several contentious issues surrounding the provision of air services worldwide, including LCCs, airline alliances, the role of regulation (particularly bilateral ASAs) and environmental implications. Within the framework of networks, various network options were outlined, and it is important to note that no single network type is ideal in all circumstances. In Europe, for example, short flight sectors means linear networks can be made to be profitable. In other countries, hub-and-spoke domestic networks can be profitable by linking hinterland regions through spoke services.

Without question, airlines worldwide operate in 'turbulent' environments, both politically (with respect to bilateral and open skies agreements) as well as economic. Despite numerous bankruptcies since 11 September 2001, the airline sector is starting to show signs of (slow) recovery. Profits for some carriers have increased (although carriers such as Southwest and Ryanair generally prospered while network carriers, particularly in the United States, suffered immediately following 11 September 2001) and traffic levels are slowly recovering thanks to demand in China and India. Aviation fuel (also known as JET-A1) cost increases have meant an increase in the number of airlines applying fuel surcharges on top of ticket prices, leading some to suggest that the industry itself is quite unique in that very few other industries allow for the addition of surcharges to make up for operational costs. LCCs, or the LCLF, is spreading rapidly around the globe, having originated in the United States but fully expanded in Europe. The European adoption of LCLF carriers has, in part, been assisted by the dense network and strong connectivity possibilities throughout the region. In many cases, as discussed, it is difficult to distinguish between LCLF and FSAs as many FSAs have recently adopted drastic cost-cutting measures as a means of increasing profit.

SELF-REVIEW QUESTIONS

1. What is a hub-and-spoke network?

2. What are the nine freedoms of the air?

3. What are the key benefits, for both passengers and airlines, of airline alliances?

Box 6.2 Case study – The (failed?) Qantas/Air New Zealand SAA (based on Duval, 2005a).

In early December 2002, both Air New Zealand and Qantas (Air Pacific, based in Fiji, was also named on the application because Qantas held 46.32% of shares in the company; Air New Zealand held 1.9% and the Fijian government, as majority shareholder, held 51%) sought authorisation from the NZCC to ratify a 25 November 2002 SAA between the two parties that would, in effect, result in a JAO network across the Tasman Sea and, more generally, throughout the Pacific theatre. The proposed SAA included elements from all three of Tretheway and Oum's (1992) typology of airlines alliances. The applications filed by both Air New Zealand and Qantas to their respective governmental bodies were contextualised in several existing international policy environments, including:

1. the New Zealand-Australia closer economic relations trade agreement (established in 1983 and designed to harmonise trade relations between the two nations);

2. the SAM (created to foster cooperation in customs matters and ostensibly allowing commercial air carriers to designate their 'home market' as both Australia and New Zealand); and

3. the TTMRA (introduced in 1996 with the aim of allowing the free flow of goods and skilled personnel between the two countries).

The applicants (Air New Zealand and Qantas Airways Limited, 2002a, 2002b) argued that there were numerous benefits to the proposed alliance:

1. Cost savings: existing markets served under the SAM between Australia and New Zealand did not warrant the profitable operations of two FSAs.

2. Scheduling and direct services: it was argued that the alliance would effectively seek to tie customer and market bases together and allow for the management of inventory to be more efficient. One area in which customers/passengers would benefit was the code-sharing of some routes in order to allow feeder services into New Zealand from Australian domestic networks.

3. Tourism: the applicants argued that they were, quite literally, two of the largest organisations involved in the promotion and marketing of tourism for both

Australia and New Zealand. They argued that it would not be in their own best interests to jeopardise this through price collusion if the alliance were approved.

Assessing the impact of potential airline alliances is not easy. In the New Zealand context, several considerations were addressed by the NZCC:

1. What kind of impact would the proposed alliance have on domestic, trans-Tasman and trans-Pacific markets and operations; and

2. what would it mean to have Qantas own, as suggested by the Agreement, almost 25% of Air New Zealand? Under Section 67 of the New Zealand Commerce Act of 1986, Qantas was seeking to purchase 22.5% of shares in exchange for NZ$550million. There were, in effect, two options for the NZCC to consider:

(a) in the 'factual', allow Qantas to own nearly 25% of New Zealand's flag carrier, but at the same time recognise that some, if not all, of the arguments made in favour of the proposed Agreement (e.g. efficiencies, cost savings) would be passed on to the consumer; or

(b) in the 'counterfactual', allow Air New Zealand to operate on its own in an uncertain, global economic environment which, as discussed in this chapter, has already meant some airlines going bankrupt.

The decision was handed down in October 2003: the Agreement was rejected by the NZCC. This decision was appealed by Air New Zealand (on behalf of the applicants) in July 2004 to the High Court at Auckland, but the High Court endorsed the NZCC decision in September 2004. As of 2005, the airlines have not publicly discounted the possibility of attempting a merger or alliance of some sort in the future. While this case study has examined the New Zealand view of the proposed SAA, it is interesting to note that the ACCC, the Australian equivalent to the NZCC, also rejected the application on similar grounds, but the Australian Competition Tribunal, upon appeal, threw out the ACCC ruling and granted authorisation for the SAA to move forward. By this point, however, both airlines had moved on (and were still subject to the High Court ruling in New Zealand) and further options were being considered. One such option was the pursuit of an 'air share agreement' that would effectively allow both carriers to sell seats on either's aircraft.

One issue with respect to the timing of the alliance deserves some speculations. In a business environment, it is often said that one must keep a close eye on one's competition. At the time the SAA proposal was launched, the financial health of Air New Zealand was under question. It had recently been recapitalised by the New Zealand government and it was facing significant competition across the Tasman Sea by Qantas as well as numerous other 'fifth freedom' (discussed elsewhere in this chapter) carriers. Qantas had already established domestic operations along the main trunk routes in New Zealand through the operations of Jet Connect, who leased the Qantas brand. Qantas' financial health was strong. One theory that was floated in some aviation circles at the time was that the proposed SAA was one way for Air New Zealand to closely watch the operations of its major Pacific competitor. In so doing, Air New Zealand could be reasonably sure that any strategic or surgical move on the part of Qantas to the detriment of Air New Zealand (routes, fare structures) would be illogical given the pair were attempting to secure a strategic alliance. Granted, there were provisions in the SAA that would clearly benefit Qantas (i.e. easy access to the New Zealand market, greater control of the Pacific routes between New Zealand/Australia and the United States), but something to consider is whether, from the perspective of Air New Zealand, the SAA was more of an exercise or example of keeping a competitor at bay while internal re-organisation and operational efficiencies are improved.

Questions for consideration

1. Why did the NZCC claim that the SAA between Air New Zealand and Qantas was not in the public good?

2. Under what circumstances might the following react negatively against an alliance similar to the one discussed: (a) a major international airport; (b) a small regional airport?

ESSAY QUESTIONS

1. Justify the introduction of non-stop services between two places; why would an airline offer services non-stop when it could offer a segmented route that shuttles passengers between multiple destinations along a linear network?

2. Research a current bilateral agreement between two countries and determine whether there are any 'barriers to entry' for other carriers.

3. Why can it be said that many FSAs are not much different from LCCs?

KEY SOURCES

Doganis, R. (2002) *Flying off Course: The Economics of International Airlines*. London: Routledge.

This book is one of the more comprehensive and accessible examples of recent titles seeking to examine and explain the complexities of the modern global aviation system. Doganis covers several topics well, including route network planning, yield management, cost control and marketing.

Holloway, S. (2003) *Straight and Level: Practical Airline Economics*. Aldershot: Ashgate.

Holloway's text is superb in its treatment of the primary economic variables that shape airline operations. Particularly useful is his discussion on yield and revenue management, unit costs and scheduling. Although based in economics, the text is accessible to those who may not have formal training in advanced economic concepts.

International Air Transport Association – www.iata.org

IATA is an airline industry association that represents approximately 260 airlines around the world. The goal of the IATA is to essentially represent the interests of member airlines with respect to regulations, safety and security and passenger management, among others, but in a global context. The opinions from the IATA site should be compared to that of the Airports Council International (www.airports.org) site.

CHAPTER 7:

MANAGEMENT OF TRANSPORT FLOWS

LEARNING OBJECTIVES

After reading this chapter, you should be able to

1. Understand the history and importance of privatisation and liberalisation of transport infrastructure.

2. Appreciate the role of airports as hubs and nodes in the global transportation system.

3. Understand the principles of yield management in transport operations.

4. Recognise and evaluate recent advances in security and safety in relation to transport.

5. Assess the impact of external events on transport operations.

INTRODUCTION

Transport management incorporates several fields of enquiry, including, for example, transport economics (Button, 1993), international trade relationships and impacts of transport cost (Rauscher, 1997), business logistics and operations management (Gubbins, 2004), information systems, planning (Banister, 2002), and legal frameworks. While limitations in space prevent a thorough overview of each of these issues, the intent of this chapter, however, is to isolate three critical management applications in the context of tourism flows' relationship with transport.

First, the management of nodes within networks of travel are examined, with the modern airport used as an example of a node that acts as a conduit for travel and tourism. As shown, airports today function not only as hubs in order to facilitate the transfer of passengers from one place to another, but they also (and sometimes necessarily) function as centres of commerce. Second, the principle of yield management is outlined. Yield management is a complex method (yet reasonably simple in principle) by which airlines (and other transport-related industries) price their product/service. The third management application examines safety and security measures in transport. As is shown, a variety of industries associated with transport pay close attention to safety and security issues. The chapter, therefore, isolates specific examples where careful managing of transport-related procedures and management issues can overtly impact on tourism development and/or tourist flows. The selected three management functions described herein, and taken together, illustrate both the complexity of the transport/tourism relationship and the importance of understanding how tourism and travel in general can be impacted by fluctuations in the various components and transport operational factors.

PRIVATISATION AND LIBERALISATION: TRANSPORT INFRASTRUCTURE

In order to situate management implications, it is important to briefly note that, much like the actual modes of transport, the ownership and management of infrastructure, as a resource necessary for the provision of transport services, has undergone processes of liberalisation and privatisation in many countries. This is perhaps best exemplified through the increasingly popular process of liberalising and/or privatising airports, but can also apply to rail tracks once under the management of governments. Privatisation is meant to imply efficiencies and increased profitability, the latter of which generally focuses on increased commercialisation (Oum *et al.*, 2003; see also Chin, 1997). It also allows for access to more private investment to fuel growth, as in the case of the BAA, where capital expenditure before privatisation was constrained by government policy (Doganis, 1992). While the BAA often emerges as the bellwether by which the successes of airport privatisation can be identified (Doganis, 1992), throughout Europe airport competition and gradual privatisation has been heightened following the deregulation of the airline industry in the 1990s (Barrett, 2000). In Australia and New Zealand, airport privatisation was initiated at roughly the same time (Forsyth, 2002), while privatisation efforts in Asia have been comparatively slow as governments are concerned with possible monopolistic abuse (Hooper, 2002). (Their concern is not without foundation, as Gerber [2002] argues that sufficient regulation needs to pre-date any privatisation effort as it helps stimulate protection for user, balance for the various stakeholders involved and helps stimulate partnerships between government and private enterprise.)

In the United States, the privatisation of airports has not been as widespread as in Europe. Some US airlines are steadfastly opposed to such a move, largely because it would introduce competition at a level over which they may feel they have little control due to their own increasing operational costs. As well, airlines and airports in the United States have historically worked together under airport use agreements that govern the charges levied by the airport and prescribe the use entitled to the airlines (Graham, 2001). In fact, Graham (2001) notes that airlines have substantial input into future development at US airports. The argument could be raised that, in the case of major airports where one airline may account for a substantial amount of operating revenues, this kind of relationship or agreement is problematic: not only does the dominant airline have substantial control of route structures and pricing schemes, but it also may have considerable input in future development plans that may threaten its dominant status. Indeed, Doganis (1992: 30) points out that the risk of monopolies is possible as 'airport managers may reduce space for passenger and cargo shippers in order to maximize revenues from a variety of commercial activities'. The differences in liberalisation and privatisation of transport infrastructure and operations in the United States and the United Kingdom can be partially explained by the FAA mandate (and, by extension, the United States government) to ensure efficient operation of the nation's air transport system (Ellett, 2003). As well, in the UK it could be argued that there is more modal competition for travel and haulage of goods (Starkie, 2002), thus prompting some liberalisation of the transport infrastructure in order to ensure profitability.

What is the most appropriate level of ownership-management? This is difficult to state clearly, largely because there are few mechanisms in place to adequately compare the output of airports (Hooper, 2002). The situation with airports is somewhat mirrored in other transport modes. For example, Welsby and Nichols (1999) point out that the privatisation of the UK rail system perhaps overlooked future demand for infrastructure. As a result, and this situation is mirrored in New Zealand as well, governments may, in some cases, need to assume the responsibility for capacity enhancement through infrastructure maintenance and construction.

NODAL MANAGEMENT: AIRPORTS AS HUBS

Following on the discussion from Chapter 2, where the spatial distribution of transport in relation to tourism was discussed, the purpose of discussing nodal management is to highlight the issues involved in managing the flows of tourism traffic. Nodes can often act as hubs, and there are significant considerations for how transport provision is made possible (and profitable) by what occurs at these hubs. The majority of the following discussion focuses around airports, largely because they play an important role in global traffic, but also because they are integrated into the transport/tourism relationship on a number of levels, including facilitating traffic flows, impacting upon regional development, and securing other economic capital flows within the countries in which they reside.

One of the major areas in which the role of the airport contributes to tourist flows is the issue of capacity constraints, particularly as many existing airports, especially international hub airports such as London Heathrow, struggle to keep up with increasing demand.

As Starkie (1998) points out, as airports become popular (for reasons of geography) there is increasing demand for the services they offer. The magazine *Airline Business* argued in June 2005 that the crowded airports of Europe, for example, are currently facing a congestion crisis as they strive to meet increasing demand, especially where some (e.g. Airports Council International) are suggesting that demand is returning to pre-11 September levels (Airline Business, 2005a). In the United States, 35 of the nation's most crucial airports for managing the flow of air traffic are now operating at pre-2001 levels (Airline Business, 2005b). Additionally, the boom in air travel in India over the past few years is also affecting two of the country's biggest airports (New Delhi and Mumbai). The IATA has reportedly suggested that India will need to work fast to increase capacity at these hubs in order to facilitate the rapid projected growth in India's air markets (Mahapatra, 2005). What this means for tourism is that the networks that serve international tourist flows are becoming congested as the number of major nodes operating at peak capacity grows.

One way of alleviating capacity constraints and congestion at many airports is to construct new runways. The United States has opened eight new runways since 1999, with a further seven to be opened by 2009 (Airline Business, 2005b). The challenges to opening new runways, however, are enormous, despite the benefits towards alleviated congestion (adapted from McCartney, 2005):

- Atlanta's (Hartsfield Airport) fifth runway project has been in progress (planning through to the construction) for nearly 25 years. In that time, new housing developments near the airport have meant increasing community opposition.
- At Boston's Logan International Airport, a new runway specifically for turbo-props needing federal and local approval has taken almost 30 years because of community and legal battles.
- Environmental opposition to two new runways to be built on landfill at San Francisco has meant that the runway development has been scrapped.

Major international hub airports, such as JFK in New York or Heathrow in London, do not, of course, have the ability to handle unlimited traffic, even if new runways are constructed. All airports have a finite number of gates and internal (airside) capacity limits in the number of people they can service. Customs and immigration areas are also limited in their resource allocation. There are also capacity limits on the use of the runways themselves (Cao & Kanafani, 2000) as departures and take-offs need to be managed in accordance with existing international (and local) safety regulations (e.g. minimum distances of separation when landing and taking off). In short, most airports are 'capacity restricted' as well as capacity constrained, which is one way of suggesting that quick fixes are not always suitable solutions.

OPERATIONS MANAGEMENT: MANAGING CAPACITY THROUGH AIRPORT SLOT ALLOCATION

Capacity restrictions at airports mean that an airline cannot simply decide to start flying to a particular airport, arriving at a time of their choosing. Unlike ground tourist transport, which is comparatively free-flowing (i.e. bus tours, which may be governed by some time restrictions, often can show up at attractions with only a rough estimate given for arrival), air transport is highly scheduled and managed to maximise resource allocation and profit across the numerous operations that are involved (e.g. airports, airlines, ancillary services). The amount that any particular airline can utilise an airport is closely managed through the allocation of slots. Slot space is usually not allocated and governed entirely by the airports themselves. Instead, at most major airports around the world allocation is governed through a series of guidelines set out by the IATA, which functions as the global airline trade association. At the same time, however, certain 'grandfather' rights are recognised, so if a particular airline has historically (i.e. perpetually over a significant period of time) utilised a landing slot at a particular airport which has, very recently, suddenly become popular and in demand by other airlines, that airline has some rights to declare its use of the newly in-demand slot(s) on the basis of its grandfather rights. In some situations, 'historical' use of airport (or, more properly, runway) slots is based on use in the previous equivalent season (Starkie, 1998).

In order to secure slots at particular airports, many airlines regularly engage in slot trading at either one of the biannual conferences established by IATA (although it should be noted that not all world airlines are members of IATA). When slot space becomes available at a certain capacity-restricted airport, airlines can often bid large sums of money to acquire the rights/permission to fly there. In the United States, however, anti-trust regulations prohibit airlines from meeting to discuss mutually beneficial route structures and timings. As Starkie (2003: 53) outlines, market forces generally dominate the allocation of slots at US airports as

> there are few restrictions at US airports limiting the allocation of landing and take-off slots; for example, airline scheduling committees, which an important slot management outside the USA, do not operate at US airports. Airlines simply schedule their flights taking into account expected delays at the busier airport. Essentially, slots are allocated on a first come, first served basis, with the length of the queue of planes waiting to land or take-off acting to ration overall demand.

Because of capacity restrictions, airports carefully manage their operations (both airside and in general) in order to ensure that the amount of traffic coming in (and going out) is both manageable and profitable (from aeronautical charges). For example, aircraft rotations or movements (i.e. take-offs and landings) are generally relegated to daylight hours (usually as a result of zoning and noise restrictions), thus airport management can accurately plan for the most efficient use of specific resources. Baggage handling, for example,

193

can be scheduled in accordance with incoming aircraft movements, so when multiple long-haul aircraft arrive at only certain times of the day or week, staffing can be allocated appropriately. Similar arrangements can be applied to customs officials, although this is usually embedded within an arrangement with national governments as customs and immigration fall under its purview.

Capacity restrictions on operations lead to slots or times that are inherently more valuable to airlines than others. Suppose an airline wishes to capture the business travel market with a flight that departs London Heathrow at 0800 and arrives at Charles de Gaulle (Paris) at 0910, a useful arrival time for business travellers who are perhaps only in Paris for the day. The problem, however, is that London Heathrow can be congested at this time of the morning, with numerous flights leaving for a variety of continental (and international) destinations. A slot for an 0800 departure may therefore not be available. Operating the flight at a later time in the morning, say 1100, may be possible due to available take-off slots, but it may not appeal to business travellers who may need a full day, for example, in Paris. It is not surprising, then, that capacity constrained airports, such as London Heathrow, actually act as a barrier to entry for some upstart airlines because the profitable and highly desirable slots are already occupied by established airlines. This is one of the reasons why many LCCs have moved to the use of secondary or regional airports (such as a London Luton or London Stansted), which may have slots available and may not be as capacity constrained. In general, the inability of new entrants to a particular market is made difficult by other factors, as Hanlon (1999) shows:

1. Airlines must acquire 'matching slots' at destinations. In other words, if a new upstart wishes to commence service from one particular airport, it must also acquire landing slots at the destination.

2. Slot trading is in place at several airports in the United States, where particular slots may be bought and sold. This raises the issue of the potential dominance of one or two airlines at a particular airport (sometimes called fortress hubs) thus limiting the ability of new entrants to initiate use.

3. Some incumbent or established carriers will utilise slots for the purpose of preventing new entrants even when policies regarding slot reallocation are introduced.

4. Slots that are reallocated (on the basis of non-use of existing slots or new slots) are often at times of the day that are not feasible and not as lucrative.

Not surprisingly, the value of slots, especially those at peak times, has not escaped the attention of airport operators. In April 2006, European Union transport commissioner Jacques Barrot hinted that congestion charges may be levied on airlines using European airports during the busiest periods (Minder, 2006). The real value, however, is in the slot itself. In 1993, the European Commission introduced Council Regulation 95/93 on 'common rules for the allocation of slots at community airports' (http://europa.eu.int/scadplus/leg/en/lvb/l24085.htm; accessed 31 August 2005). In the preamble to the regulation, the EEC (Council Regulation) stated:

> One of the main difficulties of the current system of slot allocation has been to find the right balance between the interests of incumbent air carriers and new entrants at congested airports so as to take due account of the fact that incumbent air carriers have already built up their position at an airport and have an interest to expand it further, while new entrants or air carriers with relatively small operations need to be able to expand their services and establish a competitive network.

To address this, the EEC (http://europa.eu.int/scadplus/leg/en/lvb/l24085.htm) established a series of procedures by which slots are allocated:

- A carrier using a time slot that has been cleared by the coordinator is entitled to claim the same slot in the next scheduling period. In a situation where all slot requests cannot be accommodated to the satisfaction of the air carriers concerned, preference is given to commercial air services and in particular to scheduled services and programmed non-scheduled air services.
- The coordinator also takes into account additional priority rules established by the air carrier industry and, if possible, the additional guidelines recommended by the coordination committee allowing for local conditions.
- If a requested slot cannot be accommodated, the coordinator informs the requesting air carrier of the reasons therefore and indicates the nearest alternative slot.
- Slots may be freely exchanged between air carriers or transferred by an air carrier from one route or type of service to another.
- A Member State may reserve certain time slots for regional services.
- The Regulation provides for the setting up of 'pools' containing newly-created time slots, unused slots and slots which have been given up by a carrier or have otherwise become available.
- Any slot not utilised is withdrawn and placed in the appropriate slot pool unless the non-utilisation can be justified by reason of the grounds of the grounding of a certain type of aircraft, or the closure of an airport or airspace or any other exceptional reason.
- Slots placed in the pools are distributed among applicant carriers. 50% of these slots are allocated to new entrants unless requests by new entrants are less than 50%.

Important within Regulation 93/95 is, as discussed earlier, the preservation of grandfather rights of incumbent airlines. Once allocated, carriers are required to utilise slots 80%

of the time in order to retain future grandfather rights. In doing so, an airline may utilise a particular slot solely for the purposes of creating a barrier to entry for a new upstart carrier and thus maximising their market share. (As discussed earlier, competitors have found a way around grandfathered slot allocations by establishing operations at secondary airports.) Indeed, Boyfield (2003: 34) has commented that many have seen the Regulation as 'antiquated' because it is actually centred around the concept of grandfather rights, so much so that '[c]arrriers have little incentive to hand back slots they hold at peak times, no matter how inefficiently they are used'.

Protests over the grandfathered slots have not gone unnoticed, however. Hanlon (1999) notes that a 1995 review of slot allocation by the UK Civil Aviation Authority recommended that all new slots or non-utilised slots should be turned over to new entrants. Although the system still exists in principle, it was not until 2004, when the EEC revised the regulations (EEC 793/2004; Commission of the European Communities, 2004) to widen the definition of new entrant airlines and established a list of priorities when allocating new or unused slots, that change was enacted. More significantly, the EEC later called for submissions exploring the possibility that commercialising the slot allocation system (through free-trading and auctions under the principle of liberalisation) might render the system more efficient (Paylor, 2005). Airlines, however, were concerned that the grandfather rights they hold to slots at major hubs might be jeopardised if such a system were introduced.

In fact, the commercialisation of slots has already begun, but it is far from transparent in many cases. A recent issue of *Airline Business* magazine characterised the 'grey' slot market at Heathrow as somewhat of a 'murky world'. In 2004 it was reported by the *Guardian* (Gow, 2004) that Qantas paid approximately £20 million each for two pairs of daily slots at Heathrow. The slots were previously held by LCC Flybe (formerly British European). Commercial transactions of slots, then, may ultimately favour larger carriers with access to large amounts of capital. With respect to the revised 2004 EEC regulations, the scarcity of slots at EU airports would be measured in monetary terms such that underutilisation of existing slots by incumbent carriers is financially compensated by local authorities, who would have the power to pull slot allocation(s) and reallocate to a new entrant who may promise to maximise utilisation. Although the slots themselves would be commercialised, financial compensation for incumbents, if they happen to underutilise, may be more preferable than allowing market forces to outright determine allocation. The reason for this is that new entrants would be restricted in their ability to acquire lucrative slots because of 1) incumbents refusing to sell slots or 2) local authorities unwilling to exercise authority (if any) of removing slots from incumbents and reallocating to new entrants.

Both the UK CAA and the UK government supported the introduction of the commercialisation of slots, noting that it would have an overall positive impact on the European aviation marketing (CAA, 2004; Department for Transport, 2004c), but this proposal to unleash slots to market forces has been concerning for some. The European Regions Airlines Association (*Flight International*, April 2004 issue) warned that such a system for slots may be detrimental for regional airports. It claimed that airlines operating out of larger hub airports (such as Heathrow, Frankfurt Main and Paris Orly) could shift their operations in favour of long-haul routes to the detriment of smaller regional routes. As well, smaller regional airlines may find servicing regional and major hub airports to be unprofitable, thus leaving some regions (and thus, travellers) without sufficient access to major hubs. Table 7.1 provides an overview of the concentration of airport activity throughout the UK. Critical within this table is the extent to which smaller regional centres such as Leeds/Bradford, Liverpool and East Midlands are proportionately smaller than larger airports such as Heathrow, Gatwick or Stansted.

Table 7.1 UK airport activity, 2002 (arrivals and departures at airports handling more than one million passengers)

	*Domestic**	*Scheduled*	*Non-scheduled*	*Total*
Newcastle	1165	626	1597	3387
Manchester	2743	6463	9412	18,618
Leeds/Bradford	477	486	563	1526
Liverpool	736	1685	414	2835
East Midlands	548	1097	1588	3233
Birmingham	1222	3762	2928	7911
Luton	1745	3665	1064	6474
Stansted	2461	12,385	1203	16,049
Heathrow	6674	56,237	124	63,035
Gatwick	3427	15,029	11,062	29,518
London City	417	1184	1	1602
Bristol	925	1194	1296	3415
Cardiff	108	344	964	1416
Aberdeen	1613	407	529	2549
Edinburgh	5079	1418	415	6911
Glasgow	4297	1157	2315	7769
Belfast City	1886	1	3	1890
Belfast International	2683	182	686	3551
Other UK airports	3877	1587	1726	7190

Source: Adapted from CAA (2005)

*Note: * Domestic traffic is counted at airports on arrival and departure.*

There are other concerns with the slot system, regardless of whether it becomes fully (or even partially) commercialised, that are of concern to the flow of passengers. Chief among these is that potential for passengers having limited choice in air services, and indeed the congestion and hoarding of slots at smaller regional airports in the European

Union, was one of the reasons the initial 1993 slot allocation procedures were reviewed. In other words, when one or two airlines control the majority of the prime slots at an airport, prices for that particular market may not be at all competitive due to a lack of competition.

For tourism, the attention given to airport slot allocation (and trading, for that matter), is important for several reasons:

1. It remains to be seen whether profitable routes can be sustained if the cost of doing business on the ground is not justifiable to carriers. In some circumstances, slot trading, as an outright cost borne by a carrier to enter a particular market, may jeopardise connectivity within a network as carriers may decide not to service a particular node if the cost of gaining access to that node is prohibitive. Similarly, in such cases, tourism destinations that may rely on service from that node may suffer in terms of accessibility. Overall, then, where some nodes carry high slot costs, or where slots are at a premium, carriers servicing those nodes will only likely only fly on routes that are highly profitable, thus artificially creating travel flows that favour some destinations and disfavour others.

2. Somewhat related, slots (and their timing) are market drivers. Carriers will naturally attempt to streamline the timing of their aircraft operations in order to maximise profitability (e.g. rapid turn around times, which is an inherent feature of LCCs), yet their ability to command suitable and profitable market share can often depend on departure and arrival times at destinations or other nodes throughout a network. As air transport is undoubtedly part and parcel of the overall travel experience, suitable arrival and departure times therefore become critical. In this way, revenue can be impacted negatively if the ability of a carrier to service a market is restricted by the timings of take-offs and landings not only at the origin node, but also at the destination node as well. In essence, slot allocations need to be considered across the network, and a profitable and operationally efficient balance is maximised when a carrier retains appropriate network-wide slot access.

3. By extension, markets located in proximity to an airport at which a carrier is dominant because of grandfathering of slots or through saturation of services under the first come, first served rule (as in the United States) may find choices rather limited when it comes to holiday destinations. The implications for this are twofold: on the one hand, the economic impact of hubbing at a particular airport or node can be significant when one carrier is dominant; but on the other hand, this dominance comes at a price with respect to destination choice, as smaller carriers may not have the frequency and desirable destinations within their network. Again, the dominance of a carrier at a particular node may artificially determine travel patterns and, to an extent, demand as there is less of an incentive for the carrier to expand its network to other nodes if existing services are already profitable.

4. Finally, if market forces of slot allocation are adopted at a global level, then users (passengers, tourists) may find that increased competition could bring lower fares and more options for destinations served, thus enhancing tourism. In many respects, some LCCs have chosen to directly avoid the slot allocation issue in utilising regional airports for domestic and some international flights, particularly in Europe (e.g. Ryanair).

The issue of slots at airports is a worldwide concern, but particularly in Europe where competition has grown significantly in the past decade. Ostensibly it is an issue where balance is required between the interests of the travelling public (set and overseen by government policy) and the interests of shareholders of airlines who have great interest in retaining access to profitable markets (and, by extension, profitable stages or routes). As Paylor (2005) notes, consultation on the new policies targeted for enactment in the EU are ongoing, it may be some time before full implementation is realised.

OPERATIONS MANAGEMENT: AIRPORT SERVICES

On the subject of nodes and how air transport provision is affected by regulatory mechanisms that govern their inclusion in a carrier's network, it is also important to understand how the services rendered at nodes or airports can be characterised. This is important from a tourism perspective as airports function as vectors through which international travel often transpires. Airport revenue emanates from aeronautical or non-aeronautical operations (Graham, 2001). Aeronautical revenue sources can include landing fees, passenger fees, fees for parking aircraft (between flights or longer term for maintenance), and various handling fees if certain services (such as movement and management) are not provided by carriers. Non-aeronautical sources of revenue can include rents for spaces in the terminal affiliated with specific carriers, concessions, sales of products/services managed by the airport, and parking fees (Graham, 2001). The importance of the split can vary from airport to airport. For example, when Delta Air Lines declared bankruptcy in September 2005, Hartsfield-Jackson International Airport was immediately concerned for its own future revenue streams. The airport had recently approved an expansion plan worth approximately US$6 billion, and future revenue from Delta, which already accounts for 19% of the airport's revenue, was deemed critical for future profitability (Tagami, 2005). Indeed, many airports around the world manage ancillary, but often significant in terms of revenue, means of non-aeronautical sources of revenue. For example, Dunedin International Airport, in the South Island of New Zealand, owns several small farms in the immediate vicinity and approximately 500 head of cattle. The reason for this is largely practical; the airport is situated in a rural area (but still within the city boundaries) and by acquiring ownership of surrounding properties it can control any future development or expansion needs.

One way of counteracting the reliance upon aeronautical revenue streams is the development of non-aeronautical activities. Rather than rely upon the services of one or more

199

airlines, airports around the world are instead turning to specific sources of revenue from passengers. For example, retail activities are increasingly becoming an important source of revenue for many airports, largely because passengers (both airside and landside) can be a captured market. The BAA's 2004 Annual Report (BAA, 2005) indicates that it generated nearly half a billion UK pounds from retail activities across the airports it manages (Heathrow, Gatwick, Stansted, Southampton, Glasgow, Edinburgh and Aberdeen). At Christchurch International Airport in New Zealand, concessions and lease rentals accounted for almost 40% of the total operating revenue in 2004, up from approximately 36% in 2003 (Christchurch International Airport, 2005). This reflects expansion activities that have resulted in many direct flights from the airport to Australian and other Pacific destinations.

Hanlon (2001) offers several reasons for the rapid growth in non-aeronautical or commercial activities at airports worldwide:

1. Commercialisation and privatisation has allowed many airports to effectively be run as commercial entities rather than publicly funded organisations managed at the local or national government level.

2. Airlines are 'exerting increasing pressure on the airport industry to control the level of aeronautical fees which are being levied' (Hanlon, 2001: 127), which means that airports must search for other means by which overall revenue and profit are realised.

3. The type of traveller utilising an airport has changed; where air travel at one time was reserved for those of high-net worth, advances in technology has meant that air travel is within reach for most people in industrialised countries, and this overall market is one which is demanding of (or at least responds to) retail opportunities.

4. Competition between airports, according to Hanlon (2001), may mean that, for some passengers, consideration of retail activities may be at least a secondary consideration for choosing air services (with the nature of those air services being the first).

Freathy and O'Connell (1998) argue that airports' concentration on non-aeronautical revenue reflects a change in the customers they service, and thus the role of the airport itself is somewhat blurred:

> There are those who remain close to the traditional view of an airport, i.e. that it exists to ensure the efficient movement of passengers between one destination and another. An alternative, and perhaps more eclectic approach, views airports within the framework of consumer change. In this context airports are seen not only as modal interfaces but also as leisure attractions and primary destinations in their own right. If airports are viewed as locations through which passengers are to be moved as quickly and as efficiently as possible, then the role of commercial activities within

airport operations will always remain limited. If however an airport is viewed as a primary leisure destination it itself, them it will remain possible to develop further the commercial opportunities within the airport. (Freathy & O'Connell, 1998: 16)

Freathy and O'Connell's comments are interesting in that they point to the challenge that airports face: on the one hand ensuring that customers/passengers are able to get to their aircraft efficiently, but on the other the realisation that, to remain viable, they must ensure revenue extraction from these same passengers is maximised. Freathy and O'Connell (1999) also proposed a typology of airport retailing: concessionaire-based retailing (the most widespread form of retail activity, and where the airport authority acts as the landlord), authority managed retailing (where the authority occupies retail space, although concessions may also be present), management contract (all operations handled by third or fourth party), and joint venture operations between multiple partners, some of whom may be overseas.

Without question, the means by which passengers are processed is critical to the overall efficiency of the airport (including customs controls and immigration), but airports can also maximise their physical layout to ensure that passengers are at the very least exposed to retail and other service opportunities. Doganis (1992) argues that there are three locational factors involved in relation to passenger flows: (1) position of shops; (2) the floor level; and (3) the split of available space between airside and landside areas. As Freathy and O'Connell (1998: 76–77) point out, there are several key factors that determine retail locations in airports:

- The logic of the passenger traffic flows: the location of shop units needs to mirror the direction in which the passengers are travelling. They should not be required to retrace their steps or go in a counter direction.
- Floor levels: retail outlets should be on the same floor as the departure gates and passengers should not have to ascend or descend stairs in order to shop.
- Distance: shops should be accessible without passengers having to traverse long distances. They should be sufficiently removed from the security and passport checks to allow the traveller to make the mental adjustment to a shopping environment.
- Visibility: before encountering the retail offer, passengers should have the retail outlets in their line of vision. This will help stimulate purchasing behaviour and possibly trigger impulse sales.

Retail activities have become increasing developed at many major global hub airports, incorporating elements of design strategies used elsewhere in retail environment planning (Crawford & Melewar, 2003; Rowley & Slack, 1999). At London Heathrow's Terminal 3, for example, the traveller is exposed to a wide range of retail services, ranging from distinctively up-market fashion to magazines and snacks. For the most part, the design of the shopping promenade in the terminal is meant to emulate a shopping mall. In fact, the design of Terminal 3 is such that, at almost any point, a clear view of the available shops is possible, a clear strategy in encouraging impulse purchasing (Crawford & Melewar, 2003). Some major international hub airports are moving beyond retail activities to include other activities aimed at passenger comfort and increasing revenue. At Changi International Airport in Singapore, the expansive nature of the passenger concourse (airside) in both Terminal 1 and Terminal 2 facilitates almost 25 million passenger movements (i.e. transferring or otherwise) per annum, and the airport features seven natural areas or gardens, a movie theatre, various napping areas, a fitness centre, a transit hotel, and over 100 retail outlets.

Many airports, especially those which have seen substantial increases in passenger traffic, fund their expansion activities through levies on passenger traffic. Whereas in the past airports have been able to fund airport expansion from the aeronautical revenues collected from carriers, many LCCs attempt to negotiate smaller landing and operational fees for aircraft handling and space rentals (Francis *et al.*, 2004), thus leaving airports with few options for revenue collection apart from retail activities. In 1997 Edmonton International Airport introduced an airport improvement fee (AIF) to fund capital expenditures and related financing costs for redevelopment and expansion of the terminal facilities. This new programme, however, has not entirely covered the cost of the redevelopment, according to the airport's most recent Annual Report (Edmonton Airports, 2005). As of 31 December 2004, total cumulative expenditures sat at CDN$357 million while the cumulative net AIF revenue was CDN$133 million.

In late August 2005, Ryanair announced that it was cutting 12 flights a week to and from Cornwall because of a recent decision by the Cornwall County Council to implement a £5 tax on departing passengers over the age of 16. Ryanair argued that it would cost the region over £10 million in lost revenue from visitors as approximately 100,000 fewer passengers per year would fly to the airport (Times Online, 2005). The airline also argued that an extra tax would be enough to dissuade people from flying to Cornwall, especially when passengers are 'price sensitive' (i.e. air travel is elastic) and may elect to chose a different destination simply because the amount of the tax, especially to an LCC, can amount to a substantial proportion of the cost of the ticket. As Graham (2001) points out, taxes levied by governments (at varying levels) are often used for airport improvements when they have some degree of ownership, but they are also used to fund destination marketing organisations and their marketing operations. For tourists, airports need to be functional as well as ergonomic (Caves & Pickard, 2001) and offer amenities and services that are both useful and efficient. The management of airport space, however, involves balancing the needs of several stakeholders: airlines want to be able to provide their transport service and make a profit, airports want to be able provide the departing (and incoming) passenger with a range of amenities to enhance their experience.

YIELD MANAGEMENT APPLICATIONS IN TRANSPORT

Related to the profitability and management issues associated with airports is how specific modes of transport manage their operations through yield management with the intent of maximising their revenue through a process of carefully managing supply and price in order to ensure that passengers pay the maximum possible price they are willing to pay. This concept relates to that of pricing. Pricing is strongly associated with economic concepts of value and scarcity, where individuals assess the value of a particular good or service in relation to its cost and/or availability. Price-sensitive consumers will react to higher prices by assigning a higher value to the money in their pocket (or for another good or service) than the good or service that it buys them. Non price-sensitive consumers may place a higher value on a good or service because it is convenient (or scarce). They are willing to pay more for that good or service. Thus, firms seek to offer their good or service at a cost that the market will bear; raising prices may or may not have an impact on sales. In the case of transport there are several markets to consider, each with its own sensitivity to price. As a result, transport operators consider how much each market segment would be willing to pay for what the transport operator is providing. This is roughly considered when discussing yield management, which is exactly that: the process by which overall yield is managed (or maximised) given different markets.

Yield management is perhaps one of the more important concepts when examining tourism and transport linkages because of its pervasiveness; both LCCs and network carriers use yield management systems, as do charter carriers, cruise lines, rail companies and car rental firms. In this section, the aviation industry is used, largely because it is the most common transport industry in which yield management practices are examined. Whereas Chapter 3 examined the network of operations and how, geographically, the extent and concentration (and connectivity) of that network needs to be carefully planned, yield management helps to understand the financial implications of providing services across a network to a diverse range of markets.

In the next chapter, the nature of the airline seat will be discussed from the perspective of what is actually marketed to the consumer. The question of concern here, however, is how an airline can adequately price a seat on an aircraft and still retain a profit. Does an airline make money on offering a seat at, for example, 10p? The short answer is 'no' as the cost of providing that seat for sale is likely more than 10p, but in the context of managing operations, achieving market share and maximising revenue, it may, in the long-term, be profitable and strategically advantageous for the seat to be sold at that price. Any business must cover its costs first and foremost before entertaining the possibility of posting a profit, and almost every business tracks their costs down to the most feasible unit. In the case of airlines, trains and cruise ships, this means a firm will know exactly how much it costs to transport someone from point A to point B.

For an airline, the total required to physically transport that seat from sector or stage (i.e. origin to destination) is what is called the seat-mile cost (O'Connor, 1978). The total revenue collected from that seat is referred to as the revenue passenger mile, which is the revenue collected transporting a passenger along a sector. A ratio between the two can be calculated as the load factor, which is the relationship between the passenger miles and seat miles (O'Connor, 1978). For example, an Air New Zealand ATR-72 aircraft, one that is used to regularly fly passengers between some of the smaller provincial centres, will hold 66 seats. The flight distance along the Dunedin and Christchurch sector is 204 miles (although this may differ slightly because of slight variations in the route due to other circumstances). This ATR-72 flight thus outputs 13,464 seat miles (66 seats × 204 miles). If, on a particular flight, 45 seats in the aircraft are filled, then the passenger miles are 9180 and the resulting load factor is thus 68% (seat miles divided by passenger miles).

Load factor calculations are not necessarily calculated by merely dividing the total seats available by the total seats occupied because airlines watch carefully the cost per seat mile, which is governed by several factors. For example, short-haul flights can often carry similar base costs to long-haul flights, such as baggage handling or check-in staff (O'Connor, 1978), yet short-haul flights are often more expensive to operate for an airline because, comparatively, landings and takeoffs consume more fuel than when cruising. As well, frequent rotations (i.e. the number of take-offs and landings) means higher costs through aeronautical charges at airports (O'Connor, 1978).

In general, airlines must carefully control their costs (which is why LCCs generally compete on the basis of lower costs per seat mile than traditional FSAs). Taneja (2003: 11) suggests that 75% of an airline's total costs are fixed costs (the actual aircraft, salaries and airport/maintenance facilities). As many airlines operate in a highly competitive environment, profit margins are thin and managing revenue is thus critical when fixed costs are high. The way an airline will manage its sales is through the process of yield management, which is a system by which airlines (and other service providers, such as hotels, cruise companies, rail companies, etc.) will attempt to match demand with supply and offer the product at a reasonable price. Doganis (2002: 283) notes that yield management

> involves the management of seat access through an airline's reservation control system in order to maximise the total passenger revenue per flight. This is not the same as ensuring the highest load factor or the highest average yield. In fact maximising revenue may in some cases mean that neither of these aims is achieved.

Doganis' point is that it is the revenue accrued from a flight that is of importance to an airline, and not whether the flight was full. The process of maximising revenue is complex and requires software that projects income based on a variety of factors (see Box 7.1). Human input is also provided as a means to inject certain relevant information (e.g. long-term marketing goals or unforeseen or non-trackable externalities) into the system that eventually determines how much the average tourist will pay for a flight from, say, Cairo

BOX 7.1 Overbooking: Managing passenger demand with restricted capacity

Holloway (2003) notes that there are two dependent areas or departments within an airline that control pricing and availability. A pricing department will 'create and administer the passenger fare and freight fare structures applicable to each market' and some larger international airlines will have several pricing departments worldwide that are charged with responding to local market competition and demand (Holloway, 2003: 113). A revenue management department will 'allocate the physical space available on each individual flight-leg (augmented by overbooking limits) between the different fare and rate bases available for sale on that leg' (Holloway, 2003: 115). The role of overbooking is really one of compensation for the operations of a particular sector or flight-leg. For example, a passenger that fails to show up at the airport can mean missed revenue opportunity if sufficient demand for the seat (in the form of late-booking or standby passengers) is low or non-existent. This is deemed as 'spoilage costs' (Holloway, 2003: 546). Similarly, if a passenger cancels a booking at a point in time where it becomes too difficult (or costly) to logistically fill that seat, lost revenue is incurred. A carrier may elect, in planning for no-shows or booking cancellation (with fare refund) in a number of ways:

1. Control of booking classes (see Chapter 6) to minimise cancellations or no-shows: for example, lower fares may carry restrictions on cancellation, ranging from non-refundable to only partially refundable, thus minimising last-minute cancellation by the passenger. Passengers booking on this fare are less likely to not show up for their flight.

2. Overbook in selected classes in order to contain lost revenue from no-shows or cancellations: for example, a flight may be 'overbooked' if there are more passengers than seats. Carriers will intentionally hedge their bets that not all passengers will show up for a particular flight, although this hedging will, to some extent, be managed by the revenue management department by alternating the fare classes sold in relation to demand in the market. For example, in a price sensitive (or elastic) market, a carrier may elect to offer more fare classes with fewer restrictions or penalties on cancellations, but at the same time overbook the flights in preparation for the historic trend of no-shows at the airport. Likewise, a carrier operating in a particularly inelastic market (or on a route where demand is particularly inelastic in relation to the city-pairs served or time of flight operation) may elect to offer fewer fare classes where cancellations are allowed without

penalty as the market will bear these fare types and demand is strong enough to warrant the minimal use of discounted fares or fare types allowed for free cancellation.

As Holloway (2003: 546) points out 'Airlines facing low yields and high break-even load factors have an incentive to overbook more aggressively', although this can be managed based on whether the carrier opts to restrict cancellation or refund opportunities. An RMS, then, is employed to minimise the spoilage costs, yet at the same time control what are known as 'denied boarding costs' (Holloway, 2003: 546). In reality, overbooking can sometimes mean more passengers with valid tickets than seats on the aircraft. This is why, and it usually happens at the gate immediately prior to boarding, airline officials will call for volunteers to wait and take a later flight to the same destination. The incentives offered to passengers willing to forfeit their seat for the flight are the denied boarding costs that a carrier must incur if their overbooking policy results in more passengers than seats. These incentives (to passengers) or denied boarding costs (to airlines) typically can include vouchers for food within the airport, certificates for use on future ticket sales (usually, if not always, with the same airline), or bonus frequent flier miles. These costs also include the cost of administration in situations of overbooking and denied boarding of some passengers, and as Holloway (2003) suggests, can even include the loss of value as held by the passenger and perhaps impact on whether or not future bookings are made with the same carrier.

In February 2005, the EU introduced new rules that compensated passengers if their flights were delayed or cancelled. If a flight is cancelled less than two weeks before flight time, passengers are to be offered a refund or re-routing to their destination, as well as meals, telephone calls and appropriate accommodation. Airlines (especially those represented by the European Low Fares Airline Association) argued that the cost of compensation (up to 600 euros) is often more than the cost of the ticket, and thus appealed the EU rules to the European Court of First Instance in late 2005. In a decision rendered in January 2006, the Court upheld the EU rules. It seems, then, that the practice of overbooking as an element of yield management may be more carefully managed in the future, at least in the EU.

Airlines are not the only transport mode to utilise overbooking procedures. Many rail operators will utilise overbooking strategies to manage supply and demand issues. For example, a large number of tickets sold for travel between cities in close proximity to one another may be last minute, but some may be sold as a 'book' of multiple-journey tickets that are valid for travel at any time. The rail operator, then, must carefully watch demand and loadings to account for the number of platform sales (sales immediately prior to departure) and advance sales. The one benefit that some rail operators have, however, is that the number of passengers can often exceed the number of seats in situations where commuter trains operate. For example, on a recent trip from Edinburgh to Glasgow, this author found himself standing for the entire journey because of unexpected demand for the service operating at the specified time. Throughout the train, passengers were standing in the aisle and the price they paid would have been the same whether or not they would have been able to get an actual seat. Although people were standing, operators still need to carefully manage occupancy rates of rail cars in order to keep within safety limits, which are usually imposed by governments. What some operators may argue, however, is that any train leaving without people standing means lost revenue, thus enforcing the idea as discussed in Chapter 8 that the 'product' on offer may not always be an actual seat.

Questions for consideration

1. To what extent can it be argued that overbooking constitutes a strategy for controlling yield?

2. How might a transport provider, accustomed to overbooking, decide to abolish this policy? What are the marketing and operational implications, including any potential impact on fares?

to London. Knowledge of the market is very important when determining fares. As Hanlon (1999: 190) notes:

> From their market research, airlines know that high-income travellers, business travellers and those travelling for urgent personal reasons (e.g. to attend a funeral) have relatively high price-inelastic demand. At the same time the airlines are aware that holidaymakers, those visiting friends and relations, students on vacation, etc. are all very sensitive in their demand to the fares charged. ... Differences between elastic and inelastic travellers in these respects are often rather wide and present airlines with good opportunities to segment the overall market by reason for travel and to use this as the basis for price discrimination.

Airlines will therefore attempt to maximise their profit by selling a seat at the right price to a particular market segment. As well, the time a ticket is purchased can also play a role, with some last-minute purchases often much higher than those tickets purchased months in advance (and vice versa as the marginal cost of adding an extra person on a flight is minimal).

Interestingly, some airlines will often price certain seats well below what it costs them to operate. As Taneja (2003: 11) points out, '[t]he high fixed costs combined with low marginal costs and the perishable aspects of the product have led managements to introduce some fare structures and levels during normal times that did not, and still do not, reflect the cost of equipment, let alone full-allocated costs'. This may seem somewhat illogical, but there are several justifiable reasons for this (Taneja, 2003). For one, and as indicated in the last chapter, aircraft costing between US$10 and US$300 million dollars do not accrue revenue if they are idle on the ground. As well, an airline may wish to acquire an increased market share. Thus, customers become familiar with the product/ service and may chose to fly with the airline at another time. Finally, as Taneja (2003) points out, in the aviation industry the fixed costs tend to be quite high, and even with low fares the airline has an opportunity to generate some cash flow, however small.

Although low fares are quite common in the airline industry today, the reality is that few airlines will allow their flights to operate at a loss. There may be some circumstances where the offer of low fares may be justified. For example, 're-positioning' flights can be run when an airline needs to move one aircraft from one location to another outside of the existing schedule. As a result, the airline may offer extremely low fares which generate both cash flow and even a small profit, especially given that the aircraft needed to be re-positioned anyway. As well, some carriers may even elect to operate a route at a loss if they know that it feeds into a much more profitable (i.e. international) route. For example, consider the network in Figure 3.2. An existing carrier may know that the market for travel from A to B is too small to warrant direct services that would be profitable, so they elect to utilise a smaller aircraft (e.g. an ATR-72) and move passengers to C first, at which point they board a 737–800 and fly to B. This carrier may elect to run the A to B flight at a slight loss if they think that they are achieving significant market penetration in A to warrant. As well, they may price their seats from B to C in such a way as to make up for the losses accrued on the A to B sector. Of course, when competition enters into this mix, downward pressure on prices between any of the sectors may hinder profitability and even more so if a new entrant decides to offer direct services from A to C.

Providing real examples of the concept of yield management is almost impossible as airlines will closely guard their costs and revenue by sector. There are, however, some basic principles by which yield management works in relation to the pricing of aircraft seating. What is important to keep in mind, however, is that quite often airline seat sales or low fares in general are designed more as a marketing tool than a means to offer seats to customers. Airlines cannot continuously price all seats at below cost else the flight will not return a profit. As Doganis (2002) points out, Philippines Airlines fares between the Philippines and the United States, in the early 1990s, were low enough to capture 70% of

Table 7.2 Yield management and fare levels, example 1

		Price per seat	
Distance (km) from A to B			1050
Capacity (number of seats on the aircraft)			150
Available seat kilometres (ASKs)			157,500
Operating cost (estimated)			$18,000
Cost per ASK (cost per available seat kilometre, or unit cost)			$0.11
Loading (seats sold)			136
'Ultra saver' fare	10 seats	$109	$1090
'Super saver' fare	110 seats	$199	$21,890
'Try 'N' save' fare	16 seats	$299	$4784
Total flight revenue			$27,764

the market. Despite the fact their load factors were high, the airline lost money on the sector.

In the example in Table 7.2, a flight from point A to B (totalling 1050 kilometres) has the capacity of 150 seats, all of which are configured for economy class. This results in 157,500 ASKs (although available seat *miles*, or ASMs, are often used). With an estimated operating costs of $15,000 (the exact currency type is irrelevant for the purposes of this example), the cost per ASK (which is the unit cost) is $0.11. This is what it costs the airline to physically transport that seat on this particular route. Suppose, then, that 136 people are on this particular flight. When looking at the fare splits for all passengers (and three broad fare types are used to illustrate the point, but there are often, in reality, many more fare classes and prices), 10 passengers paid $109 (the ultra saver fare) for their seat, 110 paid $199 (the super saver fare) and only 16 paid $299 (the no saver fare). The total revenue for the flight (based on tickets only, not including any snacks or beverages for which the airline may choose to charge) is therefore $27,764. If the load (136 people) is multiplied by the distance (1050 kilometres), the result is 142,800, which is the total passenger kilometres flown on this particular flight. This figure can then be used to calculate the yield per seat kilometre (also known as the yield), which is simply the net revenue divided by the passenger kilometres flown. In this case, it is $0.19, which is compared to the cost per ASK of $0.11.

If the airline in this example was to sell all of the seats on that sector at the ultra saver fare, the results become quite interesting. In flight represented in Table 7.3, seats were sold at $109 which, because of strong demand at that price, resulted in a full aeroplane.

Table 7.3 Yield management and fare levels, example 2

Distance (km) from A to B			1050
Capacity (number of seats on the aircraft)			150
ASKs			157,500
Operating cost (estimated)			$18,000
Cost per ASK			$0.11
Loading (seats sold)			150
		Price per seat	
'Ultra saver' fare	150 seats	$109	$16,350
'Super saver' fare	0 seats	$199	0
'Try 'N' save' fare	0 seats	$299	0
Total flight revenue			$16,350
RPKs (revenue passengers multiplied by the number of kilometres flown)			157,500
Yield per seat kilometre (yield)			$0.10

The overall profitability of the flight, however, is negative, with the yield per seat kilometre now sitting at $0.10 when the cost per ASK is still $0.11. In other words, costs exceeded yield. Of course, another way of examining the profitability of this particular flight is to note that revenue ($16,350) versus costs ($18,000), but the point here is to illustrate how an airline might maximise revenue by selectively pricing seats. To be fair, this example and associated calculations are highly simplified, but the basic principle is that airlines will generally not consciously sell all seats on a flight at a low fare such as to jeopardise profitability of that flight. As indicated above, however, an airline may decide to operate a flight at a loss for a variety of reasons, and it may be perfectly reasonable from a business standpoint to operate a flight such as the one represented in Table 7.3.

Figure 7.1 demonstrates the principle of yield management in the context of consumer surplus, where the balance is set between what a consumer is willing to pay and the price at which an airline is willing to sell a seat in the context of marginal revenue it gains from selling that particular seat. Hanlon (1999: 192–194) notes that this price/cost versus output model aptly illustrates pricing strategies adopted by airlines. A fare at P1 would maximise profits as the amount of revenue (MR) is equal to the marginal cost (MC) of providing that seat. The airline may also sell some seats at cheaper fares (Q2 up to Qm, where the MC increases because demand would require introducing an additional flight). The shaded areas represent consumer surplus, where passengers are effectively gaining a net benefit because the price they pay is effectively below the level of demand. In other words, consumer surplus is present where the price paid is below what they would be willing to pay. Of course, an airline would like nothing better than to sell all of its seats at P1, but demand for seats is not constant at that price level. Thus, different pricing levels exist that are meant to reflect demand for seats at those prices. As Hanlon (1999: 194) notes

> By discriminating between passengers the objective of the airline is to expropriate as much as possible of what would otherwise be passenger consumer surplus if all seats were sold at MC (the shaded areas). The purpose of restricting the availability of the cheaper fares is inhibit passengers trading down from more expensive fares, or to limit what in airline parlance is called 'revenue diluation'.

For the average tourist, then, the big question is often 'When should I buy my ticket'? Given Figure 7.1, it might seem that purchasing in advance always results in the cheapest fares, but an airline may discover that advance purchasers on some rates have a higher willingness to pay than those who book and purchase closer to the time of departure. Figure 7.1 is simplified because it illustrates how an airline might approach fare prices, but

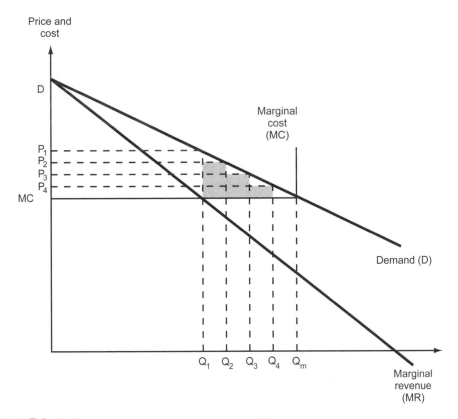

Figure 7.1 Yield management and consumer surplus

Source: Hanlon (1999) with permission from Elsevier.

in reality fares can change many times in the months or days leading up to departure. What is evident from the above discussion, however, is that the management of revenue in transport operations is paramount and often means that some travellers will pay more for a seat than other travellers, even when that physical difference between the seat itself and the service on offer is non-existent. For the passenger, the negatives of price discrimination (another term for yield management, although one with a somewhat negative overtone) can be outweighed by the value held of the good or service in question. As Botimer notes:

> Revenue management uses differentiated fare products priced at varying levels and booking limits to provide a flexible, flight-specific method of seat inventory allocation. To achieve efficiency in allocation, it is desirable to fill the aircraft with those passengers who value the service most *at departure* prompting a need to: (1) identify the passengers who value seats most; and (2) protect seats to accommodate those passengers. Airlines identify passengers by their values of willingness to pay by using price *discrimination* techniques. Airlines protect seats in their computer reservations systems for higher paying passengers through *seat inventory control* (or yield management) techniques based on forecasts of expected arrivals for different classes of service. (Botimer, 1996: 314; italics in original)

For airlines, yield management is an effective way of addressing fluctuations in demand and shifting costs. It is in use because, for the most part, it can be highly effective. For example, Belobaba and Wilson (1997) ran a simulation that showed that utilising effective yield management results in an overall increase in revenue. More importantly, they also were able to demonstrate that an airline who is first to utilise yield management in a competitive situation will benefit. The fact that the vast majority of carriers worldwide utilise an RMS or some form of yield management would suggest that its importance and success is meaningful.

SAFETY AND SECURITY MANAGEMENT

In many ways, safety and security are two separate issues, yet they remain inextricably linked. A secure transport environment is not necessarily a safe transport environment and vice versa. One could even argue that a totally secure transport environment is largely unattainable. For tourism, the perception of being safe and secure is often the price of entry for most destinations (Hall *et al.*, 2003). The need to incorporate safety and security measures in transport can be rationalised in several ways:

1. The perceived level of safety of a mode of transport can have significant impacts on a destination; for example, in January 2006 the EC established a blacklist of airlines (most of which are African) that were no longer permitted to fly into EU airspace because they lacked the ability to maintain aircraft to accepted international standards (Table 7.4). Such an action can have severe implications, not only for the airline(s) but also for those passengers who arrive in the EU as tourists and for those EU residents who might wish to utilise a blacklisted airline to travel to a foreign country.

2. Security levels have generally increased worldwide as a result of a combination of reasons: increased traffic overall, increased threats to security and an increased number of destinations served. Security issues are generally associated with airlines, but other forms of transport are involved in monitoring and establishing security and safety requirements. For example, British mainline railway stations are likely to soon introduce security screening procedures (x-ray scanning and frisking by security personnel) not unlike those found in airports (The Age, 2005).

3. Security is enhanced and made relevant both prior to and during transport. As such, it often falls under the purview of different government organisations as well as private enterprise. Following attacks on cruise ships by modern pirates around the Horn of Africa in 2005, there have been calls for regional governments to get involved in securing the waters and help the existing United Nations and Unites States patrolling warships (ABC News, 2005).

Table 7.4 List of carriers whose activities are banned within the EU

Legal name on air operating certificate (trading name, if applicable)	Operating state
Air Koryo	DPRK
Air Service Comores	Comores
Ariana Afghan Airlines2	Afghanistan
BGB Air	Kazakhstan
GST Aero Air Company	Kazakhstan
Phoenix Aviation	Kyrghizstan
Phuket Airlines	Thailand
Reem Air	Kyrghizstan
Silverback Cargo Freighters	Rwanda
Africa One	RDC
African Company Airlines	RDC
Aigle Aviation	RDC
Air Boyoma	RDC
Air Kasai	RDC
Air Navette	RDC

Legal name on air operating certificate (trading name, if applicable)	Operating state
Air Tropiques s.p.r.l.	RDC
ATO – Air Transport Office	RDC
Blue Airlines	RDC
Business Aviation s.p.r.l.	RDC
Butembo Airlines	RDC
CAA – Compagnie Africaine d'Aviation	RDC
Cargo Bull Aviation	RDC
Central Air Express	RDC
Cetraca Aviation Service	RDC
Chc Stelavia	RDC
Comair	RDC
Compagnie Africaine D'aviation	RDC
Co–za Airways	RDC
Das Airlines	RDC
Doren Aircargo	RDC
Enterprise World Airways	RDC
Filair	RDC
Free Airlines	RDC
Galaxy Corporation	RDC
Gr Aviation	RDC
Global Airways	RDC
Goma Express	RDC
Great Lake Business Company	RDC
I.T.A.B. – International Trans Air Business	RDC
JETAIR – Jet Aero Services, s.p.r.l.	RDC
Kinshasa Airways, s.p.r.l	RDC
Kivu Air	RDC
LAC – Lignes Aériennes Congolaises	RDC
Malu Aviation	RDC
Malila Airlift	RDC

Legal name on air operating certificate (trading name, if applicable)	Operating state
Mango Mat	RDC
Rwabika 'Bushi Express'	RDC
Safari Logistics	RDC
Services Air	RDC
Tembo Air Services	RDC
Thom's Airways	RDC
Tmk Air Commuter	RDC
Tracep	RDC
Trans Air Cargo Services	RDC
TRANSPORTS AERIENNES CONGOLAIS (TRACO)	RDC
Uhuru Airlines	RDC
Virunga Air Charter	RDC
Waltair Aviation	RDC
Wimbi Diri Airways	RDC
Air Consul SA	Equatorial Guinea
Avirex Guinee Equatoriale	Equatorial Guinea
COAGE – Compagnie Aeree de Guinee Equatorial	Equatorial Guinea
Ecuato Guineana de Aviacion	Equatorial Guinea
Ecuatorial Cargo	Equatorial Guinea
GEASA – Guinea Ecuatorial Airlines SA	Equatorial Guinea
GETRA – Guinea Ecuatorial de Transportes Ae5eos	Equatorial Guinea
Jetline Inc.	Equatorial Guinea
KNG Transavia Cargo	Equatorial Guinea
Prompt Air GE SA	Equatorial Guinea
UTAGE – Union de Transport Aereo de Guinea Ecuatorial	Equatorial Guinea
International Air Services	Liberia
Satgur Air Transport, Corp.	Liberia
Weasua Air Transport, Co. Ltd	Liberia

Legal name on air operating certificate (trading name, if applicable)	Operating state
All air carriers certified by the authorities with responsibility for regulatory oversight of Sierra Leone, including,	Sierra Leone
Aerolift, Co. Ltd	Sierra Leone
Afrik Air Links	Sierra Leone
Air Leone, Ltd	Sierra Leone
Air Rum, Ltd	Sierra Leone
Air Salone, Ltd	Sierra Leone
Air Universal, Ltd	Sierra Leone
Destiny Air Services, Ltd	Sierra Leone
First Line Air (SL), Ltd	Sierra Leone
Heavylift Cargo	Sierra Leone
Paramount Airlines, Ltd	Sierra Leone
Star Air, Ltd	Sierra Leone
Teebah Airways	Sierra Leone
West Coast Airways Ltd.	Sierra Leone
African International Airways, (Pty) Ltd	Swaziland
Airlink Swaziland, Ltd	Swaziland
Jet Africa	Swaziland
Northeast Airlines, (Pty) Ltd	Swaziland
Scan Air Charter, Ltd	Swaziland
Swazi Express Airways	Swaziland

Source: Adapted from http://europa.eu.int/comm/transport/air/safety/doc/flywell/2006_03_22_flywell_list_en.pdf (accessed 25 March 2006)

With respect to marine transport, Lois *et al.* (2004: 104) note that there are several points along the operational phase of a cruise ship in which passengers and crew are at risk: passenger boarding (fire in the terminal, gangway collapse, noise), leaving port (passengers falling overboard, collision), at cruise (collision, grounding, fire, engine failure), at dock (hard docking, fire) and disembarkation (slips and falls). Of course, marine transport is not the only mode to have safety risks. Janic (2000: 44) discusses the nature of risk and security in aviation and outlines five features of airline accidents that distinguish them from the impact of accidents involving other modes of travel:

1. Because flying may take place over long distances, accidents may occur at any point in time or space. Hence, there is exposure to individual and global hazard.

2. Passengers and aircraft crews are primal target groups exposed to risk of an accident but there are individuals on the ground who may be exposed to the same accidents albeit at a lower probability.

3. Although being a rare event in an absolute sense, aircraft accidents can have severe implications.

4. Conditionally, any aircraft movement is an inherently risky event, then, according to probability theory, aircraft accidents may be classified as highly unlikely (although possible) events.

5. With respect to time dependency, risk is always present during given time and space horizons (i.e. whenever a fight takes place). The effect is non-cumulative and particularly related to the separate exposures of the people on board.

Health safety on cruises has been the subject of much media attention, with the problems experienced by passengers ranging from 'simple' food poisoning to more series viral outbreaks. Not all incidents, however, can be definitively linked to poor sanitation efforts. In early October 2005, reports emerged suggesting that warming ocean waters in and around Alaska are thought to be responsible for the contamination of local shellfish with bacteria. These shellfish are subsequently consumed on-board ships cruising from the west coast of the United States up into the Alaskan region. Previously, the same waters were thought cold enough to thwart the proliferation of harmful bacteria in shellfish. Nonetheless, illness outbreaks on cruise ships do occur (see Table 7.5).

Despite air accidents frequently being shown on evening newscasts (to some extent, because they are relatively rare [Motevalli & Stough, 2004]), the ICAO reports that the fatality rate in aircraft accidents actually decreased between 1970 and 1993, even though the number of passengers increased some 500% (Janic, 2000). In fact, air travel is comparatively safe:

1. New aircraft designs have incorporated numerous safety measures, largely by virtue of international agreed specifications, but also because safer aircraft can be marketed by the manufacturer to potential customers (e.g. airlines).

2. Navigation systems have become sophisticated to the point where human error in the past can be mitigated through active on-board monitoring of environmental conditions and nearby aircraft movements. For example, many airlines have introduced collision-avoidance systems in some aircraft to minimise the risk of mid-air collisions (Kayton & Fried, 1997). Improvements in ground radar systems at major airports have also helped to mitigate ground incursions.

Table 7.5 Selected illness outbreaks on cruises for Q4 2004

December 16	*Golden Princess* (Princess Cruises)	A passenger reports there was a significant outbreak of illness (hundreds of passengers affected) on the Caribbean cruise, December 12–19. According to CDC [centre for disease control]: the ship reported that an unusual number of passengers and some crew members were experiencing GI [gastrointestinal] illness. The ship's medical staff reported that 84 of 2742 (3.06%) passengers and 15 of 1100 (1.36%) crew members were ill. Predominant symptoms are diarrhea and vomiting. The cruise ends in San Juan on December 19.
December 13	*Silver Shadow* (Silversea Cruises)	The ship reported that an unusual number of passengers and some crew members were experiencing GI illness. The ship's medical staff reported that 11 of 275 (4%) passengers and 20 of 291 (6.87%) crew members were ill. Predominant symptoms are diarrhea and vomiting. The cruise ends in San Diego on December 14.
December 10	*Norwegian Dream* (Norwegian Cruise Line)	The ship reported that an unusual number of passengers and some crew members were experiencing GI illness. The ship's medical staff reported that 73 of 1749 (9.37%) passengers and 9 of 744 (1.21%) crew members were ill. Predominant symptoms are diarrhea and vomiting. The cruise ends in New Orleans on December 12. According to a passenger, the numbers were significantly higher: 'Once the passengers realized that reporting gastrointestinal illnesses would result in a 48 hour quarantine, many just suffered and gutted it out (pun intended) in their rooms.'
December 10	*Norwegian Sea* (Norwegian Cruise Line)	The ship reported that an unusual number of passengers and some crew members were experiencing GI illness. The ship's medical staff reported that 44 of 1531 (2.87%) passengers and 4 of 683 (0.58%) crew members were ill. Predominant symptoms are diarrhea and vomiting. The cruise ends in Houston on December 11.
December 7	*M/V Explorer*	The ship reported that an unusual number of passengers and some crew members were experiencing GI illness. The ship's medical staff reported that 71 of 758 (9.37%) passengers and 1 of 195 (0.51%) crew members were ill. Predominant symptoms are diarrhea and vomiting. The cruise ends in Fort Lauderdale on December 8.

| November 2 | *Sun Princess* (Princess Cruises) | The ten-day cruise was struck with an outbreak of illness. According to one passenger: 'I was confined (quarenteened) to my cabin for 5 days with the Norwalk Virus. I got it 4 days into the sailing and at the 1st port we arrived at. I never left the ship. We were later told that the service in the dining room was so poor because everyone was sick. Then we heard from the entertainer that she had the flu and a fever and the whole crew was given a vaccine (flu shot).' Representatives of the CDC boarded the ship, but no information has yet been released. |
| October 3 | *Veendam* (Holland America Line) | The ship reported that an unusual number of passengers and some crew members were experiencing GI illness. The ship's medical staff reported that 39 of 1230 (3.17%) passengers and 2 of 557 (0.36%) crew members were ill. Predominant symptoms are diarrhea and vomiting. The cruise began in San Diego and will end October 14. |

Source: Reproduced with permission from www.cruisejunkie.com

3. Air traffic control systems have undergone substantial technological improvements in the past several decades (Brooker, 2003), thus rendering aircraft movements in the sky and on the ground safer now than at any point in the history of civil aviation, despite rapid growth in the industry. Concern still exists, however, for the organisation and operation of aircraft in high-density air space, such as continental Europe where air traffic has increased over the past several decades and where individual airports often have their own traffic control centres. The concern generally relates to the rate at which aircraft are 'handed off' from one control centre to another when crossing multiple countries and control areas. Traffic across the Atlantic Ocean, from eastern North America to Europe, is handled by the OACC (operated by National Air Traffic Services, UK), which covers some 633,000 square miles:

> As radar only has a range of some 200 miles, controllers in the Oceanic Area Control Centre [OACC] use position reports and estimates passed from pilots to ensure aircraft are safely separated. This information is provided using high frequency radio and is transmitted and received via a radio communications station at Ballygireen, near Shannon in Ireland. To take account of passenger demands, time zone differences and airport noise restrictions, most North Atlantic flights operate in two time blocks – westbound in the late morning and afternoon and east bound during the night and early morning. Because of this and the limited height band for economical jet operations the airspace is comparatively congested. To safely accommodate as many aircraft a possible an organised track structure is created every 12 hours. This allows aircraft to be handled in an orderly and efficient manner. The OACC is respon-

sible for the day track system and the Gander Centre in Canada provides the night track system. After entering oceanic airspace, pilots are required to make position reports at every ten degrees of longitude. These reports are fed into the controllers' flight data processing system (FDPS) which automatically updates the controllers' traffic information displays. The FDPS alerts the controller if a report is overdue, or if the situation is other than that expected. Based on the time of the last reported position and the forecast winds the computer will work out an estimate for the next position. It will also warn the controller if the separation between any two flights is approaching the minimum allowed. (http://www.nats.co.uk/services/soacc1.html; accessed 27 September 2005)

4. Various international organisations have implemented safety monitoring programmes that cover technical and manpower elements. These include the United States' IASA, the Safety of Foreign Aircraft Programme from the European Civil Aviation Conference, and the Universal Safety Oversight Audit Programme established by the ICAO. The adherence to these, and many other, safety protocols is enshrined in international agreements and is often required before bilateral or multilateral agreements are signed and enacted. In general, organisations such as the ICAO have limited powers of enforcement (Button *et al.*, 2004). Instead, programmes such as IASA in the United States are designed to evaluate airline compliance with the codes established by ICAO.

5. Passengers are more aware of safety regulations and advice. Most, if not all, major international airlines feature on-board safety videos and demonstrations before take-off. They may also include safety information affixed to the back of seats or on cards in seat pockets. Like guests at hotels, passengers are often asked to note clearly the location of nearby exits in case of emergency.

An excellent example of the importance of passenger understanding of safety measures on board an aircraft was the crash of Air France Flight 358 on 2 August 2005 at Toronto's Pearson International Airport (IATA: YYZ). The aircraft involved was an Airbus A340–300 with 309 passengers and crew onboard. Preliminary accident analysis indicated that it may have skidded in wet conditions on the runway, possibly due to a microburst (a sudden downward wind blast), resulting in the aircraft coming to a rest in a small ravine and catching fire. Miraculously, no passengers were killed. The on-duty flight attendants successfully evacuated all passengers within minutes, and this obviously contributed to the lack of fatalities. Most modern aircraft are designed to be evacuated in as little as 90 seconds, regardless of the passenger loading. Likewise, crew are exhaustively trained in safety and evacuation procedures. Airport emergency services are also trained to respond to an accident within minutes, as was the case with AF358. Shortly after the crash, Transport Canada indicated that it would be reviewing safety procedures at airports, specifically new technologies that are designed to minimise aircraft damage and death and injury. One such measure is the use of soft end-point material at the end of runways to prevent massive overshooting such as the case with AF358.

POST-11 SEPTEMBER

The events of 11 September 2001 spawned radical changes in the securing of the infrastructure associated with transport around the world. In the United States, the Federal government instituted the Department of Homeland Security, the purview of which was the safety and security of the entire country but whose focus was publicly associated with the guardianship of air transport facilities. This meant that, almost overnight, the screening of air passengers was shifted out of the hands of private contractors at most airports in favour of the new TSA. For many travellers, the frequent consequence of the increased security was lengthy delays in security screenings at major airports, and a very detailed list of what could and could not be brought aboard an aircraft (see www.tsa.gov for the most recent list). Since its inception, the TSA has come under increasing scrutiny by the travelling public. On many online bulletin boards (e.g. www.flyertalk.com), travellers post stories about inconsistent screening procedures, failure to find and confiscate prohibited items and rude treatment by TSA staff.

In Australia, news reports in late 2005 suggested that the Australian national government is looking to spend upwards of AUS$200 million on upgrades to airport security programmes across the country. This follows a report, commissioned by the government to identify the security arrangements at major airports, which suggested that policing at the country's airports is largely inadequate and dysfunctional. As Australia has been involved in recent global conflicts and has been the victim of terrorist attacks overseas, the report is a severe indictment of the current state of affairs. Opposition parties in Australia have also been calling for a similar government department to oversee security throughout the country, not unlike the United States' Department of Homeland Security. Using PNRs, the United States acquires passenger information for flights entering US airspace and flights arriving from Europe. Although these data have been provided by the EU since 2003, the European Court of Justice recently indicated that handing over data to the United States government is potentially in breach of individual's privacy (RTE News, 2005).

The events of 11 September, and even the events that have followed (Bali bombings, the war in Iraq) have clearly ushered in a new era in safety and security management. At one point, it was once not unheard of for passengers to carry firearms on domestic flights in the United States (with the airline of course having full knowledge as they would have been declared prior to boarding). Today, the immediate confiscation of cigarette lighters is not uncommon in some airports, and significantly more passengers are subjected to secondary screening at security checkpoints prior to boarding. In some cases, such as Auckland International Airport in New Zealand, passengers departing New Zealand for the United States (either Los Angeles or San Francisco) are subjected to two screenings: once after check-in and again before entering the boarding lounge before being seating on the aircraft. For the airport, this has meant that certain areas (airside) are cordoned off for use by flights on these routes. The cost of this additional screening is generally borne by either the airport or the carrier (and sometimes both), which can result in increased fares or the application of security fees on top of fares.

OTHER SECURITY THREATS TO TRANSPORT

Between 1990 and 2003, over 52,000 animal strikes, 97% of which were bird strikes, were reported at United States airports. Aircraft components such as the nose (or radome in larger aircraft, where sensitive radar equipment is housed), windshields and engines bear the most brunt, and the results can often be catastrophic. Many bird species are attracted to airports as they are expansive and flat sections of land. As well, migratory birds such as Canada Geese were a factor in over 800 strikes at US airports between 1990 and 2003 (FAA, 2004). Canada Geese are not small, and as such they can do tremendous damage to aircraft components, but even smaller birds can cause engine failure and thus force pilots to abort either a landing or take-off and re-direct or re-attempt a landing or take-off. After an Aero Perlas de Havilland DCH-6 (Twin Otter) aircraft was affected by a bird strike in Central America, the aircraft was forced to ditch at sea and 20 people lost their lives.

Managing wildlife within airport boundaries can be challenging. At Dunedin International Airport in New Zealand it is not uncommon to see a vehicle speeding up and down the runway searching for birds in the immediately vicinity only minutes before an aircraft departs or arrives. If small groups of birds are found, the driver fires a small pistol (with blank bullets) in order to drive them away. At one time, the vehicle used to light up its orange safety lights which used to be enough to rid the runway area of birds. Over time, however, the birds became used to the lights and it was necessary to utilise loud noises to encourage them to move along. At Anchorage, Alaska, the problem is not so much with bird strikes but with moose wandering on to the runway. These mammals are by no means insignificant and can do incredible damage to an aircraft as they can typically weigh over 450 kilogrammes and stand almost 2 metres in height. In October 2005, this problem seems to have been solved with the introduction of a electric strip along the runway which gives wandering moose a short electric jolt (along with a loud snapping sound) if they are about to cross onto the runway itself (Reuters, 2005).

Other safety and security issues relate to those modes of transport operated by tourists themselves. Tourists travelling to foreign countries may only need an International Driving Permit (IPD) to legally operate a motor vehicle. The IPD was established by the 1949 UN Convention on Road Traffic (after an earlier Convention in 1926) that allowed holders of valid driving licences in their home country to receive the IDP without any further testing, although in many countries there are restrictions in place for holders of IDPs. In New Zealand, many tourists elect to rent smaller motor homes (called 'Maui vans') and tour around the country. Some of them may hold IDPs while others may hold permits from their country of origin that may be recognised as valid in New Zealand. The problem, however, is that it is incumbent upon tourists to learn the rules of the road whilst visiting. For North American visitors, rental agencies constantly remind tourists to 'keep left', yet on more than one occasion this author has encountered Maui vans pulling into the wrong lane onto a major highway (in New Zealand, a major highway is often a bi-directional, two-lane highway). This illustrates that there are safety issues with managing tourist flows where they utilise personal transport in their mobility. It is therefore incumbent upon destination marketing organisations to work closely with automobile associations, rental agencies and even publishers of guidebooks worldwide to ensure that rules of the road are available to tourists.

While transport is the vector for passengers and tourists, it can also be the prime vector for the spread of disease. The speed with which passengers can move across the globe is ultimately the speed at which viruses and disease can spread. Governments, airlines and the public can be forgiven for being scared at the potential economic and social upheaval the threat of airborne particulates can bring from overseas destination. Kenyon *et al.* (1996) found that passengers seated on an aircraft within approximately two rows of an infected person with tuberculosis were in danger of getting infected themselves. The SARS scare in Asia in 2003 wreaked havoc with many airlines, including American Airlines and Singapore Airlines, both of which had extensive route networks that included many Asian destinations. Passengers were either simply afraid to fly or local governments ordered flights cancelled. On 11 April 2003, a 48-year-old Hong Kong businessman tested positive for the SARS coronavirus after a flight from Hong Kong. Before confirmation of SARS was established, this particular passenger travelled on seven flights within continental Europe. Using passenger manifests of the seven flights on which the patient flew, Breugelmans *et al.* (2004) tested the capacity of SARS to be transmitted within an aircraft cabin by conducting tests on a sample of passengers (although all passengers were contacted, not all agreed to participate in the study). None of the passengers tested were infected with the SARS coronavirus, and the researchers concluded that efficient transmission in an aircraft cabin was not likely. Breugelmans *et al.* (2004) note that their results are consistent with other studies that had similar conclusions.

Although SARS is not easily vectored in closed environments such as aircraft cabins, the public perception was that it was dangerous to board aircraft to and from the Asian region (Mason *et al.*, 2005). Almost immediately after the presence of SARS was announced, the impact on tourism in the Asian region was swift and, unfortunately, catastrophic. According to the WTTC (2003), tourism arrivals fell by at least 70% across the Asian region. McKercher and Chon (2004, referencing a news article that appeared in the *London Evening Standard* newspaper) note that media coverage showing airline passengers falling ill had severe repercussions of how the Asian region was perceived as a tourism destination.

More recently, attention has been directed at the potential impacts of the H5N1 subtype of the avian bird influenza virus. Once again, airlines and certainly other transport firms, are carefully watching the situation and considering the impact in several ways. The impact on their operations, largely in the form of considerable drop in demand, could be significant if a global flu epidemic strikes, especially if it is found that the virus can easily spread in, for example, an aircraft cabin (unlike SARS). In late 2005, some major international hub airports, for example Los Angeles International Airport, were already developing plans for quarantining passengers if necessary (Mercury News, 2005). The lessons learned from SARS, and even from H5N1 (it is still unclear as to the real danger it poses), suggest that modern transport in all its efficiencies of mobility that it provides can be responsible for the rapid spread of disease and viruses. What is necessarily is for organisations such as the ICAO and the IATA to carefully consider any new policies of disinfecting and cleaning that may be required when future epidemics or pandemics occur.

CHAPTER SUMMARY

The shape of management of transport operations can have significant impacts on tourism. For example, the decision of one or two airlines to cease operations out of a particular airport can have negative implications for the destinations served by the carriers addressing these markets; on the other hand, in the short term there may be an increase utilisation of coach or other forms of ground transport (perhaps even personal cars) for the purposes of recreation or tourism closer to home. Revisiting Prideaux's transport cost model (2000a, 2004; see also Chapter 2), the form of transport can, in some situations, dictate travel flows. Similarly, pricing can be quite erratic in some markets due to competition and efforts by a transport provider to stimulate demand. In some cases, the decision might be made to 'open up' certain fare classes and/or utilise yield management to ensure maximised revenue.

This chapter has utilised air transport heavily in its explanation of yield management largely because the stability of air transport is, at the time of writing, a concern that is worldwide. Finally, safety and security is paramount not only in the destination but also in the form of transport. In some preliminary research conducted by this author in New Zealand across a representative sample of residents, safety and security was found to be one of the top three travel motivators that determine whether people choose to fly or not. In effect, safety and security are the price of entry. In many situations, countries are no longer willing to risk their very public reputations as safe destinations when competition for global tourist traffic is immense. This may be one reason why several countries around the world are closely examining the safety records of both airlines and manufacturers.

Box 7.2 Case study – Airport expansion: Heathrow's Terminal 5

The ongoing construction of the new terminal at London's Heathrow airport is seen as necessary by many because London functions as one of busiest international airports in the world. In 2004, 67 million passengers passed through the airport, representing an increase of over 6% from the previous year (ATW, 2005b). Terminal 5's construction was approved in 2001. From May 1995 until March 1999, the longest public inquiry in the UK's history was held, totalling some 525 days. The inquiry focused on a number of subjects, including the overall economic need of the new terminal, land use policies, access to the terminal itself, and noise and air quality concerns. The project is being funded by the BAA (with some help from British Airways) at a cost of £4.2 billion. Construction started in September 2002 and the first phase of the project is to be completed by April 2008. The new terminal will undoubtedly be a marvel of mobility, with fully integrated road and rail transport (including tunnels for rail service along the Piccadilly Line and the Heathrow Express Line) and a full automated people mover to shuttle passengers from the main terminal to the satellite buildings. Other facts about the new terminal:

- The site of the new terminal is 260 acres, roughly the size of Hyde Park in London.
- The terminal building will be 400 m long and have five full floors.
- It has provision for an additional 60 aircraft stands, with approximately 15 of these designed to be used by the new Airbus A380.
- It has the capacity to serve 30 million passengers per year, which will mean Heathrow overall will serve some 90 million per year (www.baa.com).
- Two rivers have already been diverted around the new development.

The planning and development of Terminal 5, however, has been somewhat controversial. Many opponents to the new terminal point to the likely increase in noise and air pollution, and suggest that congestion at the airport and the immediate area will be substantial. As the airport is located in the London Borough of Hillingdon, residents there have continued to voice opposition to the development. Friends of the Earth argue that Heathrow already causes pollution in the South East from air and road traffic emissions, and the noise levels from existing flight operations already 'disrupt the sleep of half a million people' (Friends of the Earth, www.foe.co.uk/campaigns/transport/case_studies/ heathrow.html). The controversy surrounding the expansion of Heathrow will likely continue, however, as there have been recent calls for the construction of a new runway (which will mean demolition of certain areas) and a future Terminal 6. This announcement (from 2003) was met with similar consternation. The HACAN recently suggested that the expansion will be a detriment to the communities surrounding the airport (Trivedi, 2003). As BBC News reported (Trivedi, 2003), the government has several 'conditions' that the new terminal will need to address with respect to the overall sustainability of an additional terminal beyond Terminal 5, including the use of clean fuels for service vehicles and the introduction of charges for driving to the airport.

Questions for consideration

1. Make a list of all stakeholders that would need to be consulted regarding the expansion of the largest airport in the country in which you live.

2. To what extent could it be argued that airports constitute a 'public good' and thus need to be allowed to operate without intrusion from external interests?

SELF-REVIEW QUESTIONS

1. Describe, in your own words, RPKs and ASKs.

2. Define yield management in relation to consumer surplus.

3. What are the differences between aeronautical and non-aeronautical charges in the context of airport management?

ESSAY QUESTIONS

1. What impact can terrorism have on travel flows (both domestic and international) and the operations of airlines or other forms of transport?

2. Is a slot a piece of real estate available for an airport to sell (or for an airline to trade) or is it a right of access to which governments must attribute rules and regulations?

3. Why has there been a difference in approaches to privatisation and liberalisation of transport infrastructure and operations in the United States and the UK?

KEY SOURCES

Doganis, R. (1992) *The Airport Business.* London: Routledge.

Graham, A. (2001) *Managing Airports: An International Perspective.* Oxford: Butterworth-Heinemann.

These two titles offer excellent explorations of the structure of modern airports. It is useful to juxtapose them to understand how the study and analysis of airports has changed somewhat over the nearly a decade that separates their publication. Where Doganis' volume concentrates on overall structure, including cost structures, scheduling, operations and relationships with customers (e.g. airlines), Graham's text captures the privatisation and liberalisation of modern airports in depth (owing to nature of the operational environment of many airports during which the text was written) and outlines the move towards non-aeronautical revenue streams.

ACI – www.airports.org

ACI is an international organisation representing the interests of airports worldwide. The website provides various publications and position papers relating to issues in global air transport and airport management. To compliment the views from ACI, the IATA website should be visited as well (www.iata.org).

CHAPTER 8:

TRANSPORT AND TOURISM MARKETING

LEARNING OBJECTIVES

After reading this chapter, you should be able to

1. Appreciate the importance and complexity of transport marketing in relation to tourism.

2. Understand and explain the differences and similarities between treating transport as a service and as a product.

3. Recognise that transport marketing can often involve the marketing of destinations.

4. Appreciate the various forms of strategic marketing in transport and their relationship to tourism marketing and development.

5. Understand how service quality is assessed in transport operations.

INTRODUCTION

One of the more intriguing aspects of transport marketing is that, quite often, the marketing of transport is critical to the overall development and viability of attractions, destinations and countries. In other words, it could be argued that transport marketing in association with tourism fulfils a dual role: 1) that of marketing the product/service (a distinction to be debated later in this chapter) in its own right in order to drive business; and 2) the secondary destination marketing that transport providers ultimately provide, such that a transport operation may elect to promote and advertise a destination itself for the purpose of driving its own business or operations. The second point is critical as it demonstrates how the marketing of transport operations is often closely allied with other

228

marketing efforts as established or implemented by destinations or attractions. Thus, while distribution systems feature strongly in the sales of transport products/services, perhaps more consistent and meaningful characterisation would rest with the multi-line, often integrated marketing efforts where a single marketing strategy is adopted yet ultimately serves the purposes of more than one organisation or enterprise.

The purpose of this chapter is to dissect the nature of transport marketing as it relates to tourism. The chapter does not address strictly 'advertising/promotion' elements in the marketing of transport, although these are considered, but rather considers the numerous factors that determine the markets for transport (e.g. segmentation) and the consequential marketing efforts in a Maslowian sense designed to address the needs, wants and desires of these markets. Just as important in understanding the markets for transport, what also must be discussed is the nature of the transport product. More specifically, the question remains as to whether transport, especially the type relating to tourism, is a product or a service.

Marketing is a key ingredient in any tourism experience as it links desire and latent demand with conversion into actual holiday experiences. Thus, markets and marketing are key areas where transport providers need to be as aware as possible, especially when part or much of their core business is involved in the servicing of passengers who are embarking on travel for personal, pleasure or business reasons. The purpose of this chapter, then, is to draw attention to the complex nature of marketing efforts designed by transport enterprises around the world and position these in the context of the consequences this can have for tourism development. As will be shown, transport firms have engaged in multiple strategic and tactical marketing programmes designed to entice passengers, not least of which include pricing strategies in relation to yield management strategies as discussed in the last chapter, frequent flyer programmes as established by many airlines, alliances (including mode-specific and multi-mode integration programmes) and traditional advertising.

MARKETING 101: RELATIONSHIPS TO TRANSPORT

Traditional marketing texts, and even those written for the purpose of isolating key strategies in use by the global tourism sector, tend to isolate several key factors when discussing marketing and tourism. Seaton and Bennett (1996), for example, discuss the nature of the marketing mix in tourism, pointing out that tourism is comprised of multiple sectors, including accommodation, transport, attractions and even the wider destination. Each of these sectors features core, tangible and augmented tourism products. For example, Seaton and Bennett (1996: 122, Table 5.4) argue that the core product of an airline is simply transport, and the basic need function is to transport customers from point to point. The tangible products, then, include the airline name, lounges, the quality of in-flight service and perhaps the comfort of the seats. Tangible products are 'the specific features and benefits residing in the product itself; styling, quality, brand name, design, etc' (Seaton & Bennett, 1996: 121). Shaw (2004) argues that the aviation industry's product is intangible and is thus 'instantly perishable and cannot be stored'. Middleton and Clarke (2001: 373), however, suggest that

> most airline marketing focused on product augmentation, corporate images and the quality of service provided by staff. Apart from obvious distinctions between first-class, business-class and economy-class, and with limited but important exceptions such as Concorde, the traditional approach to marketing airline products was rather sterile and unimaginative. Seats on transport are just commodities in the eyes of most consumers.

The augmented product of an airline, according to Seaton and Bennett, could be elements such as the chauffeur to the airport, any add-ons to the flights, such as hotels or other car services, and frequent flyer programmes. These augmented products represent 'add-ons that are extrinsic to the product itself but which may influence the decision to purchase' (Seaton & Bennett, 1996: 121). An airline, in an attempt to understand the specific product that its customers may require, really has numerous levels of product to consider. Augmented products may, for some customers, be a primary influence on the decision to purchase, perhaps even more so than tangible products. For example, rather than chose an airline on the basis of tangible elements such as seat pitch or seat quality, a frequent traveller may elect to travel with a certain air carrier over others because of the ability to collect points towards future air travel. In other words, and depending on the market segment, some tangible elements may not be part of the decision-set criteria when selecting an airline. Graham (2001) notes that the composite nature of airports almost precludes the concept of the augmented product from being applied, and this may perhaps be extended to airline operations as well.

The core, tangible and augmented product differentiation can also be applied to other modes of transport. Cruise companies, for example, may offer, by virtue of external linkages to other services and products, ground transport at specific destinations. Such augmented products can often determine itineraries and market demand. Tangible products relating to cruise tourism can include the décor of the ship itself, with more and more ships, as noted in Chapter 5, being outfitted as luxury vessels in an effort to segment the cruise tourism market. With cruise tourism, even the core product can be questioned. For example, using the blurry distinctions discussed in Chapter 2, one question that arises is whether a cruise company's core product is even transport. Cruise tourism is complex from the perspective of transport in that the ship itself functions as a mode of transport as well as an attraction. With more and more cruise companies offering on-board shopping and other amenities, the fact that the ship moves from port to port is, to some extent irrelevant. Cruise companies would prefer passengers spend their money on-board as opposed to the duty-free areas in port at the next destination. Thus, perhaps the core product of many cruise operations is not transport, but rather service and amenities relating to the experience on offer.

PACKAGING

Middleton and Clarke (2001: 372) describe what they call 'passenger transport bundles', which are comprised of:

- Service availability and convenience (reflecting routes offered, schedules and capacity).
- Cost in comparison with competitors on the same routes.
- The design and performance of the vehicle (comfort and speed).
- Comfort, seating, ambience and any services offered during the journey.
- Passenger handling at terminals and car parks.
- Convenience of booking and ticketing arrangements.
- Contact with staff and their roles in contact with customers.
- Image and positioning of each operator. (Middleton & Clarke, 2001: 372)

The importance of these bundles is that they highlight what is available to marketers. As the core product remains largely undifferentiated, it is the tangible and even augmented elements that are promoted in order to help potential customers differentiate one firm offering passenger transport from another. There are, of course, significant relationships between core, tangible and augmented products relating to a particular mode of transport, as shifts in one can have significant impacts on the other. For cruise tourism, it could be argued, the tangible benefits on offer for cruise tourists have slowly become core products for some cruise tourists. The same could be said for airlines. At the beginning of the aviation revolution, which saw the world conceptually shrink in time and space (Hall, 2005), airlines provided a basic core product of transport, but some might argue that stiff global competition has resulted in the core product of an airline shifting from transport to service. This raises the issue as to whether transport itself, particularly in relation to tourism, is a product or a service.

PRODUCT OR SERVICE? POSITIONING TOURISM TRANSPORT

Exploring the distinction between tourism transport as a product or service necessarily involves examining the concepts of tangibility, heterogeneity, perishability and separability (Bitner et al., 1993; Hartman & Lindgren, 1993). These differences are critical in understanding exactly what is being marketed when considering the relationship between transport and tourism, but also because it calls into question the nature of the transport experience and where it is situated in the wider tourism experience.

231

1. *Tangibility*. Products are often physical entities; that is, one is able to handle and inspect them before purchase. Services, on the other hand, cannot be directly handled, but they can be inspected. For instance, an individual is able, if they wish, to observe the behaviour of restaurant staff before deciding to dine there, on the assumption that the information gleaned from observation will provide an indication of the level of service on offer. Services are therefore intangible, and must adopt affiliated marketing schemes that should create a sense of what that service is like in the mind of the customer (Seaton & Bennett, 1996). In relation to transport, the question is whether a seat purchased from British Rail from Exeter to London constitutes the purchase of product (in this case, transport from one place to another) or a service. When one purchases a ticket, they are not purchasing the seat itself. The sale, rather, involves a contract that states that the passenger has permission to board the train, sit in an assigned seat (or not, depending on the level of service purchased – and the irony is noted here), and disembark at the appropriate end point. In other words, there is very little that is tangible about the journey.

2. *Separability*. Unlike most products, services are effectively consumed and produced at the same time (Onkvisit & Shaw, 1991). The journey is happening as you experience it, and for this reason, it can be argued that the production and consumption of transport is, for the most part, inseparable.

3. *Perishability*. As a service, transport cannot necessarily be stored or carried forward for consumption at a later date, although some purchase options may allow this to happen. Generally, empty seats or berths represent lost revenue opportunity.

4. *Heterogeneity*. For the most part, services are difficult to standardise simply because they are fluid offerings that differ from one encounter to the next due to the variable nature of human behaviour. Unlike physical products, the service encounter is highly erratic (Onkvisit & Shaw, 1991). Once again the question is raised whether tourist-related transport has as its core product the simple notion of being transported from one place to another or whether it is the tangible products that form the transport experience and are the elements by which potential customers are converted into actual customers. There will exist a standard level of service on a train from, from example, Exeter to London: passengers will be offered coffee, they will be allocated to a seat which is roughly similar to the other seats in the same class, and they can expect a degree of service quality at either end of the journey in the form of, for example, check-in and baggage retrieval. In some situations, delivery of standard service may not be possible. Coffee services may be withdrawn if technical problems arise or baggage may be lost. More importantly, as standard as this service offering may be, passengers will experience them differently and thus hold differing levels of satisfaction. One passenger may find the seats uncomfortable, while another, perhaps a businessperson, may find that the constant motion of the train itself unsuitable for reading or working on his/her laptop. As standard as the service offering may appear, how that service is perceived and received is highly variable.

STRATEGIC MARKETING IN TOURISM TRANSPORT

Strategic marketing focuses on the available processes designed to introduce new consumers to existing products or develop new products for existing consumers. It also incorporates the mission and goals of a corporation or firm and how these can be reoriented towards proving the right product or service to the right markets (Kerin & Peterson, 1998). Transport operators, like many other companies, can engage in various means of strategic marketing, including market penetration strategies, market development strategies and product development strategies. Each of these is examined below.

MARKET PENETRATION STRATEGIES

A typical market penetration strategy 'dictates that an organization seek to gain greater dominance in a market in which it already has an offering' (Karin & Peterson, 1998: 7). A common means of achieving this is through promotion or advertising, but particularly increasing brand recognition. In fact, many modes of transport utilise indirect forms designed to increase awareness of a particular brand. For example, Dragonair, an airline based in Hong Kong, painted new 'livery' (decorative elements on the exterior of aircraft) to celebrate it's twentieth year of operation. The livery design featured Chinese motifs on one side of the aircraft and Hong Kong motifs on the other. Ryanair is known for some of the more 'creative' advertising in the airline industry. Shortly after the bombings on the London Underground in July 2005, the airline developed a print campaign that ran in the Independent and the Telegraph and featured a likeness of Winston Churchill calling for people for visit London regardless of the potential for renewed attacks. In a deliberate play on the words of Churchill's famous speech during the Second World War, the ad features the byline: 'We shall fly them to the beaches; we shall fly them to the hills; we shall fly them to London.' The ASA in Britain received 318 complaints about the ad on the basis that 1) 'the advertisement was offensive and distressing, because it sought to use the recent terrorist attacks in London for commercial advantage'; and 2) 'the use of Winston Churchill's image and the parody of his famous speech were offensive and disrespectful to the memory of Winston Churchill and to those who lost their lives during World War two' (ASA, 2005). In their adjudication, the ASA chose not to uphold the complaints:

> The Authority acknowledged that the complainants had found the approach extremely tasteless, but noted many media commentators had, in the days that immediately followed the terrorist attacks, commented on the positive and determined response of Londoners to continue with life as normal. Because it noted that response was a source of strength and pride to many, and the advertisers had restricted the theme of the advertisement to the stoical response that followed the attacks, the Authority concluded that the advertisement stopped short of causing serious or widespread offence or promoting further distress... The advertisers believed the use of Winston

Churchill's image and the parody of his speech were neither offensive nor disrespect-ful; they pointed out that his response to the bombings during World War 2 was one of defiance and 'business as usual'. The advertisers believed they had captured the spirit of Churchill and the Blitz mentality of Londoners in the days that followed the terrorist attacks. (ASA, 2005)

Peattie and Peattie (1996) note that price-based promotions are the most popular form of promotions in travel and tourism. This is largely because the use of specific modes of transport for the purposes of leisure and tourism is largely elastic; that is, the higher the cost of transport, the less likely a customer is to utilise that mode of transport without making sacrifices for the nature and quality of their overall experience assuming a fixed budget.

Segmenting the market

As some passengers are more price sensitive than others, different segments of markets can be isolated and targeted. Segmentation is 'the process of portioning markets into seg-ments of potential customers with similar characteristics who are likely to exhibit similar purchase behavior' (Weinstein, 1987: 4). It builds on the nature of the decision-making process that customers undertake, which involves innate needs and motivations, percep-tions or ontologies of the world, demographic characteristics (e.g. lifestyle attributes), the overall awareness of the product, and purchasing behaviour (Weinstein, 1987). If a firm is able to understand this particular segment properly, it can use this to its advantage over its competitors by tailoring and promoting a product or service that is targeted towards this segment. Segmentation, then, is a powerful means by which firms come to 'know their market' and excel in a highly competitive market.

Transport firms offering services to tourists, including business tourists, utilise segmen-tation to ensure that their product and service offers are relevant and in demand. Airlines, for example, use segmentation analysis as a means of relationship marketing processes to target specific travellers and thus increase market share. For example, those who fly on business may do so more frequently than those who travel for the purposes of leisure. As a result, business travellers may have an affinity towards more comfortable environments whilst travelling and 'in transit', hence the use of business class and first class, as well as airport lounges, targeted towards this type of traveller. Leisure travellers, however, may be less demanding of such amenities, and may respond more favourably to cheaper fares, although recent research by Dresner (2006) noted that business and leisure passengers expected similar levels of service at airports. Mason (2000) found price to be the most important determinant in business travellers' decisions to fly with particular low-cost airlines.

Some cruise companies target specific segments of the population in an effort to build market share. Stelios Hadji–Ioannou, the founder of easyJet in Europe, announced easy-

Cruise in March 2005, which, according to Hadji–Ioannou, will target 'younger crowds, in their 20s and 30s, rather than wealthy older people who like more traditional cruises' (Carassava, 2005). What is interesting about easyCruise is that the emphasis is on the destination rather than the on-board amenities that most cruise companies emphasise. In targeting younger markets, easyCruise is making various ports in the French and Italian Riviera the primary attractions. At each port, passengers will disembark and enjoy the nightlife 'scene' before re-boarding and moving on to the next destination (Carassava, 2005). Other cruise companies, such as Princess Cruises, are utilising advertising to target people who have never taken a cruise. Carnival Cruises is doing the same, with its tagline, introduced in late 2002, of 'So much fun. So many places' (Griswold, 2002).

Rail companies are also keen to tap into new markets in an attempt to re-package their produce/service and lure people away from the popular low-cost airline alternative to transport. The American Orient Express was developed to combine luxury ambience with a heritage tour of national parks such as Grand Canyon, Grant Telton and Yellowstone. The rail cars used for the trip are refurbished 1940s and 1950s models (Mulrine, 2002). As Mulrine (2002) reports, interest in the American Orient Express surged after the 11 September 2001 attacks on the United States, with the owner noting that 'we began getting more calls from people who wanted to travel but didn't want to get on a plane' (Multine, 2002: np).

With the popularity of, and often exceptional value provided by, low-cost air travel across Europe, it has been increasingly difficult for rail companies to compete. Virgin Trains recently launched an aggressive campaign across England in order to increase modal market share. Television ads featured cinematic music (from the 62-piece London Metropolitan Orchestra) against modern visual shots of passenger rail cars. Intriguingly, the passengers in the ads feature modern travellers juxtaposed with some of Hollywood's famous celebrities from the past:

> Margaret Lockwood and May Witty order a pot of tea, asking the waiter to make sure the water's boiling. Cary Grant tries to chat up Eva Marie Saint. And (rather strangely, since the train has already been racing through the fields) Tony Curtis and Jack Lemmon hobble down a Euston platform in high heels and make it on to the train just in time. The advert – a full minute long – reaches its climax (to the kind of orchestral soundtrack one might expect from a Hollywood epic) with Cary Grant declaring: 'Beats flying, doesn't it.' (BBC, 2004d)

Clearly, an ad as such as this was designed to be interesting to the viewer from an historical perspective, but it is also follows many of the key criteria that Morgan and Pritchard (2001: 46) argue constitute good advertising: simplicity, relevance, uncomplicated and long term. On the BBC website, however, readers were invited to post their comments, and one reader challenged the veracity of what was being promoted in the television spot:

I absolutely cannot fault this very well crafted short film. Unfortunately, it doesn't work as an advert, because it is rather let down by the reality it promises to deliver. If only the trains in the UK were really this nice! It is a fantasy film therefore, to be enjoyed on only that level, and not to be regarded as reflecting the reality of train travel in the UK today anymore than 'The Lord of the Rings' films reflected the reality of life in New Zealand. (BBC, 2004d)

Air New Zealand in early 2005 announced a revamped long-haul 'product', the advertising for which had substantial reach and frequency throughout New Zealand. Print ads (Figures 8.1 and 8.2) were run in major daily newspapers (thus reaching most of the country) and television adverts were featured in key advertising time periods (for example, television prime time, from roughly 7 pm to 9 or 10 pm). With this particular advertisement, what is being promoted are the many improvements in the service intangibles and produce tangibles associated with flying long-haul routes on the airline. New Zealand as a country is somewhat geographically peripheral to most of the world (save for perhaps Australia), and as such Air New Zealand must be seen to offer a higher degree of comfort on its long-haul flights versus its competitors. In this particular print ad, what is

THE AMAZING NEW
THINGS YOU'LL
DISCOVER ON
AIR NEW ZEALAND.

NEW LONG-HAUL.

Figure 8.1 Print run example (2005) from Air New Zealand showing new business premier service (reproduced with permission)

Figure 8.2 Print run example (2005) from Air New Zealand (typical third of a page newspaper run (reproduced with permission)

shown is the upgrading of the 'in-flight' experience for all classes of travel. The re-vamped long-haul product being advertised is intended to be a value driver; that is, short of purchasing or leasing yet-to-be-developed faster aircraft, many airlines now focus on particular elements of the in-flight experience that help drive the perception of value in the mind of the customer/passenger. Contrast this with Figure 8.2, where the objective is to highlight fares and destinations. Indeed, it is an example of the tight integration that airlines have with tourism, and how airlines themselves have a keen interest in ensuring the viability of destinations worldwide (consider, for example, the line in the ad in Figure 8.2: 'So whether you want to admire the iconic Mt Rushmore, experience theme park thrills in California, or relax on the sun kissed beaches of Hawai'i, the best way to see the great sights is with our great prices.'). What is not mentioned in the ad in Figure 8.3 is the services and conveniences with flying Air New Zealand, even though this advertisement ran in local newspapers throughout New Zealand. It contains several critical elements for ensuring that potential travellers have several ways of purchasing tickets: a toll-free 'Freephone' number (NZ only), mention of the Air New Zealand Travelcentre (with locations throughout the country) and the main website. What is clear is that the airline is interested in having a strong amount of control over how its service/product is distributed.

MARKET DEVELOPMENT STRATEGIES

Market development strategies generally involve attempts at securing market share in markets not currently being served (Kerin & Peterson, 1998). For example, LCCs (or VBAs) have often sought to enter new markets using price as the means of generating business. Mason (2001) discusses how some VBAs (or LCLF carriers, from Chapter 6) in the European Union have begun targeting business travellers. Mason compared the travel habits, but particularly the importance placed on specific attributes or tangible products, of business travellers that originated from London's Luton Airport and London Heathrow. Mason's study attempted to determine whether there was a difference between those business travellers who utilise a VBA out of Luton and those who utilise a network carrier (or FSA) out of London Heathrow. Mason's hypothesis was that business travellers who utilise a VBA fundamentally act as a different market segment than those who use a FSA. The results of Mason's research suggest that, among short-haul business travellers, there are not enough commonalities between those who utilise VBAs and those who use FSAs:

> The profile of business travellers using low-cost airlines seems to be different in the size of the company that they work for, the booking process used, the channel used to book flights, and the importance placed on the price of travel, in-flight service, frequent flier schemes, and business lounges. While there are differences between the two groups there was also much common ground, particularly in the proportion of travellers who select their own flights, the opinion travellers have about the value for money of business class products, and the importance of airline punctuality, flight frequency and ticket flexibility. (Mason, 2001: 108)

Mason found that price sensitivity was paramount for those business travellers utilising both VBAs and FSAs. He notes that business travellers working for small- or medium-sized companies were more likely to utilise VBAs because 1) 'people working for smaller companies are likely to see a clearer delineation from travel costs to the profits made in their businesses than travellers working for larger companies' and 2) larger companies may be in a better position to negotiate more lucrative travel deals based on potential volume of sales (Mason, 2001: 109; see also Mason, 2000).

In some instances, LCLF carriers have sought to unseat traditional full service or network carriers by competing directly at the price point, but also in the development of new markets. The case of Ryanair (IATA: FR) and Aer Lingus (IATA: EI), two Irish airlines, is particularly interesting as it demonstrates how an upstart VBA can generate new business for a variety of destinations through market development. Ryanair generally succeeded as a VBA for two reasons (Kangis & O'Reilly, 2003). First, significant competition in the beginning was primarily Air Lingus, which operated out of Dublin Airport and targeted specific niche markets which, according to Kangis and O'Reilly (2003) exposed the airline to the vagaries of the international aviation environment. Air Lingus' operations were essentially value-added (Kangis & O'Reilly, 2003: 106), which entailed providing seamless service throughout the experience of flying with the airline. Ryanair, on the other hand, adopted a 'focus on core activities', which became its key strategy for success (Kangis & O'Reilly, 2003: 106). This entailed providing a low-cost, low-frills service in order to maximise sales (Kangis & O'Reilly, 2003: 106). The result has been an almost 20-year 'battle for the skies' that has seen Ryanair become one of the largest, if not the largest, VBA in Europe. In order to achieve this, Ryanair adopted a business model (use of secondary airport and a generally a point-to-point or linear network configuration, see Chapter 6) that features cautious expansion principles without the risk of jeopardising inflated costs.

These two elements combined meant that Ryanair was able to introduce low-costs flights to new markets. As a result, '[t]he Ryanair policy is one of market creation where fares and frequencies are attractive to contingent or fringe travellers, such as those who would not normally make the journey or who would travel by another form of transport' (Kangis & O'Reilly, 2003: 107). Hanlon (1999: 179) has argued that while point-to-point travel and a directive of reduction in costs may have allowed for increase travel, 'successful operation of point-to-point routes owes at least something to the unusually high proportion of ethnic demand in the particular city pair markets serviced and also to the fact that the only surface competition comes from a sea crossing'. As well, Ryanair generally paid lower commissions to travel agents, preferring instead to sell seats to customers through Ryanair Direct (Barrett, 2000). More recently, the airline has primarily sought to sell tickets through its website. It also does not have a frequent flyer programme, nor does it make use of extra frills at the airport such as lounges. Air Lingus, on the other hand,

> employs traditional arrangements with travel agents and pays normal rates of commission. Aer Lingus markets on the strength of its image, and has tried to create brand loyalty. It operates a 'gold circle' club, providing luxury lounges at all the airports that it is servicing. It also has a frequent flier programme for its most loyal customers. (Kangis & O'Reilly, 2003: 108)

PRODUCT DEVELOPMENT STRATEGIES

Product development strategies involve the introduction of new products to existing (and with the potential of development of new) market segments. These efforts usually extend beyond the addition of value-added components, which are designed to enhance existing products; instead, the focus is on the recognition of a need for new offerings. Such strategies have been prevalent in cruise tourism in the past few decades with the increase in global competition. In many ways, cruise ships have ceased being modes of transport and, as discussed in Chapter 5, now feature a range of on-board amenities designed to appeal to numerous market segments. For example, on the *Voyager of the Seas*, a passenger can go ice skating, practice their rock climbing skills, and be entertained in movie theatres. The president of the CLIA in 2002 remarked that '[y]ou could spend a whole week on a ship and you wouldn't know you're on a ship' (Elliot & Silver, 2002). What this marks is a fundamental shift in the product that is purchased. Cruise ships have effectively diversified their product offering and, as a result, tapped into the all-inclusive market segment in countries such as the United States.

New product development is somewhat more limited when it comes to aviation, but the industry is not without examples. In mid-2005, two new airlines were launched offering business-class travel between the United States and the UK. Eos is a privately held company hoping to compete directly with high-quality and well-received business-class offerings from Virgin Atlantic and British Airways. The airline will use Boeing 757 aircraft with 48 seats rather than the usual 200 in double-class (economy and business) configuration. The second, MAXjet, is also a privately held company and will offer return services from New York to London Stansted using Boeing 767-200 aircraft with 102 seats (whereas the aircraft is normally configured in approximately 200-seat configurations). On the website for MAXjet, it states that the airline 'was founded to bring the low-cost carrier revolution to the international market. It will be the first.' This may not be entirely correct, depending on how 'international' is defined. Ryanair has been flying internationally (or transnationally) for quite some time, but MAXjet was the first to offer trans-Atlantic business class-only services in an LCC format. Both Eos and MAXjet will offer standard business-class amenities: large seat-pitch (the distance between the front of one seat to the same point on the seat ahead) and full meals and beverage services. The product that MAXjet and Eos have each developed is a business class-only airline, and the intent is to use the produce to target business class travellers who generally prefer comfort and amenities to cheap prices, but this is an interesting example in product development in that the airline industry normally attempts to offer two or three class service offerings on most flights, thus catering to multiple markets on a single flight. Over time, the test will be to see if Eos and MAXjet are able to secure enough of a market share to ensure ongoing financial viability. There may be several factors that may hinder profitable performance, however:

1. Downturns in economies (either in the United States or the UK) may mean a reduction in corporate spending on travel. Traditional multi-class carriers may not be as vulnerable in these situations, but a strictly business class airline will need to offer price points that are competitive as perhaps the average economy class fares of its competitors. On the website, however, MAXjet is promising fares that they claim are often what one would pay to fly economy (and use the example of £599 one-way).

2. Some business travel may be seasonal (for example, generally less travel during Christmas holidays), so it is not clear whether sufficient transactions will be realised in order to be profitable during the off-peak periods.

3. As evidenced by some online travel discussion boards (e.g. www.flyertalk.com), some business travellers undoubtedly select airlines on the basis of accruing frequent flyer miles and privileges. These two business class-only airlines, for now, will not be part of a formal global alliance, nor is it clear whether they will offer any type of frequent flyer reward programme.

4. The use of London Stansted as the primary airport reflects the restrictions on slot space (discussed in Chapter 7) at larger airports such as London Heathrow, and it remains to be seen whether passengers on business place a higher premium on using an airport closer to the central city to which they are travelling.

Alliances as marketing tools

As discussed in Chapter 6, code-sharing and schedule coordination is a common alliance format for airlines. This type of tactical alliance (or 'mode-specific alliance' as the concentration of market benefits sit within a specific mode of transport) often has significant benefits for marketing because it allows airlines to tap into new markets that would have otherwise necessitated the introduction of their own services and, therefore, consuming substantial amounts of capital investment (Glisson *et al.*, 1996). Table 8.1 highlights the strength of the major airline alliances operating in the summer of 2006, including the extent to which each alliance is in competition with another as measured by the duplicate destinations served. The table also indicates that over one-half of all global destinations are served by at least one airline alliance using some 60 billion ASKs.

Global airline alliances are of course not the only means by which different airlines link networks. Oum *et al.* (1996: 187) note that code-sharing agreements 'are used to enhance

Table 8.1 Comparison of alliance network strength based on weekly scheduled operations in (northern hemisphere) summer 2006

	Total Destinations	Duplicate Destinations	Total countries served	Capacity (available seat kilometres)	Proportional share
Star Alliance	873	318	147	22.6 billion	20.6%
Oneworld	591	163	128	15.9 billion	14.5%
SkyTeam	730	366	141	20.4 billion	18.6%
Total				58.9 billion	53.7%

Source: Adapted from Airline Business (2006b)

services and to create a marketing advantage of on-line connecting services'. Code-shares are the means by which one airline can be seen to offer 'seamless' service across a substantial network that it otherwise would not have been able to service itself due to the prohibitive cost (Beyhoff, 1995):

> Codesharing is a marketing arrangement between two airlines whereby one airline's designator code is shown on flights operated by its partner airline. Two letter designation codes are provided by the International Civil Aviation Organisation (ICAO) to identify the airlines on passenger tickets, computer reservation systems (CRS), airline guides and airport information boards. Under a codesharing agreement the connecting flights being operated by two separate airlines may be listed as bring a 'single carrier' service, which gives substantial marketing advantages. (Oum *et al.*, 1996: 188)

The effects of codesharing have been scrutinised by Brueckner (2001), who found that passengers from interhubs (i.e. those city-pairs that serve a major hub) are somewhat financially disadvantaged when a code-share alliance is established for the purpose of serving two or more major hubs. However, Brueckner (2001) also found that there was an overall gain in consumer and total surplus following the formation of a code-share alliance. What this suggests is when selecting a potential code-share partner, an airline would think carefully about the potential market gain from servicing major hubs using such an arrangement, and thus sharing the revenues in an interline market, versus the total costs for operating in the network alone.

RELATIONSHIP MARKETING

Relationship marketing can come in a variety of forms, from direct sales marketing to the establishment of an online presence designed to remove distributors from the total transaction. Within the last decade, many transport firms have utilised IT to form ongoing relationships with their customers (Bejou & Palmer, 1998). Many airlines, for example, employ direct marketing in the form of regular (or erratic) email announcements to members of their frequent flyer programme. These can be highly effective in targeting specific and multiple segments, as Roberts and Berger (1999: 167) argue:

> [T]he airline can manipulate the information it already has to help it better understand the behavioral patterns of its frequent customers. It can also collect and store additional information, which may be attitudinal as well as behavioral. This additional information can either be general (request of all or most of the frequent fliers) or specific (requested of only those who have not flown the airline for the past three months, for example).... It can then promote these programmes directly to the fliers who have a high probability of responding instead of wasting its promotional dollars on those who have a low probability of response.

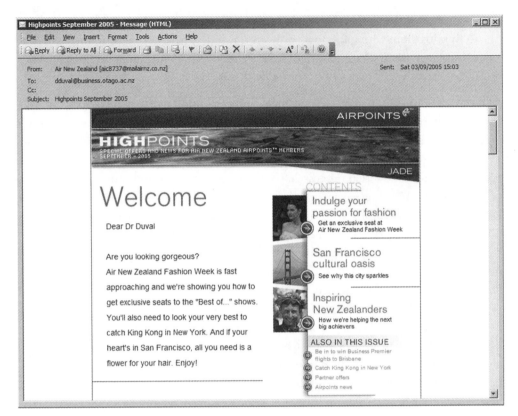

Figure 8.3 Direct marketing gone slightly askew or lack of segmentation?

For example, in a bid to enhance its relationship with this author, Air New Zealand sent an email (Figure 8.3) in early September 2005 that at once exemplifies non-directed relationship marketing. Everyone listed as a frequent flyer member for Air Points is likely to have received the email, which illustrates how direct marketing should perhaps be segmented. Of course, the intent of the email was to showcase the range of destinations that one might be interested in, but by providing it via this medium (the email included links to the airline's online portal and booking engine) it is attempting to form a relationship. Relationship marketing is a means by which firms put as much effort into retaining existing customers as they do seeking new ones. In doing so, firms seek to gain a competitive advantage and foster a long-lasting relationship with a customer (Gilbert, 1996). What is involved is rather simple in principle, but somewhat difficult given strong competition, as Shaw outlines in the context of airlines:

> Once someone has decided on a particular airline, that airline only has the task of persuading them to continue to make the same choice. This should be relatively easy, because it will be a question of reinforcing in their mind the correctness of

their original decision. If on the other hand, a carrier needs to persuade someone to fly with them who until now has been traveling with a competitor, they will need to convince them that their original choice was wrong. Human nature being what it is, few of us readily admit our mistakes. (Shaw, 2004: 229)

In reality, relationship marketing is far more complex, and involves a wide range of sub-markets (internal, referral, supplier, recruitment and influence markets) and involves establishing plans and strategies to shift the focus of a particular enterprise (Peck *et al.*, 1999). It is at once concerned with service issues and customer satisfaction, but it is also concerned with the internal processes that are often just as critical. While the full scope of relationship marketing is vast and thus only briefly discussed here, it is raised because it has significant analytical and explanatory value when it comes to the marketing of transport in relation to tourism (Bejou & Palmer, 1998; Long & Schiffman, 2000).

Although there are other means of securing relationships with customers, one of the more prevalent examples of relationship marketing in tourism transport is an administrative tracking programme designed to reward loyalty. Loyalty is not an unusual means of relationship marketing as many businesses utilise specific practices to increase retention in purchase behaviour (Hennig-Thurau & Hansen, 2000). Celebrity Cruises, for example, offers a Captain's Club programme where frequent cruisers can enjoy specific rewards and upgrades depending on the number of times they have cruised. 'Classic' status members are those who have accumulated 1–5 'cruise credits', and can enjoy a 'one-category upgrade' or have access to an 'exclusive' Captain's Club party on their next cruise. Those who have 11 or more cruises to their name are known as 'Elite' status members and can enjoy, among other things, private shipboard departure lounge access and an invitation to the Captain's or Officer's table for dining.

Cruise companies are not the only form of transport to offer a loyalty-based programme that rewards frequent use of a service. Via Rail Canada runs their VIA Préférence programme, where a member earns one point for every dollar spent on rail travel along VIA's network across the country and on some services to the United States. As with most other programmes, as a passenger spends more money on services within a particular qualifying period (usually a 12-month period), they receive higher status in the programme, which consequently brings more privileges and benefits.

Most airlines offer some form of mileage accrual programme to its customers (see Long *et al.*, 2003; Yang & Liu, 2003). These programmes are relatively simple in their construction: the more a passenger flies, the more 'miles' or 'points' become accrued that can be used at a future date for fare-free/'reward travel' or other services (as of late, some airlines offer redemption of points or miles in exchange for new electronic goods or full holidays). The points accrued through travel can be used in the future when a certain threshold has been reached, for reward travel. The amount of miles accrued is largely dependent on the level of service (e.g. economy, business, first) and, to some extent, on

BOX 8.1 Tax implications for loyalty programmes/relationship marketing

The frequent flyer miles that some businesspeople accumulate on business trips are often used (and not always legally) for personal travel. Some national governments, however, would suggest that using miles in this manner constitutes a taxable benefit, a move not unsurprising given that, as a global currency, frequent flyer miles have a value which far surpasses many of the world's major monetary currencies (Clark, 2005). In Canada, the CRA has made it explicit under the Employees' Fringe Benefits section of the Income Tax Act that such benefits are indeed taxable. According to Section 14 under Part A (Amounts To Be Included In Income):

> Under this program, which is usually sponsored by an airline, a frequent air traveller can accumulate credits which may be exchanged for additional air travel or other benefits. Where an employee accumulates such credits while travelling on employer-paid business trips and uses them to obtain air travel or other benefits for the personal use of the employee or the employee's family, the fair market value of such air travel or other benefits must be included in the employee's income. Where an employer does not control the credits accumulated in a frequent flyer program by an employee while travelling on employer-paid business trips, the comments in Paragraph 3 above will not apply and it will be the responsibility of the employee to determine and include in income the fair market value of any benefits received or enjoyed. (http://www.cra-arg.gc.ca/E/pub/tp/it470r-consolid/it470r-consolid-e.html#P128_12793; accessed 10 August 2005)

Upon seeing this, some business travellers may think twice before cashing in their frequent flyer miles for a trip to MedSun destination with their family. This guideline, however, introduces several issues:

1. What is the cost-benefit for the CRA to undertake regular audits of business travellers to determine whether there is failure to report additional income based on the 'fair market value' of the points accumulated? Presumably, however, cases of tax evasion of this sort are sorted during regular audits.

2. What is 'fair market value' when, as discussed in Chapter 6, airfares are volatile and extremely dynamic? Would a capacity-dump situation, where airlines unload excess capacity on certain flights, count? Or, would a mean fare value be more appropriate? If so, what period of time should be established in order to generate a mean baseline fare (months, weeks, days)?

3. On reward travel, the customer normally must pay the taxes and surcharges, and it is unclear whether these can be deducted from the fair market value of the reward flight(s) given that the value of the ticket itself is to be treated as income which, eventually, is subjected to certain levels of tax.

Questions for consideration

1. What are the ethical implications for booking a reward flight for personal travel using frequent flyer miles generated from travel for business?

2. To what extent is the root of this problem based on the fact that frequent flyer miles are bestowed on the individual on the plane as opposed to the individual or corporation responsible for paying for the ticket? How might this be rectified yet still retain the sense of 'loyalty' claimed by airlines?

the particular fare class. For example, some airlines will not allow accrual on low-fare tickets (e.g. N fare class, but this designator is not uniform across all airlines) but will on full-fare or mid-fare tickets (e.g. Q or Y fare class). Some airlines even offer up to 1.5 times as many miles if a passenger purchases a business- (e.g. J class) or first-class (i.e. F class) fare.

Gilbert argues that airlines need to consider three critical elements to the micro-marketing features inherent in frequent flyer programmes:

(1) There must be a greater emphasis on the database information held so as to slice and portion it to provide improved aspects of individual or group service delivery.

(2) The essence of relationship marketing needs to be the provision of differentiated communication and services. These have to be customized based on the researched characteristics of potential and current customers.

(3) There is a need to track and monitor each member of FFPs to ensure there is an assessment of the lifetime value and retention history of the individual. (Gilbert, 1996: 582)

Gilbert is advocating the use of frequent flyer programmes to offer members' personalised services. In one sense, this accurately represents a key reason for the establishment of almost any loyalty programme: the company has access to the purchasing habits and service/product usage of as many programme members that are in its database. This is a powerful tool for tailoring marketing and promotional activities. For example, an airline may discover that residents of a particular city regularly undertake travel to another regional centre for the purposes of onward travel to a foreign destination. Specific marketing (such as ad buys in newspapers or television spots) can be created for distribution in that particular market. Put another way, the existence of loyalty programmes such as those outlined above are examples of a means by which customer-specific data are acquired for the eventual purpose of market segmentation and the specific database marketing (Robinson & Kearney, 1994) examples mentioned.

Although some may argue that these programmes are designed to induce loyalty to a particular airline, with the introduction of global alliances between airlines (see Chapter 6), the value of loyalty has, in some cases, shifted from the individual airline towards a global alliance that recognises frequent flyer benefits across a number of airlines (although Bruning, 1997, notes that, on the whole, price followed by an airline's country-of-origin were key purchase determinants in his study of Canadian travellers). Passengers flying on any Star Alliance member carrier can earn (and burn) points on any carrier that is also a member of the alliance. First, however, it is important to note that each Star Alliance member programme has its own status tiers (often three or more). The higher tiers correspond to the Star Alliance-branded higher tiers recognised across the Star Alliance network, thus resulting in an almost global recognition of status and access to benefits and privileges. For example, Air Canada's Aeroplan has three tiers: Basic, Prestige (Star Alliance Silver) and Elite (Star Alliance Gold). Air New Zealand's Airports Programme has three levels as well: Silver (Star Alliance Silver), Gold (Star Alliance Gold) and Gold Elite (Star Alliance Gold as well). Those who achieve Gold Elite on Air New Zealand's programme benefit from Air New Zealand-specific benefits, such as valet parking and hotel or rental car vouchers. The general benefits of Gold status, as indicated on the Star Alliance website, generally include:

- Priority Reservations Waitlisting – when there aren't any seats left on your preferred flight, gives you priority should a seat become available.
- Priority Airport Standby – gives you the flexibility to change plans at the last minute when you don't have a reservation.
- Priority Airport Check-in – avoid long lines regardless of paid fare or class of service.
- Priority Baggage Handling – like you, your bags get priority treatment and are among the first to be unloaded.
- Extra Baggage Allowance – an additional 20kg (44 pounds) or one additional piece of luggage which means you can check in three bags instead of two.
- Priority Boarding – enjoy the freedom of boarding at your convenience along with First and Business Class passengers.

- Airport Lounge Access – available worldwide for you and a friend when you travel with any Star Alliance airline regardless of your class of travel. (http://www.staralliance.com/star_alliance/star/content/status.html; access 25 July 2005)

Like most loyalty programmes, status is awarded based on the number of flights in a qualifying period. Passengers achieve status through an annual measure of the amount of miles accrued during a particular qualifying period on any Star Alliance airline, yet the amount of miles or points required to achieve status recognition by the Star Alliance network varies from programme to programme. Air Canada, for example, will award Gold status if a passenger accrues 35,000 miles through eligible flights operated by any Star Alliance carrier. With United Airlines' Mileage Plus Programme, however, 50,000 miles are needed to reach Premier Executive status, which brings with it Gold status across the Star Alliance network. What this discrepancy demonstrates is that, despite the nature of a global alliance designed to foster loyalty to a global group of airlines, the means by which that loyalty is awarded and recognised is not entirely uniform.

A further aspect of global alliances in the airline industry is that the type of customer that they have effectively 'produced' has almost erased the need for individual airline frequent flyer programmes. As such, the global nature of frequent flyer programmes, coupled with the alliance networks that have been formed, now means that one's place of residence is almost irrelevant. This has interesting implications for individual airline loyalty programmes. For example, the author of this book, as a resident of New Zealand but Canadian citizen, has chosen to accrue miles through the Air Canada Aeroplan Programme (although he is also a member of several other airline programmes for 'research purposes', including the discovery in 2002 that, in fact, one flight can be credited to two separate programmes, a revelation that came as quite a surprise to both airlines concerned when it was pointed out to them). However, at present the vast majority of his flights are on Air New Zealand and Singapore Airlines aircraft. As a result, this author has accumulated a substantial number of miles that have been credited to his Aeroplan account, but less than 30% were actually accumulated on Air Canada metal. Put another way, in this particular case miles are being rewarded onto Air Canada's frequent flyer programme yet the bulk of revenue has gone to Air New Zealand and Singapore Airlines. However, Aeroplan carries the contingent liability (i.e. the possibility that Air Canada may be asked by this author for redemption of those miles in the form of a reward ticket for travel on an Air Canada-operated flight). Contingent liabilities are rarely reported (Bennett, 1996), which means it is difficult to determine how many outstanding points are being held by customers. However, because Air Canada is part of the Star Alliance, this author's miles can be redeemed on any Star Alliance airline. As members of this particular alliance, the individual airlines would hope that the network-wide redemption eventually balances out any contingent liability of its own programme, which means that the entire network has contingent liability for all frequent flyer miles/points accrued by all member airlines.

A commonly held belief is that the contingent liability of existing air miles held in frequent flyer programmes is problematic for many airlines in that they may not be able to provide seats to match demand without realising a significant drop in marginal revenue. The reality, however, is somewhat different. First, the manner in which air miles are collected by the public is often not from flying. The most common means of collecting miles is through credit card purchases. In this arrangement, a bank will reward air miles to a customer's frequent flyer programme according to a set ratio (charges:miles awarded). To do this, banks purchase miles from airlines or the often separate companies that airlines have established in order to manage their frequent flyer programmes. According to the *Economist* (2005), this arrangement is worth some US$10 billion to airlines around the world. The airline's marginal cost to provide seats to those who request a reward seat is relatively small. As a result, given that more frequent flyer miles are being accumulated from non-flight activities (Economist, 2005), and that airlines are being paid a considerable amount for miles by banks in order to allow the latter to use them for promotional purposes, airlines actually make money from their frequent flyer programmes, and sometimes even when the airline itself may be running at a loss. The second reason why air miles, as a contingent liability, are not necessarily a threat to airline solvency rests with their ability to alter the value of reward miles themselves. In other words, an airline may elect to introduce certain blackout periods where reward travel is not possible or increase the number of miles or points needed to redeem reward seats.

Airlines regularly use global network alliances and frequent flyer programmes as means of promotion. Each may feature specific offers for vacation packages, upgrades and hotel discounts to its members. For some passengers, particularly business tourists, the value of the specific frequent flyer programmes is high enough such that the selection of the airline they chose to fly can be based almost purely on whether reward miles can be accrued or status levels can be maintained. For example, a businessperson wishing to travel from Chicago to Auckland may elect to purchase a ticket with a carrier who is aligned with a particular global alliance such as One World or Star Alliance. The businessperson may chose to fly from Chicago to Los Angeles with United Airlines, and then Air New Zealand for the flight from Los Angeles to Auckland. Using these two carriers guarantees (again, depending on the fare type, but most international flights will accrue some miles) mileage accrual. As well, if the businessperson travels enough on an annual basis to be considered Gold in status, he or she will be able to utilise specified lounges in Chicago, Los Angeles and Auckland. Specified Star Alliance Gold Lounges are frequented by business travellers because they offer free alcohol and other beverages, free food (in some lounges), free wireless internet access, showers and a comfortable atmosphere. In many lounges, small conference rooms and business centres are available.

Shaw (2004) points to several lingering problems with the use of alliances to differentiate products in a strongly competitive environment. For one, alliances represent a particular 'mindset' by which airlines, in the face of competition, will form marketing and service alliances with other airlines rather than engage in competitive behaviour:

> When faced with a tough competitor, it has nearly always been the airlines' instinct to form a collusive, rather than competitive relationship... Without the challenge of strong competition, airlines do not work sufficiently hard to control their costs, particularly their labour costs. They therefore quickly eat up all the guaranteed revenues which collusive behaviour gives them, leading to calls for a further diminution of competition. (Shaw, 2004: 111)

Another problem that Shaw points to is the fact that the financial performance of many air carriers in alliances varies considerably, and points to the example of Lufthansa (a carrier that enjoys comparatively strong financial health) and United Airlines (which has been in financial trouble for years). This leads to some serious questions about the uniformity of product/service delivery across any alliance. For example, and as pointed out by Shaw (2004), how can United be expected to finance future service availability (or, rather, limit the erosion of service quality) on its own network when it belongs to an alliance where partners such as Lufthansa are capable of retaining existing service levels? To this end, Shaw (2004: 112) asks: '[H]ow can the Star Alliance carriers generally engage in strategic discussions about developing their long-term co-operation, when they must all know the possibility exists that United will not survive?'

SERVICE QUALITY CONSIDERATION IN TRANSPORT AND TOURISM

Regardless of whether transport is considered a product or a service, the quality of delivery must be considered. In transport research, considerable attention has been devoted to the issue of service quality and how it is to be measured (see, for example, Chen & Chang, 2005; Gursoy et al., 2005; Tsaur et al., 2002). Tripp and Drea (2002) found, in a study of over 2500 Amtrak passengers in the United States, on-board conditions, café car conditions and on-time performance were considerations in both overall attitudes towards Amtrak and whether passengers intended to re-purchase. In air transport, service quality has been examined with respect to the impact it can have on passenger behaviour (e.g. Park et al., 2004), passenger expectations (Gilbert & Wong, 2003) and safety perceptions (Rhoades & Waguespack, 2000). Suzuki et al. (2001) and Suzuki and Tyworth (1998) have modelled the important relationship between service quality and airline market share utilising a weighted average of on-time performance, over-sales, mishandled baggage and in-flight food quality when measuring an air carrier's overall attractiveness. Notably, Suzuki et al. (2001: 786) argue that 'if an airline's service quality falls below the market reference point (median service quality of the industry at time t), the airline's market share will decrease significantly, but the service increase from the reference point may not increase an airline's market share'.

One of the more popular measures of service quality is the gap method introduced by Parasuraman *et al.* (1985). In essence, this method measured the difference between customer's expectations and perceptions. Based on a specific instrument (SERVQUAL), Parasuraman *et al.* (1988) measured service perceptions in various organisations (bank, credit card company, repair and maintenance company, and a telephone company) and arrived empirically at a list of service dimensions (Parasuraman *et al.*, 1988: 23, 1991), which I paraphrase here:

1. Tangibles, including facilities, buildings and staff appearance.

2. Reliability, which relates to the ability of the service to be performed consistently and accurately.

3. Responsiveness, or the ability of the organisation or service provider to respond to customer needs.

4. Assurance, or the extent to which staff exude confidence and trust.

5. Empathy, which refers to the ability of the organisation to be caring and attentive when necessary.

Competition on the basis of service levels is a feature of some modes of transport. Airlines and many rail services (particularly in the UK) position themselves (via business models) on the basis of service. As seen in Chapter 6, the LCLF and FSA concepts differ on a number of elements, including service provision. Interestingly, with the lack of service provision generally attributed to LCLF carriers (in an effort to reduce cost), one might be inclined to think that this would have a negative impact on revenue and profits. O'Connell and Williams (2005), however, found that passengers travelling on LCLF carriers generally placed price ahead of other considerations, but those travelling on FSAs may tolerate higher fares in return for some service provision. Even more interesting was that O'Connell and Williams (2005) found that preference for the LCLF concept of low fare and minimal service was international as they surveyed passengers in both Asia and Europe.

Service level considerations extend beyond actual modes of transport. Tam and Lam (2004), for example, explore passenger wayfinding in airports in the context of service provision. In the context of airport design, such considerations are critical as airports are increasingly relying upon non-aeronautical revenue (see Chapter 7). Overall, the service encounter (Bitner, 1990) in the context of transport and tourism can occur at several points:

1. Pre-travel interactions between the traveller and the transport provider, including booking (online or through direct call centres).

2. During travel, including customer evaluation and experience of core and ancillary products and services.

3. Post-travel, including quality of post-sales service and attention (e.g. lost baggage, correct credit of frequent flyer miles).

CHAPTER SUMMARY

The marketing environment for transportation is constantly changing. Although these changes can be said to reflect the dynamic (some may even argue turbulent) environment in which competition for patronage is intense, many of the more significant changes are in the media through which advertisements and promotions are delivered. The example of direct marketing using email has already been mentioned above, but it is perhaps the internet that is providing the means through which substantial interest in marketing has taken place. As discussed, Virgin's internet advertising campaign for its UK train service was extensive and utilised the latest in imbedded advertising on web pages. Google searches are often peppered with 'google ads', which are meant to bear resemblance to the search term you inputted. Those involved in advertising are always conscious of where people get their information, and transport operations are no different. Where television advertising used to mean a captive audience, the introduction of hard disk recorders (such as TIVO in the United States) now means that consumers can time-shift their television programmes automatically and even omit commercial advertisements.

The marketing of transport operations in relation to tourism is more comprehensive, of course, than the means by which the product is advertised. What is important is how consumers perceive the advantages of the type and level of service being offered. This is particularly paramount in air transport, especially in a domestic environment where, quite often, LCCs offer cut-throat fares in a bid to drive traffic and enhance their presence in a market. With the increasing concerns over the cost of oil, however, it remains to be seen whether air travel, in some situations, will be seen as economically feasible for most travellers (see, for example, the discussion of Prideaux's transport cost model [2000a, 2004] discussed in Chapter 2).

Box 8.2 Self-check-in: New service experiences in airports

The *Airline Business* IT trends survey from 2006 reported that an average 27% of all passengers are using self-service kiosks at check-in (Airline Business, 2006a). Roughly one-half of all airports around the world employ some form of self check-in kiosk, and this is predicted to rise to 75% over the next two years (SITA, 2006). The importance of these data cannot be understated: passenger growth is predicted to rise exponentially in many regions around the world. The UNWTO's World Tourism Barometer reported in July 2006 that preliminary numbers indicate an increase of 10 million more tourist arrivals worldwide for the first four months of 2006 over the similar period in 2005. Managing these flows is becoming critical, and airports and airlines are both seeking ways of managing passenger throughput more efficiently. Whereas many airports will see a rationale for improving customer 'throughput' during peak hours of a given day, airlines may wish to utilise IT to enhance the passenger experience or as a means to reduce costs. Regardless of the strategic rationale, technological advances have helped with this process.

Passenger activities in airports generally consist of check-in, customs and security, shopping and then boarding. Although there are variants at many airports (including the introduction of enhanced retail and service activities, as discussed in Chapter 7), many airports, and airlines, have sought ways to speed up the process of passenger check-in. Check-in represents a significant time when passengers queue, although in an increasingly security conscious world, queues can also be found at customs and security checkpoints (of which there may be more than one). Many airlines and airports have sought to decrease the time spent checking-in for most passengers.

Self-check-in kiosks are becoming increasingly common at many airports (Figure 8.4). The benefits for airports are clear: they improve passenger movement and could facilitate growth in passenger numbers without a corresponding increase in foot congestion. A downside for airports, however, is that faster throughput may mean less time actually in the airport. Given the increasing reliance on non-aeronautical revenue, some airports may be reluctant to move towards more kiosks. For airlines, self-check-in kiosks can be financially lucrative. Ground staff numbers can be reduced, and thus ground operational costs shrink. For passengers, having access to multiple kiosks as opposed to proportionately fewer ground check-in staff can mean faster processing of travel documents and acquiring boarding passes more quickly. A few examples from airline press releases:

> Continental Airlines (NYSE: CAL) announced on Tuesday (18 April) that it has installed its 1,000th Self Check-in Kiosk in Bogota, Colombia. According to Continental the carrier now offers more kiosks per customer than any other airline. The kiosks, which offer service in English, Spanish, French and German, enable travellers to print their boarding passes at the airport. In addition domestic passengers can select or change their seat, upgrade or

Figure 8.4 Service efficiency: Self-check-in kiosks at Dunedin International Airport, New Zealand

stand by for First Class, verify or input their OnePass number, and change their flight for same-day travel by using a Self Check-in Kiosk, the airline said. (http://www.findarticles.com/P/articles/mi_m0CWU/is_2006_April _19/ai_n16126991)

easyJet is transforming the airline industry once again with its plans to introduce self-check-in across all airports, a process which is currently in place at Nottingham East Midlands, Geneva and Berlin airports. Some airlines have offered self-check-in for passengers with hand luggage for a number of years and a very small number allow self check-in for those with hold baggage. However, all these airlines allow passengers to check-in using either kiosks or traditional methods. Keeping two channels open is an unnecessary duplication and increases, rather than decreases, costs. Ultimately dual processes keep fares unnecessarily high. This major initiative is part of easyJet's drive to reduce costs by using technology to simplify processes and reduce complexity, which in turn will help keep your fares low. It will also help to increase airport capacity by using space more efficiently. You should also find that the check-in process is much faster, with less time spent queuing! (http://www.easyjet.com/ EN/Flying/leapcheckininfo.html)

Emirates ensures that each step of your journey is a unique experience in itself. The 24-hour Self Check-in facility at Dubai International Airport helps make checking in faster and more convenient than before. So whether you are travelling light with hand baggage only, or off on a longer journey with baggage to check in, this is the easiest way to fly. This service allows you to check in, select your seat, register your Skyward Miles and even obtain your First or Business Class lounge invites. (http://www.emirates.com/TravellerInformation/airport/SelfCheckin/selfcheckin.asp)

The caveat with self-check-in kiosks, however, is that passengers still be required to queue to 'drop off' check-in baggage. In other words, queue congestion is transferred from the check-in area to other areas, although it could be argued that is more efficient overall in terms of moving passengers throughout a terminal. For business travellers and those who travel only with carry-on luggage, kiosks can dramatically save time.

Other improvements in the check-in process have been made by several airlines. Many airlines are turning to the internet to facilitate the check-in process. Singapore Airlines, for example, offers passengers with a valid KrisFlyer (their frequent flyer programme) membership account to check-in for most international flights online. During online check-in, the passenger has the ability to change his or her seat. They are also able to confirm details such as passport identification and preferred frequent flyer programmes for mileage credit. In some instances, a boarding pass can even be printed. Such a service helps the passenger in that less time is spent at the airport check-in counter. It also helps the airline because it confirms who is meant to show up for a flight, thus allowing for planning of last-minute sales.

Overall, self-check-in represents another example of self-service trends in many economic sectors. Supermarkets, petrol stations and online shopping are good examples of how the customer's interactions with the service provider or manufacturer can be minimised. In the case of air transport, this trend in self-service is yet another means of reducing costs when margins are getting thinner and thinner. In many respects, this is ironic in that, through relationship marketing, many airlines are making conscious efforts to form relationships with customers and potential customers, yet the self-service paradigm ensures that customers are far removed from interactions with staff.

> **Questions for consideration**
>
> **1.** What problems are associated with self-check-in procedures for air travel?
>
> **2.** What other forms of transport have begun to utilise self-check-in?

SELF-REVIEW QUESTIONS

1. What are frequent flyer programmes and to whom are they targeted?

2. How is tangibility defined in the context of transport products and services?

3. Define core, tangible and augmented products in relation to tourism and transport.

ESSAY QUESTIONS

1. When your purchase an airline ticket from Copenhagen to London, what are the possible core, tangible and augmented products that you are purchasing?

2. Why do airlines engage in frequent flyer programmes? Compare the programme structure, qualification levels and reward procedures for two airlines within the same alliance. Why are they both different *and* similar?

3. How has the market for cruise tourism shifted? What new product/service offerings have been introduced, and why?

KEY SOURCES

Shaw, S. (2004) *Airline Marketing and Management* (5th edn). Aldershot: Ashgate.

This is an excellent introductory text for understanding the nuances and complexities with airline marketing. Shaw links concepts from the marketing literature and ties these together with the uniqueness of airline marketing activities. The chapter on relationship marketing is particularly strong, but the author also discusses distribution models, revenue management, and advertising and promotional strategies.

Bennett, M.M. (1996) Airline marketing. In A.V. Seaton and M.M. Bennett *The Marketing of Tourism: Concepts, Issues and Cases* (pp. 377–398). London: International Thomson Business Press.

Bennett's contribution to the Seaton and Bennett volume is particularly useful in that it shows how concepts discussed in the wider volume can be applied to airline environments. While the text itself is set as an undergraduate/graduate level offering, Bennett's discussion of products and strategic marketing is excellent, if not slightly dated given recent changes.

CHAPTER 9:

FUTURE TRENDS IN TOURISM AND TRANSPORT

LEARNING OBJECTIVES

After reading this chapter, you should be able to

1. Understand the complexity of sustainable transport in relation to taxation and regulation.

2. Understand the relationship between the Kyoto Protocol and transport.

3. Explain the relationship between the peak oil hypothesis and transport and tourism.

4. Evaluate the development of space tourism as a new mode of transport for tourism.

5. Critically assess future trends in transport that may have significant impacts on tourist flows and tourism development worldwide.

INTRODUCTION

Having considered the salient issues in the scope of tourism/transport relationships over the past eight chapters, attention can now be devoted to potential future trends. It is important to begin with broad trends that resist mode-specific or industry-specific 'predictions'. One of the more salient topics that has received significant media attention internationally since the early part of this century is the nature and feasibility of sustainable transport provision. With the growing threat of global warming, critics are pointing to the transport sector (but in particular aviation) as contributing significant emissions. Unfortunately, global capital flows, and indeed some countries' GDP strength, rely on air transport networks. This makes the question of sustainable tourism not merely one of protecting the environment but recognising the economic importance of transport. This is discussed in the context of juxtaposing command-and-control emission control measures versus natural competitive frameworks. Embedded within this is how 'global' agreements such as Kyoto impact upon on transport and, by extension, tourism.

One other significant trend in the development and relationship between transport and tourism is space tourism. As discussed, there are regulatory and management implications for such activities. Even the environmental implications are, as of the time of writing, still somewhat unknown given that space tourism as a viable commercial means of tourism transport has not yet been extensive due to logistical and financial realities. The chapter concludes with several key issues relating broadly to future trends in transport provision and tourism.

SUSTAINABLE TRANSPORT?

Transport has been extensively considered in the context of environmental impacts in the academic literature (e.g. Hayashi *et al.*, 1999). Lumsdon (2000: 372) outlines a series of stages when developing a sustainable tourism transport network:

> The first stage involves an analysis of existing policy frameworks, in relation to an audit of existing infrastructure and available data on the market. In the second stage, it will be necessary to re-appraise existing land use and assess future proposals for tourism development against, for example, core sustainability indicators. In terms of the likely criteria to be adopted, proposals for new tourism attractions, for example, would be assessed in relation to access on foot and by cycle in contrast to the current trend towards extensive car park provision. The third stage would include a synthesis of work undertaken in stage two, in the form of policy guidance or documentation. The aim would be to secure an appropriate balance in the tourism transport system, which might be different according to local conditions at each destination. In devising a tourism transport network, priority would be given to modes of travel which enhance the visitor experience, but the process would involve a weighting of this gain in relation to social and environmental impacts on residents. The final stage involves implementation and continuous monitoring in terms of both software and hardware requirements of the tourism transport network.

Lumdson's characterisation of sustainable transport planning is useful in that it highlights the extent of integration between transport and tourism. In some situations, sharp integration may be needed where transport provision is necessary for economically feasible tourism development. In other cases, less integration may be more appropriate given overall tourism development plans and policies.

The issue of sustainable transport is reaching the agendas of intergovernmental commissions and committees struck to gauge future impacts. The First International Conference on Tourism and Climate Change (http://www.world-tourism.org/sustainable/climate/brochure.htm), held in Tunisia in 2003 with representatives from IOC,

UNESCO, IPCC, UNCCD, UNEP, UNFCCC, WMO, and UNWTO, concluded with a Declaration that addresses key considerations in the balance of climate change with global tourism. The Declaration outlines specific levels of agreement relating to transport. For example:

- To encourage the tourism industry, including transport companies, hoteliers, tour operators, travel agents and tourist guides, to adjust their activities, using more energy efficient and cleaner technologies and logistics, in order to minimize as much as possible their contribution to climate change.
- To call upon governments to encourage the use of renewable energy sources in tourism and transport companies and activities, by facilitating technical assistance and using fiscal and other incentives. (Djerba Declaration on Tourism and Climate Change, http://www.world-tourism.org/sustainable/climate/brochure.htm)

The conference paid close attention to the role of transport in affecting change to climate, noting that

[w]hile concern about tourism's polluting effects covers all aspects of a tourist's activity, there was a consensus that the primary issue relates to travellers' consumption of transport services, notably road and air transport. In the former case, there is clear evidence from major tourism destinations such as France, that the use of road transport by travellers contributes significantly to greenhouse gas (GHG) emissions. The conference clearly felt that the tourism industry shares some of the responsibility for road transport pollution and thus also shares a responsibility to minimise harmful emissions by encouraging sustainable, carbon-neutral road transport solutions.

Air transport, although currently contributing substantially lower levels of GHGs than road transport, was also raised as a cause for concern. The proportionate contribution made by air transport to total GHG emissions was agreed to be rising rapidly. Schemes to achieve carbon neutral air transport by the introduction of voluntary levies have already been described. There is evidence from countries such as the UK and New Zealand that carbon taxes of one kind or another are increasingly being placed on the political and environmental policy agenda. It seems inevitable that, at some future date, serious consideration will be given to additional environmental taxes or levies targeting the air transport sector specifically. The conference was concerned that the tourism industry acknowledge the polluting effects of air transport and take steps to minimise its impact. When considering the control of the air transport sector for its emissions, the socio-economic impacts of the control measures on destinations should also be examined, as it can affect local economies especially in long-haul destinations in developing countries. (WTO, 2003b)

Broadly, the concern with sustainable transport can be sketched in two ways: 1) the control and regulation of emissions; and 2) the availability of alternative means of power in light of the peak oil problem.

EMISSIONS CONTROL: KYOTO AND BEYOND?

The Kyoto Protocol (negotiated in 1997, but enacted in 2005) was/is designed to produce mechanisms and incentives to cut the production of GhGs produced as a result of human productivity and industries whose by-products reduce the health of the atmosphere and, as a result, the biosphere. It is important to note, however, that not every country in the world ratified the Kyoto Protocol (although over 140 did): the United States and Australia (both of which signed the original Protocol but have not ratified the Agreement) have generally refused because the internationally regulated reduction of GhG emissions could have serious impacts on their respective national economies. Both have argued that national- or region-based programmes would be more efficient.

The Kyoto Protocol states that '[t]he Parties included in Annex I shall pursue limitation or reduction of emissions of greenhouse gases not controlled by the Montreal Protocol from aviation and marine bunker fuels, working through the International Civil Aviation Organization and the International Maritime Organization, respectively'. This is an important point because the international aspect of aviation is not mentioned. Article 2 of the Protocol states specifically that, in order to promote sustainable development practices, countries should '[i]mplement and/or further elaborate policies and measures in accordance with its national circumstances, such as ... [m]easures to limit and/or reduce emissions of greenhouse gases not controlled by the Montreal Protocol in the transport sector' (http://unfccc.int/resource/docs/convkp/kpeng.html). (The Montreal Protocol, signed initially in 1987, was instrumental in generating incentives towards removing several noxious elements from the stratosphere, including CFCs, carbon tetrachloride, halons and methyl chloroform by 2000.)

The Protocol states clearly that those countries who have voluntarily bound themselves to the Protocol must work towards reducing aviation emissions in their own countries, but of course aircraft operations are not restricted only to domestic operations. The transnational nature of aircraft operations makes it somewhat difficult to affix emissions violations or curtailment successes to one particular country. In fact, the closest form of an 'emissions tax' seems to be one based on, or embedded within, aeronautical charges levied by airports (as is the case with some German airports). In the absence of international agreements (and incentives) to cut emissions, however, the impact of aviation-related emissions will have difficulties with establishing meaningful management.

Although Kyoto may not have been the success that some economists and environmentalists may have wished for (particularly with respect to transport), there are examples of regional approaches to emissions reductions. For example, the European Commission introduced in July 2006 legislation proposals that address the exponential growth in air traffic and the resultant emissions. Specifically, the EU Parliament called for airlines to be brought in to the carbon trading schemes already in place in the EU as well as pay a tax on aviation fuel. The intent of these suggestions is to introduce incentives to curb emissions. Such a plan is not without its critics, however. The IATA suggests that regional schemes do little to curb emissions overall, and that a global solution is necessary. Writing in *The Times* (UK), columnist Carl Mortishead argues that, while it is admirable to want to curb emissions, all evidence points to the fact that demand for air travel is increasing in spite of fuel surcharges and that the airline industry is 'so inefficient that it can absorb more fuel costs as it eliminates chronic waste and subsidies' (Mortishead, 2006).

The EU Parliament proposal is not the only regional effort. In 2005, a new vision for pollution reduction and energy efficiencies was agreed to in principle by Australia, China, India, Japan, Korea and the United States. The Asia–Pacific Partnership on Clean Development and Climate is designed to sit next to Kyoto as opposed to outright replacing it, with the major difference being that the signatories will be allowed to set their own emissions reductions and not be penalised for missing the targets. Although non-binding, however, and with very little mention of transport issues, there may be some consolation in that the countries involved are some of the biggest developed-nation polluters in the world. The Worldwide Fund for Nature, however, cast doubt the on the pact, suggesting that '[a] deal on climate change that doesn't limit pollution is the same as a peace plan that allows guns to be fired' (BBC, 2005).

A series of meetings of countries from around the world was held in Montreal in December 2005 with the intent of paving the way for a new international focus on climate change and emissions. Some gains were made, particularly in the area of future talks that would extend climate control beyond Kyoto, although no specific mention was made with respect to transport emissions (at least not publicly nor in the final statements at the wrap-up of the conference). Importantly, however, one noteworthy event transpired at the Montreal meetings that raises the question of who is to pay for the cost of achieving sustainable transport through emissions reductions. Stephen Byers, the former environment secretary in Britain, suggested (Independent, 2005) that, if the United States would not opt in to a new agreement, regulations should be imposed on United States-registered aircraft entering and using European airspace and landing at European airports. This model of emissions control is one where absolute cuts are enforced. If the European Union enforced an absolute cut in emissions by United States-registered carriers, the cost of compliance by an American carrier could well be higher than the returns or profits realised. For some there may be little economic incentive to continue operations to Europe. As the trans-Atlantic routes are some of the most profitable for many United States major airlines, reduction in services in order to comply with absolute emissions cuts could further destabilise the United States air industry.

A popular means (as measured by support from governments and environmentalists) by which aircraft emissions may be reduced is through Pigovian taxation (Pigou, 1920). A Pigouvian tax effectively requires a polluter to internalise the cost of pollution through the imposition of a tax that represents the cost to society as a consequence of the pollution generated. Thus, if the European Union were to impose an emissions tax (per quantity of some particulate released into the air, but usually through the amount of GhG emitted) on United States-registered airlines flying to and from Europe, the cost to those carriers effectively becomes a set amount based on emissions. Thus, the airline would have to determine the cost of its operations at present with the tax versus the cost to cut emissions. In other words, the airline would need to determine what the opportunity cost is to continue operating on the route where emissions taxes would be imposed, particularly if other routes to other destinations may not attract similar taxation.

Taxation to cut emissions places the cost of reducing emissions firmly in the hands of airlines. Of course, the cost of adopting newer operational procedures designed to cut emissions will likely not be the same for all airlines, and this illustrates one problem with the option of taxing emissions on international aviation. Some airlines, for example, may be able to adopt processes that cut emissions even further than one-half of current amounts if significant tax savings are realised, while others may find that the cost in tax is cheaper than adopting more efficient operations. In essence, the principle at work here is that, for a cost-effective reduction in emissions, particularly for transport, the marginal cost of abatement for each airline (or cruise company, or car manufacturer) is the same, but the problem is setting a tax rate where the maximum cut in emissions is realised. This is difficult because the cost of reducing pollution is unknown for all companies, and will naturally vary. However, with the taxation approach, as opposed to absolute cuts, firms will cut emissions as cheaply as possible (i.e. they will select the best option where profits are maintained, and the cost to society is minimised). In fact, absolute emission cuts can be politically dangerous, as those firms whose costs rise substantially may be forced to trim workforce size (thus society may incur some cost for absolute emissions cuts). In an international context, this could lead to strained relations between trading partners.

Pigouvian taxes, abatement and Coase's Theorem

Whether Pigouvian taxes or absolute emissions cuts are enacted, an argument could be made that the compliance costs should be borne by those who are directly affected by unsustainable activities, especially when the direct and tangible benefits of those activities is important enough to maintain. This argument is derived from the Coase Theorem (Coase, 1960; see also Millimet & Slottje, 2002):

> Given free bargaining and low transaction costs, voluntary actions of individuals in the market will allocate property rights to the most highly valued and efficient use. Both parties in a dispute over property rights have an incentive to move to this posi-

tion. Such allocation occurs automatically without regard to how property rights are initially or legally assigned. Judicial action to allocate them generates a superfluous social cost and is itself a negative externality. (Cobin, 1999: 380)

Coase essentially argued that direct regulation was inefficient in the context of addressing externalities (Hillman, 2003). Coase's Theorem has already been considered roughly in an environmental context (e.g. Hanley *et al.*, 1997; Pearson, 2000), and its principle argument can thus be considered subsequently when discussing aircraft emissions. In the example above, it is a given that the presence of United States-registered carriers covering the Atlantic route is important for both tourism and overall commerce and trade in Europe. A call for an absolute reduction in emissions may adversely affect those carriers who account for a substantial amount of the capacity between the United States and Europe. Airlines, faced with mandatory restrictions on emissions, would have two primary options: 1) reduce emissions as required through potentially costly means (i.e. newer, more efficient engines); or 2) completely stop flying to Europe. There are, of course, options within these, such as modifying some aircraft for specific use to and from European airspace. The problem with option 1 is that it would likely mean the costs are off-set to passengers. Classic supply/demand curves and price elasticities would suggest that passengers, facing increased ticket costs to Europe, may elect to travel elsewhere (although some business travel would, perhaps, be less elastic). In this scenario, the result could be a hugely diminished market for travel to Europe. Opting for option 2 could mean a hugely retrenched airline, with job losses and, perhaps as well, diminished traffic to Europe because of the subsequent decrease in capacity.

In applying the Coase Theorem, however, both the airlines and the European Union (acting on behalf of the public) would negotiate any subsidies or levels of absolute emissions cuts to the point where the social cost to both parties is effectively minimised. In other words, the question becomes whether the costs involved in introducing absolute emissions cuts should be borne by those calling for the emissions cuts or by those responsible for the emissions. The most appropriate solution, of course, is to adopt absolute emissions cuts in such a way as to minimise the social cost as much as possible. The problem, however, is how social cost can be measured internationally, especially given that Carlsson (2002) noted that different markets (especially the nature of imperfect competition) and networks would undoubtedly produce differing levels of emissions. In other words, would the social costs to Europe outweigh those in the United States, or would the robustness of the totality of both economies be maintained if the social cost is borne by, for example, the United States? It is plausible that countries that rely heavily on tourist traffic arriving by air would not likely call for absolute emissions cuts from the airlines that service it because the potential social cost to their own economy (assuming airlines decide to cease operations) may be too significant in the immediate term. In other words, airlines are less likely to absolutely cut emissions if, for example, the marginal cost of doing so is prohibitive.

At its most basic level, a Coasian approach used in addressing international aviation emissions would assume that bargaining between the polluters (airlines) and the 'victims' (individual nations or supranations, acting on behalf of its citizenry) would result in a mutually agreeable level of emissions. In adopting a Coasian framework, however, a number of issues are raised (Duval, 2006):

1. *The role of aeropolitics.* Nations and supranational entities (e.g. trading blocs) negotiate ASAs that effectively allow flag carriers to establish routes and connections between several points within a service network. Bilaterals are negotiated on basis of the economic benefits brought to a country through the provision of these services, and the move towards open skies (Canada/United States, Australia/New Zealand and potentially the US/EU) demonstrates the relative importance placed in air access for the purposes of economic growth and global connectivity of markets. The imposition of a Pigouvian tax or some form of Coasian approach at a national or supranational level may call into question the reciprocal rights of airline whose initial air operating certificate is from a foreign country. In fact, a 1999 report from the European Commission considered two options for the introduction of an excise duty on aviation fuel: option A involved taxation of all routes departing any Community airport and option B focused purely on intra-European routes:

> The results show clearly that the environmental effectiveness of imposing kerosene taxes is significantly higher where all routes departing from EU airports are taxed. Moreover, the ratio between environmental effectiveness, on the one hand, and economic and competitive impact on the European airline industry, on the other hand is, from a European view, significantly better where all air carriers are taxed, at least as long as circumvention practices by means of taking fuel in third countries is not widespread. Finally, in relation to cost-benefit considerations, it is at least questionable whether a reduction in all transport-related CO2-emissions of just 0.26% (as calculated for an EU 2005 scenario with 1992 a base year on the basis of applying option B) and of NOx-emissions by 0.12% would justify considerable pressure on the competitiveness of the European aviation industry which would have to compete head-on with third country air carriers enjoying intra-Community traffic rights, as a side-effect of the cumulative effects of so-called open-sky agreements concluded by Member States.

> Consequently, any effective approach would necessitate a system that allows for taxing/charging all carriers operating out of Community airports (Option A). Such an approach, however, if applied in the field of kerosene taxation would require fundamental changes to existing policies at ICAO-level and, in particular, to existing bilateral Air Service Agreements (ASAs) that allow for the imposition of taxation only in case of a reciprocal agreement. These changes will be difficult to achieve

without considerable concessions in other fields. For these reasons, the Commission considers that the approach suggested in its 1996 report should be maintained, for the time being, pending progress in international fora. The alternative (Option B), though legally feasible, is unacceptable in the Commission's view. It would not strike the delicate balance between environmental, economic and internal market requirements which is necessary for a coherent policy in this area. The conclusion reached as to the relative attractiveness of options A and B also applies to lower tax levels even though these may reduce the economic burden for Community air carriers. (Commission of the European Communities, 1999)

2. *Should the consumer be responsible for emissions abatement?* The benefits of airlines servicing a destination can be calculated through manufactured goods (from cargo) and tourist receipts, including the magnitude of the multiplier and any leakages present in respective service sectors. The full and direct impact of emissions, however, is not as clear. As a result, an airline (or the wider industry) might argue that the full cost of emissions production should be borne by the destination, given that the *known* social benefits of its services outweigh the (generally) *unknown* social costs of its emissions (see, for example, Duval, 2006; Vlek & Vogels, 2000). If a Pigouvian tax or a command and control (CAC) measure were introduced, the airline may cease services, thus rendering not only a reduction in emissions (a social benefit) but also a social cost to the destination because the sudden drop in the financial impact of the services provided indirectly by the presence of the airline. In this instance, the question is whether the net social benefit of emissions reduction is *greater* than the social cost endured as a result of the reduction of those services.

3. *Power structures and transaction costs.* Under a Coasian framework, where both parties would negotiate and bargain towards a level of pollution acceptable to both, the social and environmental costs are evaluated against economic benefits of the indirect and direct services provided by airlines. There is, however, a significant problem with the implementation of a Coasian approach. Negotiating or bargaining might work in a constrained environment, such as a factory emitting pollution, which has as a negative effect on local residents, but international aviation presents significant problems because of its size and transnational/globalised nature. As a result, an airline and a particular nation might well negotiate an equilibrium point given the marginal net private benefits and marginal external costs, but if this is not implemented on a global scale it provides an immediate disincentive for airlines to service a particular destination such that their marginal external costs are disproportionately allied to a specific nation or region (Duval, 2006).

Where does this leave sustainable transport? In one sense, sustainable transport as a concept needs to be thought of beyond strict environmental measures; it needs to take into account financial and social sustainability, and nowhere is this more critical than with respect to international aviation and tourist travel. More importantly, the level of measurement and analysis must be considered. If global units of analysis are not feasible, regional interpretations (e.g. Haynes *et al.*, 2005) must be developed in order to capture in aggregate as much data within a wide spatial scope.

Box 9.1 Pigouvian taxes – who pays? (based on Duval, 2006)

Ryanair announced in July 2006 that services from Prestwick, Ayrshire, Luton and Stansted to various Swedish destinations would be cut if a new general aviation tax proposed by the Swedish government is to be formally implemented. Under the tax, passengers flying within Europe will pay an additional SKR94 (approximately 10 euros) and SKR188 if travelling internationally (Forbes, 2006). However, the European Commission is, at the time of writing, currently investigated the tax (The Local, 2006). Commentators have argued that the tax may not be legal on the grounds that it stifles competition, particularly when some smaller regional airports would not be required to collect the tax on the basis that it would stimulate travel to these destinations.

Ryanair argued that the tax would add an extra 30% to the average cost of the fares on offer from the airline (Evening Times, 2006). Michael Cawley, deputy chief executive of Ryanair, argued that '[t]he introduction of this tax will make Swedish tourism uncompetitive when compared with cheaper alternatives in Spain, Italy and elsewhere in Europe' (Evening Times, 2006). This example demonstrates a potential response by an incumbent carrier to the introduction of a new tax or levy. Prior to this announcement, Ryanair had been reporting healthy profits for its operations (although in August 2006 it was announced that rising fuel costs would cut into projected profits significantly), and as such the rationale for the threat of flight cuts to Sweden was not likely on the basis of shrinking financial efficiencies of its network.

With the introduction and proliferation of LCCs throughout Europe, including low cost subsidiaries of larger FSAs, many are susceptible to the introduction of additional levies because of inherent business models that favour stringent management of marginal costs. Indeed, as outlined in Chapter 6, the success of LCCs is largely attributed to the management of costs. Naturally, any new cost from a new tax or levy can be passed onto the passenger, but when other destinations are deemed as being relatively accessible and where transport costs (and thus ticket prices) are lower, the viability of the network segment with the additional tax or levy will be questioned from a strategic management perspective. Indeed, if it is accepted that airfares are somewhat elastic (see Chapter 2) when compared to other goods and services, then the cost of any tax or levy will likely be covered by the airline in the face of significant competition for other services.

Figure 9.1 provides an idealised and hypothetical model of the impact of a tax or levy upon a particular route where high degrees of elasticity are realised. The model is useful in that it demonstrates the strategy an airline may employ in making business decisions regarding the financial feasibility of a particular route within a larger network. The model does not incorporate the complicated nature of yield management (see Chapter 6), which would suggest that, when the cost base is altered, varying fares would be offered in order to cover that new cost level. The model, then, assumes a fixed price for the product on offer, or in this case a seat on an aircraft.

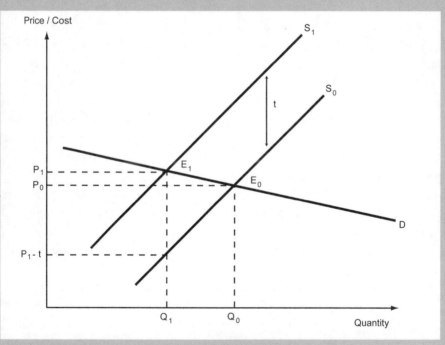

Figure 9.1 Theorised elasticity of airline seat prices (not under yield management) and impact of tax

Source: Adapted from Turner *et al.* (1993) based on Duval (2006)

In the model, the equilibrium price prior to the introduction of tax (t) is achieved at E_0, given D and S_0. The introduction of a new tax or levy shifts the supply curve to S_1, thus reflecting an increase in the cost of production, or in this case the cost of operating a particular flight. This shift results in a new equilibrium point (E_1) given that the quantity demanded subsequently falls. As a result, where P_1 to P_1-t represents the tax enforced, P_0 to P_1 represents the total cost of the new tax borne by passengers and P_0 and P_1-t the amount borne by the producer, in this case the airline. Once again, it is important to remember that this model assumes that fares are uniform and yield management systems are not in place. In reality, an airline would have the option, through yield management, of reducing the number of cheaper fares in an effort to increase their RPK. Depending on the actual monetary value of the tax itself, the airlines' ability to recover the tax from P_0 and P_1-t may not be entirely possible.

An airline facing increased costs on a particular route will undoubtedly re-evaluate the strength of the route from a return-on-investment perspective. In other words, it may decide that rather than fly that particular route to a destination that has imposed a tax on its operations, it could instead utilise its aircraft on other routes where similar taxes are not present and thus the overall operational cost base is lower. The nature of the elasticity can be used to establish how producers might respond with alternate pricing schemes. For instance, where the market for products/services (in this case, seats on an aircraft) exhibit significant inelasticity, producers will likely be able to recover increased costs as a result of an introduction of a tax or levy. Travellers who utilise air services out of necessity will bear the increase cost through higher fares. Where demand is suitably elastic, competition between producers can continue on the basis of a uniform increase in cost, but the consequential increase in price (or in this case, fares) may not be uniform. In the example above concerning Ryanair's flights into Sweden, if all airlines were charged the new tax (and indeed this is what was proposed) and if travel to Sweden on holiday was deemed to be comparatively elastic, demand for holidays to Sweden would likely fall and thus shift to other destinations. Arguably, such a simplified model does not take into account holiday cost models that feature the cost of transport as part of the overall trip cost (see Chapter 2 for a discussion of Prideaux's transport cost model). It can, however, be argued that transport cost, as part of the overall cost of travel, can be relatively inelastic while the cost of other components (such as discretionary spending, accommodation, activities, etc.) may be fully elastic. As a consequence, the imposition of the tax by Sweden, if successful, may not necessarily result in a drop in visitor numbers if those visitors were willing to sacrifice costs in other aspects of their travels.

Questions for consideration

1. What kinds of travel can be considered elastic/inelastic and how might they be affected by the introduction of new taxes that raise the cost of transport?

2. How might the cost of transport be affected if a tax or levy (or any aspect of the overall cost base) is reduced or removed? Use Turner *et al.*'s model from Figure 9.1 to explain your answer.

PEAK OIL: TOURISM TRANSPORT IMPLICATIONS

The problem of emissions may not be as much of a problem in the (near?) future if the peak oil hypothesis, which has been at the forefront of many environmental discussions in the popular press since the early part of the current century, becomes reality. According to the IEA (2002), transport is the fastest-growing economic sector in terms of energy consumption. The rapid growth in mobility of people and cargo worldwide has been supported in the past two to three decades by both the increasing liberalisation of transport markets and the provision of cheap oil (Greene & Wegener, 1997), although some might argue that this 'bubble' may be set to burst with the increasing recognition of the reality that oil, upon which this exponential growth in mobility is based, is definitely finite and, more importantly, may be at its peak in terms of efficient production. In fact, geologists' claims (which are not recent) that the world may be approaching a peak in oil production are starting to receive considerable attention by mainstream media, and thus more attention by people and governments worldwide. Peak oil could have severe implications for countries that have a high reliance on oil for base forms of energy. As the vast majority of transport (and not just tourist transport) relies upon cheap, refined oil/petroleum, the impact of peak production could very well be staggering.

If the remaining 50% of the world oil reserves are difficult to extract and refine, the cost of petrol, diesel and aviation fuel will likely increase significantly, thus transforming the nature of mobility (both domestically and internationally) as the cost of transport could well be out of reach for many individuals. The issue is not so much the lack of oil, but how much the remaining oil will cost to refine. As the cost of refinement rises, the cost of fuel product will also rise to maintain existing margins. As a result, it is entirely unclear what the exact repercussions might be for tourism as the cost of refining oil rises after peak production, but several arguments can be made:

1. If the cost of transport increases substantially and thus occupies a higher proportion to personal income, shifts in mobility patterns are likely. Some destinations may become inaccessible because of the high cost of transport. Others' network positions, and thus accessibility, could default to positions where transport costs are the least likely to dissuade markets. Some long-haul destinations, for example, could see falling arrivals if the cost of transport from their key markets becomes too great.

2. Transport networks may spatially constrict because the fixed and marginal cost of operations precludes any possibility of securing market share, particularly if wages do not increase to match proportional increases in transport costs. Transport modes may opt to offer services only where sufficient demand (and net personal benefit) is maximised. As such, the connectivity of existing networks may shrink, with cuts to services or frequencies due to a combination of higher fuel costs and depressed demand.

3. Non-essential travel may be at risk if the cost of transport is too high. Indeed, current ITs already exist that may soften the economic implications of rising transport costs. For example, some business travel, perhaps deemed non-essential, could be replaced by internet telephony (such as VOIP).

The problem is much larger than tourism, and it is of course necessary to position tourism within a wide global system of input/output processes that may be affected adversely from peak oil. While recent increases in the price of crude oil (Figure 9.2) have helped publicise the peak oil hypothesis (although it is more a fact than a hypothesis, but what is hypothetical is when the 50% mark will be reached), there have been steps taken to reduce global dependency on oil and fuel in the transport sector. To avoid rising aviation fuel costs, or at least their wide variability, some airlines have adopted operations policies that are more sustainable (e.g. use of more fuel efficient engine technologies, reduction of noise pollution), even though many are currently (as of December 2005) charging an extra fuel surcharge on top of ticket prices (e.g. Malaysian Air and Singapore Airlines, as of 1 November 2005, were both applying a $50 one-way fuel surcharge). The question, however, now becomes: what alternative modes of transport are available that do not rely upon oil?

Greene and Wegener (1997) note that there are three fields of policy and planning that can influence sustainable transport: technology, supply and demand. Technological innovation has brought several alternatives to motorised transport, particularly in the area of powerplants:

1. The introduction (and constant improvement) of fuel-efficient jet turbine engines, particularly in concert with new aircraft from Boeing and Airbus. Additionally, GE introduced its GEnz engines for use in the Boeing 787 (Dreamliner), the 747 Advanced, and the new Airbus A350. New materials and design help increase overall efficiency, but GE

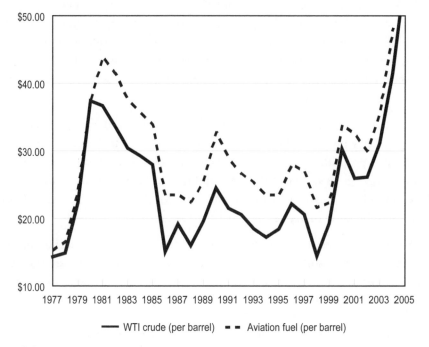

Figure 9.2 Crude oil costs and aviation fuel costs

Source: http://www.airlines.org/econ/d.aspx?nid=9336; accessed 6 January 2006

Note: Crude oil WTI, and aviation fuel costs represented by major, national and large regional United States passenger and cargo carriers.

claims that the engine is more fuel efficient (by 15%) and will reduce emissions significantly. According to the company's website (http://ge.ecomagination.com/@ v=312005_0548@/index.html): 'If all of today's fleet of 200–300 passenger aircraft had GEnx engines, annual carbon dioxide emissions would be reduced by an amount equal to removing more than 800,000 cars from the road for a year, or the carbon dioxide absorbed by 1.2 million acres of forest in a year.' Further innovations may include so-called 'zero-emissions' aircraft fueled by hydrogen and under development/study by NASA (Becken & Simmons, 2005), but these are likely several decades away from being adopted into regular use. Indeed, not everyone believes that increased air traffic and travel will result in irreversible changes to the global climate. The UK aviation technology group Greener by Design suggested recently that aviation emissions will actually be quite a bit less in 2050 than currently thought, largely because of advances in aircraft operational efficiencies, such as reducing drag coefficients (Szodruch, 2001) and new engine technologies (Airline Business, 2005c).

2. The introduction of hybrid automobiles (battery-powered/petrol combinations) in some countries (including the United States and Europe) offers owners an opportunity to engage in short-distance trips without burning petrol. The question, of course, then becomes how the power used to charge the batteries is supplied (and whether its ecological footprint favours its use).

Other aspects of technological innovation include new forms of transport that harness alternative forms of mobility, including:

1. Solar-powered ground transport: for the most part, still in the testing phase, and to some extent hampered by the small size of the vehicles (usually one occupant).

2. Electric-powered ground transport: with the introduction of hybrid automobiles, if the source of electricity is, for example, nuclear, an argument could be made the ecological footprint is still impacted, even though reliance upon a non-renewable source of energy such as oil is reduced. If it is hydro-electric, increased demand on power grids can have serious implications for a country or region's ability to meet demand (as in the case of New Zealand and parts of California).

As Holden and Høyer (2005) argue, the potential for reducing the ecological footprint of transport will soon be based on the use of both conventional and alternative fuel technologies, however this path may not be easy to follow:

> Hydropower has a very low ecological footprint, but is not a global resource with sufficient volumes to support the ever increasing transport systems. Natural gas also has a low ecological footprint but does not have these resource volume limitations, at least not for several decades. But it is not a renewable energy resource, and it does not fulfil the long-term requirements of a sustainable energy system. Biomass is globally available in large volumes and is a renewable resource, but would lead to unacceptable increases in ecological footprints if used extensively. Therefore, it seems that only a combination of more efficient use of resources, substitution to less environmentally harmful fuels, and reduced transportation will meet long-term sustainability objectives.

Demand for transport provision is ostensibly geared for existing forms of available transport. As Greene and Wegener (1997: 182) suggest:

> Medieval cities were built for walking, and this required that living and working were close together. The railway made spatial division of labour possible and so opened the way for the growth of cities. Rapid transit and the cars have facilitated the expansion of metropolitan areas over wider and wider territories with the consequence of ever longer trips and greater volumes of traffic with all their problems of congestion, traffic accidents, energy use, pollution and land consumption.

Tourism, then, has ridden this wave of expansion of transport networks fuelled by the availability of cheap oil. Shifts in the ability to provide transport due to the problem of 'peak oil' will undoubtedly be exacerbated by the layout of the existing network as they were planned and constructed at a time when concerns over peak oil had not yet reached mainstream media. If the era of cheap oil is truly in the past, then the spatial realm of people and goods may be severely constricted in the future if the expense of transport is outside the ability to rationalise its necessity. In aggregate (i.e. globally), oil prices have not had the same impact on global tourism traffic as events such as 11 September 2001 or conflict in the Middle East, but over time the cost of travel may become too burdensome, and the ability of suppliers to rationalise lower costs may be limited without having an impact of profitability.

The sustainability principle in tourism has certainly achieved strong consideration in the academic literature, and has also garnered considerable attention from government and agencies such as the UNWTO. Measurement units have been studied and disseminated in units such as MJ (energy use), emissions (such as GhGs) and carbon dioxide equivalents (CO2–e). As well, the displacement of land can also be measured. Rather than adopting site-specific or micro-level measurements of sustainable practices in tourism (e.g. certification schemes showcasing compliance or minimising environmental impacts through select visitor management techniques), one benefit of examining the sustainability principle of tourism through the lens of transport is that a wider view of sustainability is afforded. Gössling et al.'s comments (2005: 418) serve as a useful starting point to adopting a broader approach to sustainability:

> First, whether using energy consumption, greenhouse gas emissions or area-equivalents as basis for calculations, a substantial share of tourism needs to be seen as unsustainable. Second, the use of fossil fuels and related emissions of greenhouse gases is, from a global point of view, the most pressing environmental problem related to tourism (cf. Graβl et al., 2003; Sala et al., 2000; Thomas et al., 2004). Third, transport contributes overproportionally to the overall environmental impact of leisure-tourism; this is between 60% and 95% at the journey level, and including local transport, accommodation, and activities.

What this means is that efforts at achieving sustainability practices 'on the ground', while useful, do little to address the larger issue of the global climate change, and thus radical shifts in quality of life in some areas of the globe. Gössling and Hall (2006: 315) perhaps best address this fundamental misguided attempt to parley sustainability into a 'feel-good' action:

> [T]he reality is that concentrating on tourism alone, and by that we mean the tendency to focus just on what is happening at the destination, is one of the great problems with sustainable tourism. For tourism to really contribute towards security

and sustainable development it needs to be placed within the bigger picture of human mobility, lifestyle, consumption and production. The consumption and production system that seeks to use 'pro-poor tourism' by those from the developed countries to help those in the developing world is the same consumption and production system that has often led to the situations that have contributed to inadequate development practices and poverty in the first place. The most sustainable forms of tourism in many cases may well be no tourism at all, rather focusing on other dimensions of development and a full consideration of alternatives.

Becken (2002) noted that one-way travel to New Zealand by all international tourists resulted in 1900 kilotonnes of CO2 being emitted into the atmosphere. Put into context, it seems somewhat futile for a country such as New Zealand, to portray itself as '100% pure' (the current marketing slogan) when the very act of travelling to New Zealand is surprisingly (or not, to some) environmentally harmful.

SPACE TOURISM: NEW FRONTIER(S)?

One might be inclined to think that space travel is perhaps the best way of minimising the environmental impacts of commercial air transport, but even space transport comes with its own set of environmental issues. First, however, it is necessary to position space tourism development in the wider context of transport. As of the end of 2005, three individuals have already been in earth orbit as 'passengers' (i.e. not as members of a scientific crew): Dennis Tito became the first in 2001 thanks to a Russian Soyuz capsule that docked with the International Space Station (ISS), Mark Shuttleworth became the second in 2002 with a similar itinerary (and both each paid US$20 million for the privilege), and Greg Olsen also visited the ISS in late 2005. The future of tourism and transport, it seems, involves space. In fact, it may even involve other planetary bodies. In September 2005, the United States announced that it will once again go to the moon, and a few months later reports emerged that China will head to the moon by 2017. Budgeted at US$104 billion, something The *Economist* dubbed as 'suspiciously precise', NASA plan's include the use of a Crew Launch Vehicle (CLV), which will dock with the ISS, as will a second vehicle, the Crew Exploration Vehicle (CEV). The CEV will be tasked with propelling its occupants directly to the moon. All of this, including a replacement shuttle vehicle due in 2010, is targeted for 2018.

Outside of the scientific realm, space tourism is quickly becoming the latest niche product involving transport. Richard Branson's Virgin Galactic will transport tourists in from origin to destination in record times using sub-orbit flight trajectories, much higher than traditional aircraft. Significantly, some 38,000 people have already paid a deposit for travel on some of the first flights (Wardell, 2005). This is not surprising considering polls from the late 1990s suggest that space travel would be welcomed by the public (CNN, 1998). Both NASA and the Space Transport Association have long suggested that space as a frontier for travel could be developed, and a series of recommendations from 1998 established the importance of joint industry–government partnerships and support:

- National space policy should be examined with an eye toward encouraging the creation of space tourism.
- The expansion of space camps, space-themed parks and other land-based space tourism should be encouraged.
- The federal government should cooperate with private business to reduce the technological, operational and market risks – much as it has done with aviation and satellite communications.
- The government should sponsor research and development to dramatically lower the cost of space travel and demonstrate ways to reduce the effects of space sickness. (CNN, 1998)

Crouch (2001) argues that there are several forms of space tourism:

1. *Near-earth orbit:* This has yet to reach the tourist masses, although several 'ordinary' citizens have already paid for the privilege. Dennis Tito was the first 'tourist' in space, spending some US$20 million for the opportunity to ride a Russian Soyuz rocket and capsule that eventually docked with the ISS. Critics of this endeavour were quick to point out that only individuals of high-net worth could afford such adventures, and it will likely be some time before weekend trips around the world in space are realised.

2. *Low-earth orbit, high-speed aircraft:* Boeing announced its Sonic Cruiser, designed to fly at supersonic speeds, in March 2001 (see http://www.boeing.com/news/feature/concept/background.html). By flying in the upper limits of the atmosphere, and much higher than 'traditional' jumbos such as the Airbus A340 or the Boeing family of large jets, the aircraft would minimise the time spent travelling from origin to destination, and thus further compressing the time-based perception of distance first seen with the development of long-haul aircraft in the mid-19th century. Ultimately, Boeing opted out of fully producing the Sonic Cruise and instead poured its energy into the development of the Boeing 787, or Dreamliner, which favoured efficiency in operations over speed. As noted by Duval (2005b), there is potential for future development of low-orbit or sub-orbit aircraft, but travellers would need to consider the overall transport cost of the trip. For some essential travel, such as business trips, the cost-benefit could be acceptable. For leisure travel, however, unless a trend of 'massification' is realised, as it was with long-haul travel in the later part of the last century, sub-orbit travel will remain available for those individuals of high-net worth. Further still, and in relation to the sustainability of sub-orbit transport, further studies will need to consider the increased environmental damage to sensitive atmospheric layers (see also Chapter 6 for a discussion of air transport and environment).

3. *Terrestrial-based space tourism activities*: Including attractions and events such as space shuttle launches and tours to locations around the world where recent solar eclipses would be visible. Carlson Wagonlit (through www.eclipsetours.com), refers to such activities as 'astro-tourism'.

At the end of 2005, the FAA within the United States government released a series of proposed rules to regulate space tourism (www.faa.gov/regulations_policies/rulemaking/recently_published/media/ai57.pdf). Among the more notable regulations issued by the FAA (ratified in December 2006) includes the requirement that companies inform potential 'space flight participants' of the risks involved with space travel. It is also suggested that companies require tourists sign a written consent form and be fully trained in safety procedures. Technological developments in vehicles designed to travel to and from space will undoubtedly push space tourism to the point where many companies and governments are involved. As discussed, however, the management of this new mode of transport will need to be carefully considered (Duval, 2005b) (see also Box 9.3).

FUTURE TRENDS AND ISSUES IN TOURISM AND TRANSPORT

Identifying future trends in an industry as varied and dynamic as this is somewhat problematic. To some extent, the prevalence of a mode of transport in the current media can be assessed somewhat on the basis of its presence in mainstream media (and in the informal media such as blogs and discussion forums). The airline industry seems to change on an almost daily basis (reports of bankruptcies, new routes being introduced, new open skies agreements being debated, security screening systems introduced), while rail transport tends to move slightly slower (with the exception of the concerns raised over troubled firms such as Amtrak). A quick search of Google News on 8 December 2005 revealed, at the very top of the list, a report (from Newsday, New York) of family who is considering suing Royal Caribbean because their newly married relative disappeared without a trace whilst on holiday in the Mediterranean (see http://tinyurl.com/8qvhr). Outside of the current focus of the global media, several trends and issues can be identified that can help position where transport and tourism might evolve.

LCCS: EXPANDING INTERNATIONAL NETWORKS

LCCs (or LCLF carriers, see Chapter 6) could well begin servicing international markets within five to seven years, particularly as new fuel-efficient aircraft (e.g. Airbus A350 or Boeing's 787) enter service (Tretheway & Mak, 2006). To some extent, some low-frill airlines already serve destinations within a five- to six-hour catchment (e.g. JetBlue serving the continental United States or Freedom Air serving Australia from New Zealand across the Tasman Sea), but it remains to be seen whether a truly low-cost format will be utilised between major hubs internationally. Already, there have been suggestions (based on hints given by its president in the April 2005 edition of *Airline Business* magazine) that Emirates may be looking at establishing an Emirates Express brand that would utilise the new Airbus A380 in an all-economy, high-density layout (Airline Business, 2005d). LCCs will also likely begin to grow in the Middle East, with National Air Services already planning an LCC out of Saudi Arabia sometime in 2006 (Arab News, 2005).

Although the networks of LCLF airlines may shift, low-cost as a concept will likely come to dominate the airline industry. In many ways, this is a safe forecast as, for the most part, most air carriers worldwide already focus on cutting costs as a means to maximise revenue. This may have, though, little impact on destinations and tourism development as the costs offset by removing frills and ancillary services only covers increases in operating costs relating to, for example, fuel. In other words, rising aviation fuel prices, and any corresponding increase in ticket prices, may result in stagnant or slightly depressed demand for travel. This is especially the case for short-haul travel, where fuel efficiencies may not be as great. In the United States, many traditional network carriers (or FSAs) seem to have significant problems with achieving profitable operations (Table 9.1). In 2004, for example, several large network carriers reported significant losses, while LCCs such as Southwest, AirTran and JetBlue reported positive profit margins (Table 9.2). The United States BTS reported that fourth quarter results of airline performance show regional domestics carriers (as opposed to the larger LCCs and the larger network carriers) generally performing well (BTS, 2005b).

Table 9.1 Operating profit/loss of US network carriers as a percentage of total operating revenue

	Q1 2004	*Q2 2004*	*Q3 2004*	*Q4 2004*
US Airways	−11	2	−14.3	−10.1
Northwest	−2.9	4.3	1.6	−11.7
Alaska	−11.2	1	6.9	−11.7
American	−8.3	−4.6	−8.2	−14
Continental	−9.9	−4.4	−7	−15.4
Delta	−13.1	−6.3	−13	−17.7
United	−12.2	−4.7	−7.7	−22.7

Source: Adapted from US BTS, Form 41, Schedule P1.2

RAIL TRAVEL: LIMITED TO NICHE MARKETS?

Rail travel will likely continue to appeal to specialist or niche markets, but may face difficulties with any attempt to re-establish itself as a prime mode of transport for tourists

Table 9.2 Operating profit/loss of US LCCs as a percentage of total operating revenue

	Q1 2004	*Q2 2004*	*Q3 2004*	*Q4 2004*
Southwest	3.1	11.5	11.4	7.2
JetBlue	11.3	14.1	7.1	3.7
AirTran	4.3	11.3	−4.9	1.3
America West	2	2.6	−4.7	−6.9
Frontier	−8	−3.8	−0.3	−7.3
Spirit	2	−3	−13.3	−31.6
ATA	−14.8	−9.1	−12.5	−187

Source: Adapted from US BTS, Form 41, Schedule P1.2

between origin and destination. The presence of air service is often comparable in price or, if it is not, has the benefit of time in its favour. There is, however, some room for consideration that smaller, regional services may be profitable and attract some tourist traffic. Virgin Trains, for example, has developed advertising campaigns designed to shift the perception of train travel in the UK. Such services could be utilised by weekend recreationalists, thus avoiding congested roads.

SHIFTING DEMOGRAPHICS

The Population Division of the Department of Economic and Social Affairs of the United Nations recently (2005) released projections regarding aging populations worldwide (Table 9.3). The implications for rapidly ageing populations are certainly economic, including shrinking GDP, loss of economic influence and general market malaise. Hall (2005: 285) noted that

> the demographic shift in Western countries is also accompanied by changed employment patterns, particularly the increase in part-time, casual and contract employment in relation to permanent employment. Such changes in social structures have a number of flow-on effects in the tourism and leisure sectors, including the increase of short breaks as opposed to longer holidays.

As tourist flows often take place between 'advanced' or Western countries (Ryan & Trauer, 2004), some (e.g. Dwyer, 2004: 540) suggest that 'many older people wish to enjoy the activities and entertainment that they enjoyed in their youth, and they have more disposable income to spend on those activities'. As a result, it is quite likely that the heterogeneity found in current tourist flows with respect to desired experiences and holiday locations will continue, although Hall (2005) rightly notes that the spatial scale of those travels may be more limited. Coupled with ageing are overall population levels. In less developed countries (Africa, Asia [excluding Japan], Latin America and the Caribbean plus Melanesia,

Micronesia and Polynesia), population levels are predicted to rise at a much faster rate than more developed countries (Europe, Northern America, Australia/New Zealand and Japan) by 2050 (DESA, 2005b) (Figure 9.3). Given the concentration of international traffic within and between developed countries, this could change dramatically over the next 50 years (DESA, 2005b). Tourist and migration traffic between less developed countries and more developed countries will likely increase, and traffic between more developed countries may be relatively less robust.

Table 9.3 Actual and projected population growth of 65+ years age category by percentage

	1950	1975	2000	2025	2050
Africa	3.2	3.1	3.3	4.1	6.9
Asia	4.1	4.2	5.9	10	16.7
Europe	8.2	11.4	14.7	21.5	29.2
Latin American and the Caribbean	3.7	4.3	5.4	9.6	16.9
Northern America	8.2	10.3	12.3	18.7	21.4
Oceania	7.4	7.4	9.9	14.4	18

Source: DESA (2005a)

CRUISE TOURISM: RIDING THE WAVE OF SHIFTING DEMOGRAPHICS

Cruise tourism will likely continue to enjoy growth (whether moderate or high will depend on the economic conditions in the markets served), but as noted in Chapter 5, and coupled with shifting demographics in some countries as noted above, changing trends in cruise markets have meant some companies have targeted specific markets, as in the case of easyCruise and younger passengers. The major growth areas in cruise tourism will likely not be traditional regions such as the Mediterranean or the Caribbean, however. Rather, growth in Asia and the Gulf region (e.g. India, Bahrain) will likely surpass these traditional cruising regions within the next decade. Reports from late 2005 suggest 2006 will see some 11 million cruise passengers, and one of the new efforts at attracting market share will be branding exercises that cruise companies form with other companies (e.g. Royal Caribbean's partnership with Fisher-Price toys) (Asbury Park Press, 2005).

NON-WESTERN DOMESTIC TOURISM AND TRANSPORT GROWTH

In recent years several countries around the world have capitalised on global economic growth by enhancing internal (and outward-connecting) transport modes. In the past few years, and as noted in Chapter 6, China's domestic and regional aviation market has grown rapidly on the back of substantial domestic economic growth in the past five years. This

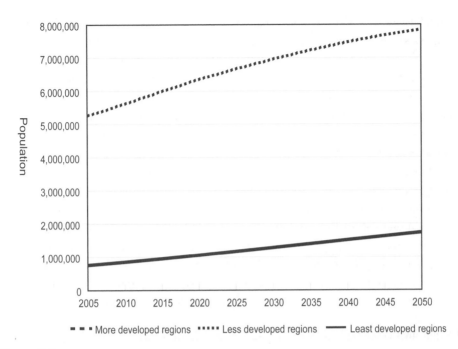

Figure 9.3 Global population forecast

Source: DESA (2005b)

has meant increases in the number of outbound Chinese tourists and a general push towards transport infrastructure renewal. Inbound tourism is also expected to grow at a healthy rate. China's integration into the global economic system is not only benefiting airlines. In 2004, for example, the Chinese government awarded contracts to three foreign firms to upgrade the country's crowded rail system (BBC, 2004c). Although SARS had an impact on China's internal travel demand in 2003, travel volumes rebounded in 2004 and 2005. By one estimate, from the General Administration of Civil Aviation of China, some 130 million Chinese people travelled on Chinese airlines in 2005, second only to the United States in terms of outbound travel. The rebound has resulted in airlines capitalising their operations with new aircraft. China Southern Airlines, in 2004, agreed to purchase 15 A320 and 4 A319 Airbus aircraft, adding to its fleet of 125 aircraft. The new planes will likely serve domestic markets, but also regional Asian centres, particularly Japan (BBC, 2004b). Overall, another 500 to 600 passenger planes (in addition to the existing 900 already in service) will be introduced by 2010, and that number is expected to reach 3000 by 2020 (People's Daily Online, 2005). There are some concerns over possible over-capacity given several major carriers plan to expand their fleets: in late December 2005 the Shanghai Securities News reported significantly slower growth (15%, compared to 33% in 2004 and 20% in 2005) in China's travel market for 2006 (Airwise News, 2005b).

Not unlike China, India's aviation market is in the middle of a remarkable growth period: domestic air travel is expected to grow 25% over the next five years (it grew almost as much in 2005 [Manorama Online, 2005]) following the Indian government allowing private companies to provide air services as 'air taxis' (Bhaumik, 2002). As well, Boeing is predicting deliveries of almost 500 aircraft to Indian carriers and the Indian government is investing in huge upgrades to existing infrastructure to accommodate the huge amount of mobility planned (Unnikrishnan, 2005). Carriers in India, particularly Air India, are also hoping for increased demand for travel following the April 2005 open skies agreement between India and the United States. Such agreements of course have the potential to stimulate demand for leisure-based travel, but they may also have the impact of stimulating demand for expatriate Indians to travel back to India. Another consequence is the potential for India to become a major hub in the region, although Sri Lanka (through Sri Lankan Airlines) is attempting to position its main airport at Columbo as a major regional hub (Airwise News, 2005c). One of the key issues facing India's rapid growth, however, is the provision of adequate infrastructure, and it is unclear whether growth in infrastructure will match growth in demand for air transport.

OPEN SKIES AND LIBERALISATION OF THE GLOBAL AIR TRANSPORT INDUSTRY

With an increasingly globalised economic climate, open skies agreements will continue to increase in number, with liberalisation efforts extending beyond simple scheduling and capacity elements to include ownership and effective control. Holloway (2003) notes that the liberalisation of existing bilateral agreements and the move towards open skies reflects general dissatisfaction with the current international regulatory regime in aviation traffic rights. Holloway (2003) also suggests that a move towards multilateral systems will continue to grow. Morocco, for example, announced in early 2006 an open skies pact with the European Union in an effort to boost its international tourism arrivals and receipts. It is likely that other countries close to the EU will adopt similar policies with the hope of capitalising on their proximity and ease of network integration. Similarly, strategic alliances of business operations (e.g. code-sharing or commercial alliances) will continue, particularly Asia and the Middle East (e.g. Qatar and Saudi Arabian Airlines announced a code-share agreement at the end of 2005).

CAPACITY CONCERNS

Capacity is likely to be a key issue in the economic sustainability of international transport (Tarry, 2004). Growing capacity without permanent reduction in costs, as Tarry (2004) suggests, may result in future capital losses for the air transport industry as investors see fewer returns. What this can mean in the future for tourism mobility is not clear due to unforeseen externalities, but for the time being global airline passenger traffic is increasing: in 2005, global RPKs were 7.5% above 2004 levels, and ASKs (capacity) only slightly increased. Together, this means that load factors were healthier in 2005 than they were in 2004 (ATW Online, 2005). Manufacturers are responding with vigour, with two the main equipment firms, Boeing and Airbus, expanding their fleet choice options (see Box 9.2).

BOX 9.2 Battle of the jumbos: The business of air transport equipment provision

Operating global networks requires significant equipment. Given that one of the largest sunk costs in establishing an airline is the cost of the aircraft, it is important to understand what is happening in the global arena of aircraft manufacturing. Our concern here is primarily with large, 'jumbo jet' aircraft because it is these that are responsible for shuttling passengers across the globe. There are two primary manufacturers for large civilian aircraft: Boeing and Airbus. Each has their own strategy and view of how civil aviation will evolve over the next few decades. Airbus recently began test flights of its new A380 superjumbo. With a maximum passenger load of 840 persons (although most options will be fitted for around 500–600 passengers in a multiple-class layout), four engine producing 77,000 lbs of thrust, and orders from airlines including Lufthansa, Singapore Airlines, Air France, Emirates, Etihad Airways, Qantas, the A380 is designed primarily as a aircraft to service international hubs. The announcement of the A380 (or A3XX as it was known during initial project design phases) ushered in a phase of excited competition between Airbus and Boeing, the world's two largest aircraft manufacturers. In one sense, the A380 was a response to the Boeing 747, which has been a global workhorse for numerous airlines around the world. Like the 747, it has been suggested that the A380 will revolutionise air travel, particularly because of the low cost per seat that could be realised by transporting more people on a single aircraft. The A380, then, represents efficiencies of scale in the air. Critics point to the problems that passengers could face: 1) queues for boarding and disembarkation; 2) costs incurred by airports in order to ensure their services are 'A380 compliant'; and 3) safety concerns with respect the time it may take to evacuate 550+ passengers (although separate evacuation tests in 2006, where 853 passengers and 20 crew left the aircraft in 78 seconds, meant the plane received approval from the European Aviation and Safety Agency and the United States FAA to carry its maximum of 853 passengers). The last point was addressed in late March 2006 when over 850 passengers, during a trial run, exited the aeroplane in less than 90 seconds with only one half of the 16 available emergency exits available, although approximately 30 passengers were injured (BBC, 2006).

On the one hand, the Boeing's response has been to introduce a replacement for the '76' family of aircraft and one that would compete with Airbus' A330 aircraft. Most importantly, Boeing wanted to develop an aircraft that was fuel efficient and generally more environmentally friendly than existing aircraft. The result was the introduction of Boeing's Dreamliner aircraft, also known as the 7E7 and, more recently, and officially (as of April 2005), the Boeing 787. Current orders for the aircraft are strong (Table 9.4).

The 787 was only part of Boeing's late 1990s/early 2000s strategy for new aircraft development. Introduced in March 2001, the 'Sonic Cruiser' was a unique aircraft designed primarily for speed. According to Boeing, it would hold '200 to 250 passengers, fly between 6,000 and 9,000 nautical miles, and travel at a speed between Mach 0.95 and Mach 0.98 – 15 to 20 percent faster than what currently is possible' (http://www.boeing.com/news/feature/concept/background.html; accessed 26 October 2005). This was distinctly different from the proposed A380 project by Airbus, which was ostensibly designed to transport more people at similar speeds to existing aircraft. According to the Boeing website:

> With its huge speed advantage, the new airplane will cut travel times by approximately 20 percent, or one hour for every 3,000 miles travelled. And because it can travel at altitudes well above today's commercial airplanes, the airlines will be less impacted by traffic congestion, an emerging industry concern. Environmental performance is very important to the Sonic Cruiser program. Because it is being designed for the environment from the earliest stages, the Sonic Cruiser will be able to deliver its speed advantage with about the same fuel burn per passenger as conventional aircraft with similar seating capacity – a level of performance that was previously thought impossible. (http://www.boeing.com/news/feature/concept/background.html; accessed 26 October 2005)

Boeing estimated travel times hitherto unheard of with conventional aircraft: Singapore–Los Angeles (2.5 hrs); Los Angeles–Sydney (2 hrs); Tokyo–Los Angeles (2 hrs). Ultimately, Boeing scrapped the Sonic Cruiser programme in December 2002 (due to lack of interest from major carriers) and concentrated development activities on the 787 discussed above.

Table 9.4 Cumulative orders for the Boeing 787 Dreamliner (January to December 2006)

Airline	Country	Order date	Total
Aeromexico	Mexico	August 2006	2
Air Pacific	Fiji	April 2006	5
Boeing Business Jet	USA USA	August 2006	1
Boeing Business Jet	USA	November 2006	1
Boeing Business Jet	USA	July 2006	2
Boeing Business Jet	USA	September 2006	1
C.I.T. Leasing Corporation	USA	September 2006	5
Continental Airlines	USA	June 2006	13
First Choice Airways	United Kingdom	September 2006	2
Icelandair	Iceland	March 2006	2
ILFC	USA	July 2006	2
Jet Airways	India	December 2006	10
Kenya Airways	Kenya	March 2006	6
Kenya Airways	Kenya	December 2006	3
Monarch Airlines	United Kingdom	August 2006	6
Nakash	USA	December 2006	2
Pegasus Aviation Finance	USA	July 2006	2
Qantas	Australia	March 2006	15
Qantas	Australia	March 2006	30
Singapore Airlines	Singapore	October 2006	20
Unidentified customer	Unidentified	July 2006	2
Unidentified customer	Unidentified	September 2006	11
Unidentified customer	Unidentified	September 2006	6
Unidentified customer	Unidentified	October 2006	10
Unidentified customer	Unidentified	March 2006	1

Source: Adapted from http://active.boeing.com/commercial/orders/userdefinedselection.cfm

More recently, the Boeing 747ADV (Advanced), or the 747-800 as it has become to be called, was was introduced at the Paris Air Show in 2003. It is set to become the new form of the highly successful 747 (itself introduced in the 1970s) that will seat 440 passengers in a standard three-class configuration. Using the existing 747 schematics, an additional two to three rows (roughly 30 seats) will be added, as will tips to the wings, which help improve fuel efficiency. The aircraft will likely have a range of 14,800 km or 8000 nautical miles.

Not to be outdone, Airbus recently introduced the new widebody A350 (officially known as the A350 XWB) after an earlier design for the same model did not find favour with major customers. With a range of 8500 nautical miles, the 'new design' long-haul, twin-engine A350 is due to begin service in 2010 and will compete specifically against Boeing's 777 and 787 aircraft. It will be produced in three variants (A350-800, A350-900 and A350-1000) and will seat 270, 314 and 350 passengers respectively. In 2006, however, Airbus faced significant challenges with its A380 programme. Initial delays on the installation of cabin wiring pushed the delivery schedule back some six months, and later in the same year the schedule was pushed back a full year due to other production problems. Many industry observers have suggested the Airbus is in financial trouble as a result of these delays, and launch customers Singapore, Qantas and Emirates may need to search for other solutions to the problem of securing suitable equipment for their operations.

Clearly, the battle of the jumbos rests in the hands of two manufacturers, Boeing and Airbus, and the success (and design) of the aircraft produced over the next few decades will shape the trajectory of air transport as well as tourism. Equally important for consideration, however, is the fact that these companies operate in global environments and face significant externalities in much the same way that their customers do.

Questions for consideration

1. Although airlines (both passenger and cargo) are generally seen as the true 'customers' for Airbus and Boeing (as well as other manufacturers), to what extent could it be argued that the traveller is also a key stakeholder? How might each company target those customers?

2. Other than the major airlines listed above, what other airlines or routes might benefit from the introduction of the A380, regardless of whether the ground-level infrastructure (i.e. airports) is capable of handling such an aircraft?

INFORMATION TECHNOLOGY

IT trends in the transport industry will continue to evolve to the benefit of the passenger. Already, IT infrastructure has altered distribution channels within all modes of transport (particularly through GDS), and this will undoubtedly extend towards providing wireless internet access during a journey. Already, some airlines and cruise ships have installed wired or wireless internet access for on-board passengers, and with the evolution of long-range wireless networks such as WiMAX, many types of ground transport could well offer similar services. The more common IT initiatives in transport generally revolve around service provisions for passengers (e.g. web check-in or purchase, on-board entertainment) but there are other initiatives that are slowly becoming pervasive: supply management applications, customer relationship management software/applications, and knowledge or content management initiatives.

CONCLUSION

Transport will undoubtedly continue to be a central feature in tourism. Changing trends in travel will continue to have a significant impact on transport provision. Increased air accessibility may often be the impetus for securing greater market shares, especially for those destinations that enjoy limited market share or are peripheral to existing key markets. In the case of Bermuda, air carriers such as USA 3000 and Spirit Airways already provide service from the United States, but government and industry representatives on the island are hopeful that JetBlue will initiate (low-cost) services sometime in 2006. As well, flights to Bermuda from London (Gatwick) on British Airways are set to increase to seven times a week in March 2006. This has coincided with a planned marketing push in the UK market of Bermuda as a holiday destination, with the hope of securing a larger portion of the Caribbean-bound UK travel market (Walters, 2005). The scope of travel may change, due to peak oil, emissions standards that limit long-haul travel or any number of other externalities, but movement and mobility will always feature in travel and tourism as activities that take place outside of familiar spaces and places. Cheaper airfares, renewed interest in alternative forms of transport (such as heritage), and the tight integration that transport has with international travel and tourist experiences would suggest that demand will not abate through shifts in market forces alone.

What may hamper demand, however, and as discussed above, is the proportion of fuel and oil costs to tourism expenditure. It has already been suggested above that, with continuing decreases in the efficiency of oil extraction and refinement, the cost of transport could continue to rise, thus some forms of transport could be reserved for individuals of high net worth, or at least those who can afford. Some might argue that the increase in oil prices in the past four to five years has resulted in a new interest in alternative modes of transport, but to some extent this may ignore a fundamental principle international economics and finance. International firms, such as airlines, and especially those operating as privately held companies, are responsible to shareholders and thus any desire to increase overall efficiencies will not be allowed to transplant potentials increases in profits. There is a social benefit to introducing sustainable means of transport, but the question remains as to who will be responsible for encouraging companies to follow this path.

The nature of tourism, and wider human mobility, naturally depends on many modes and types of transport. As well, transport is inherently linked within the business environment of tourism, and thus has an interest in the marketing of destinations, how tourism is managed and the overall viability of tourism development. As tourism development is very much concerned with issues of accessibility, and given that transport provides the networks through which tourists flow, it is integral that each pays close attention to the wider economic, social and biophysical environments within which each operates.

Box 9.3 Case study – Tourist spaceports: Environmental considerations

In Chapter 7, we considered how airports act as hubs and vectors of air transport worldwide. In this chapter, the increasing interest shown in future types of space transport was demonstrated. Indeed, it was even suggested that space tourism, or more properly extra-terrestrial vehicles, could be considered as a fourth mode of transport. While there are environmental and operational considerations of future space tourism endeavours, the ground-level support and infrastructure network is worthy of similar attention.

Several examples of the future development of spaceports have been proposed since mid-2005. Virgin Galactic entered into a partnership agreement with the State of New Mexico to construct a spaceport in the state at a cost of US$225 million. The Southwest Regional Spaceport was the first purpose-built example of a spaceport (see www.edd.state.nm.us/index.php?/about/category/locate_at_new_mexicos_spaceport/). Space Adventures (www.spaceadventures.com) announced plans to construct a commercial spaceport in the United Arab Emirates (and thus close to Dubai, which enjoys substantial air traffic from around the world), and is apparently planning further ports in Singapore and North America (Aljazeera, 2006). Currently, Space Adventures operates training facilities at the Gromov Flight Research Institute, located at the Zhukovsky Air Base near Moscow. Rocketplane (www.rocketplane.com) is closely eying the use of the former Clinton–Sherman Air Force Base (now the Clinton–Sherman Industrial Airpark) in Oklahoma that was recently granted permission from the FAA to operate as a spaceport. There are several other spaceports that have ties to space tourism. Spaceport Systems International (www.calspace.com) operates a spaceport at Vandenberg Air Force Base in California. The Mojave Airport and Civilian Aerospace Test Center (www.mojaveairport.com) operates as a spaceport and was the site of the test of SpaceShipOne, designed by Burt Rutan, which completed the first privately funded human flight into space in June 2004. Not unlike other private aerospace companies, Blue Origin (www.blueorigin. com), owned by Amazon.com-founder Jeff Bezos, operates a testing centre in Texas, although unmanned test flights have not been planned until late 2006.

Spaceports face significant challenges that are not unlike modern commercial airports designed for commercial aircraft. For example, a proposed commercial spaceport along the Gulf Coast in Texas was met with concerns from local residents over noise, potential harm to local wildlife and general loss of quality of life (Lozano, 2006). As seen in Box 7.2 these concerns are similar to those expressed in the (re)development of airports. In the case of the Clinton–Sherman Industrial Airpark, a website dedicated to the environmental impact assessment has been established (www.okspaceporteis.com).

The Southwest Regional Spaceport under construction for use by Virgin Galactic was subjected to a number of environmental considerations. According to a poster prepared by the FAA (2006) in the United States and posted on the EIS website, the agency considered several variables:

- Air Quality – launch emissions; levels of criteria pollutants; emissions from potential accidents.
- Airspace – use of airspace; impacts to other operations; changes to flight routes.
- Biological Resources – critical habitat and protected areas; wildlife movement corridors; threatened, endangered, or sensitive species; wetlands.
- Cultural Resources – historic and cultural structures; archaeological sites; Native American sites.
- Geology and Soils – mineral depletion; soil erosion; geologic hazards; hydrology and drainage patterns.
- Hazardous Materials and Waste – releases, transport, storage, use, and disposal of hazardous materials and waste.
- Land Use and Section 4(f) – conversion of farmland; existing land use plans; parks and recreational facilities.
- Noise – existing ambient noise levels; exposure of sensitive receptors to excessive noise; sonic booms.
- Health and Safety – launch operations hazards; spills, fires, and explosions; traffic and mechanical accidents.

- Socioeconomics and Environmental Justice – employment; demographics; disproportionate and adverse effects on minority and low-income populations.
- Transportation and Infrastructure – vehicular traffic and roadways; air traffic patterns; parking; utilities.
- Visual Resources – scenic vistas and resources; aesthetic quality of sites and sources of light and glare.
- Water Resources – surface water; ground water; waste water treatment facilities; floodplains; riparian issues.
- Cumulative Impacts – past, present, and foreseeable actions with potential for environmental impacts.
- Induced Impacts – impacts that could arise from activities associated with proposed action and alternatives.

It is almost certain that extra-terrestrial vehicles as a mode of transport will be developed in congruence with space tourism, but what is not certain is the extent to which terrestrial infrastructure, such as spaceports, will act in similar capacities as modern airports. One question remains as to whether the operation of spaceports will follow similar guidelines as airports themselves. While many industry-level airport associations exist worldwide, national governments impose detailed security and operational measures. It is not clear at this point whether spaceports will be governed in much the same way.

Questions for consideration

1. In thinking about the activities in modern airports, which of these may or may not be easily ported over to the operation of spaceports?

2. In addition to noise and impacts on wildlife, what other environmental hazards might need to be considered in the operation of a spaceport?

SELF-REVIEW QUESTIONS

1. What is peak oil and why is it important to tourism and transport?

2. Define Pigouvian taxes and provide an example of how they are used in transport.

3. What are some of the key issues for the future of tourism and transport?

ESSAY QUESTIONS

1. You are the manager of a major international airline and one of your destinations has just imposed a surcharge or tax of US$100 for every passenger you bring to that destination. Write a one-page briefing document to your shareholders that outlines the purpose of such taxes/surcharges.

2. Write an essay that compares the safety considerations of space travel with the three other modes of transport discussed in this book.

3. Write an essay that considers the increasing cost of fuel and the impact that may have on peripheral tourism destinations around the world.

KEY SOURCES

Gössling, S., Peeters, P., Ceron, J.-P., Dubois, G., Patterson, T. and Richardson, R.B. (2005) The eco-efficiency of tourism. *Ecological Economics* 54, 417–434.

An excellent overview of the ecological footprint of tourism activities, this article is a must-read in order to appreciate the environmental implications for tourism.

Gössling, S. and Hall, C.M. (eds) (2006) *Tourism and Global Environmental Change: Ecological, Social, Economic and Political Interrelationships.* London: Routledge

An excellent (and recent) overview of environment change concerns stemming from tourism activities. The book is unique in that it introduces the political underpinning for (in)action relating to climate change. The concluding chapter by the authors is particularly insightful and should serve as a starting point for those interested in the topic.

Flight International – www.flightglobal.com

Flight International is a leading weekly trade publication that often features extensive information on extra-terrestrial vehicles and the business of transport. It is also an excellent source of up-to-date information on trends in global air transport.

APPENDIX I:

SELECTED SOURCES OF FURTHER INFORMATION

The internet provides numerous sources of information relating to transport issues in general and in the context of tourism. Listed below are several key academic and trade journals relating to transport. As well, a selection of websites by topic is provided. These sources may be used as a starting point in searching for additional information. Most of the popular internet search engines can be used as well.

ACADEMIC JOURNALS

Journal of Air Transport Management (www.elsevier.com)
Journal of Transport Geography (www.elsevier.com)
Transportation Research (particularly Parts 'A' through 'E') (www.elsevier.com)
Journal of Transport Economics and Policy (www.jtep.org)
Transport Review (www.tandf.co.uk)

TRADE JOURNALS

Airline Business (www.airlinebusiness.com)
Airport Business (www.airportbusiness.com)
Flight International (www.flightglobal.com)
Air Transport World (www.atwonline.com)
Aerlines Magazine (www.aerlines.nl) – Free, quarterly e-zine

COMMERCIAL MAGAZINES

Airliner World (www.airlinerworld.com)
Airways (www.airwaysmag.com)
The Railway Magazine (www.countrylife.co.uk/subscriptions/railway.php)
Porthole Cruise Magazine (www.porthole.com)
Cruise Magazine (www.cruisemagazine.com)

AIR TRANSPORT-RELATED WEBSITES

www.iata.org – International Air Transport Association
www.icao.int – International Civil Aviation Organization
www.eraa.org – European Regions Airline Association
www.aea.be – Association of European Airlines
www.elfaa.com – European Low Fares Airline Association
www.raaa.com.au – Regional Aviation Association of Australia
www.ebaa.org – European Business Aviation Association
www.iapa.com – International Airline Passengers Association
www.afraa.org – African Airlines Association
www.aapairlines.org – Association of Asia Pacific Airlines
www.airlines.org – Air Transport Association of America
www.airlinebusiness.com – *Airline Business* magazine
www.airportbusiness.com – *Airport Business* magazine
www.airlinemeals.net – comprehensive listing and evaluation of on-board meals by
    various airlines
www.faa.gov –   Federal Aviation Administration (US)
www.aviation.dft.gov.uk – Department for Transport (Aviation) (UK)
www.wirelessairport.org – Wireless Airport Association
www.routesonline.com – Annual airlines and airports event/trade show
www.cate.mmu.ac.uk – Centre for Air Transport and the Environment (Manchester
    Metropolitan University, UK)
www.flyaow.com – Class availability for many global flights
www.airlinecrew.net – Bulletin board for airline crew
www.airliners.net – Bulletin board and an excellent source of copyrighted photos
www.flyertalk.com – Bulletin board focusing primarily on frequent flyer programmes and
    air travel in general
www.johnnyjet.com – An incredible amount of air transport-related links

GROUND TRANSPORT-RELATED WEBSITES

www.eriksrailnews.com – Erik's Rail News
www.railtimes.com – Rail times
www.railnews.net – Rail news
www.railcan.ca – Railway Association of Canada
www.ara.net.au – Australasian Railway Association
www.aar.org – Association of American Railroads
www.riagb.org.uk – Railway Industry Association (UK)
www.ukhrail.uel.ac.uk – Heritage Railway Association (UK)
www.irfca.org – Indian Railways Fan Club Association

www.traininc.org – Tourist Railway Association

www.railjournal.com – *International Railway Journal*

MARINE TRANSPORT-RELATED WEBSITES

www.cruising.org – Cruise Lines International Association

www.iccl.org – International Council of Cruise Lines

www.f-cca.com – Florida–Caribbean Cruise Association

www.atlanticcanadacruise.com – Atlantic Canada Cruise Association

www.medcruise.com – The Association of Mediterranean Cruise Ports

MISCELLANEOUS BLOGS, WIKIS, ORGANISATIONS AND OTHER SITES

http://wiki.jeffsandquist.com/default.aspx/AirPower/HomePage.html – AirPower Wiki
(locate power outlets at airports around the world)

www.seatguru.com – Locate the best seats on a variety of aircraft types (sorted by airline)

www.atrsworld.org – Air Transport Research Society

www.irtsociety.com – The Society of International Railway Travelers

www.thirtythousandfeet.com – Aviation directory

REFERENCES

ABC News (US) (2005) Attack on ships show pirates emboldened, 7 November 2005 (http://abcnews.go.com/International/print?id=1286496). Accessed 11 November 2005.

ABC News (Australia) (2005) Branson looks to plant waste for jet fuel, 17 November 2005 (http://www.abc.net.au/news/newsitems/200511/s1508526.htm). Accessed 18 November 2005.

Adler, N. (2001) Competition in a deregulated air transportation market. *European Journal of Operational Research* 129: 337–345.

Advertiser-Tribune (2004) Train tourism planned for Fostorias economic future, 29 February 2004 (http://www.advertisertribune.com/news/story/0229202004_new01trains2-29.asp)

The Age (2005) UK rail stations to get airport-style security, 31 October 2005 (http://www.theage.com.au/news/world/uk-rail-stations-to-get-airportstyle-security/2005/10/30/1130607151927.html#). Accessed 31 October 2005.

Airbus S.A.S. (2005) Global market forecast, 2004–2024. (http://www.airbus.com/en/myairbus/global_market_forcast.html). Accessed 30 December 2005.

Airline Business (2005a) Capacity crunch, 6 June 2005 (www.airlinebusiness.com/Articles/Article.aspx?liArticleID=198195). Accessed 7 June 2005.

Airline Business (2005b) Concrete plans, 6 June 2005 (www.airlinebusiness.com/Articles/Article.aspx?liArticleID=198193). Accessed 7 June 2005.

Airline Business (2005c) Saving the planet, 3 November 2005 (http://www.bizbuzzmedia.com/blogs/airline/archive/2005/11/03/678.aspx). Accessed 9 November 2005

Airline Business (2005d) Low-cost A380 services – imagine that, 16 December 2005 (http://www.bizbuzzmedia.com/blogs/airline/archive/2005/12/16/915.aspx). Accessed 26 December 2005.

Airline Business (2006a) IT trends survey 2006, July 2006, 42–49.

Airline Business (2006b) Stellar orbit, September 2006, 46–48.

Air New Zealand and Qantas Airways Limited (2002a) Commerce Act 1986: Restrictive Trade Practice, Section 58: Notice Seeking Authorisation, 9 December 2002. Public version accessed via New Zealand Commerce Commission website (since removed).

Air New Zealand and Qantas Airways Limited (2002b) Commerce Act 1986: Business Acquisition, Section 57: Notice Seeking Authorisation, 9 December 2002. Public version accessed via New Zealand Commerce Commission website (since removed).

Airwise News (2005a) Air Canada puts a price on comfort, 15 October 2005 (http://news.airwise.com/story/view/1129294309.html). Accessed 15 October 2005.

Airwise News (2005b) China's air traffic growth to slow in 2006, 27 December 2005 (http://news.airwise.com/story/view/1135689408.html). Accessed 27 December 2005.

Airwise News (2005c) Sri Lankan Airlines targets regional hub, 19 December 2005 (http://news.airwise.com/story/view/1135027423.html). Accessed 19 December 2005.

Aljazeera (2006) Spaceport comes to UAE, 18 February 2006 (http://english.aljazeera.net/NR/exeres/BFCD648C-FE21-498B-921A-324AC5D7EE28.htm). Accessed 28 July 2006.

Allen, W.H. (1992) Increased dangers to Caribbean marine ecosystems: Cruise ship anchors and intensified tourism threaten reefs. Bioscience 42 (5), 330–335.

Andersson, K. and Eklund, E. (1999) Tradition and innovation in coastal Finland: The transformation of the archipelago sea region. *Sociologica Ruralis* 39 (3), 377–393.

Arab News (2005) First low-cost airline planned in the Kingdom, 22 November 2005 (http://www.arabnews.com/?page=6&section=0&article=73519&d=22&m=11&y=2005). Accessed 25 November 2005.

ASA (2005) Non-broadcast adjudication. (http://www.asa.org.uk/asa/adjudications/non_broadcast/Adjudication+Details.htm?Adjudication_id=40129). Accessed 10 August 2005.

Asbury Park Press (2005) Cruise trends for 2006: Themed trips and six new ships, 12 December 2005 (http://www.app.com/apps/pbcs.dll/article?AID=/20051204/LIFE02/512040321/1006/LIFE). Accessed 12 December 2005.

ATW Online (2005) World RPKs rise 7.5% in 2005, passengers top 2 billion, 19 December 2005 (http://www.atwonline.com/news/story.html?storyID=3466). Accessed 21 December 2005.

ATW (2005a) The world's top 25 airlines 2004. July 2005, 33.

ATW (2005b) Europes biggest airports. July 2005, 120.

BAA (2005) Success through innovation: Annual Report 2004/2005. BAA.

Banister, D. (1995) *Tourism and Urban Development*. London: Spon Press.

Banister, D. (2002) *Transport Planning* (2nd edn). London: Spon Press.

Bastin, R. (1984) Small island tourism: Development or dependency? *Development Policy Review* 2 (1), 79–90.

Barrett, S.D. (2000) Airport competition in the deregulated European aviation market. *Journal of Air Transport Management* 6, 13–27.

Bayles, F. (1998) Accidents shadow whale-watching industry. *USA Today*, 21 September, 03A.

BBC (2000) Transport 2010 at a glance, 20 July 2000 (http://news.bbc.co.uk/1/hi/uk_politics/843129.stm). Accessed 7 March 2004.

BBC (2004a) South Korea launches high speed train, 1 April 2004 (http://news.bbc.co.uk/2/hi/asia-pacific/3589591.stm). Accessed 14 May 2004.

BBC (2004b) Airbus wins $1.2b China order, 12 April 2004 (http://news.bbc.co.uk/go/pr/fr/-/2/hi/business/3619353.stm). Accessed 12 April 2004.

BBC (2004c) China to modernise rail routes, 29 August 2004 (http://news.bbc.co.uk/go/pr/fr/-/1/hi/business/3610200.stm). Accessed 29 August 2004.

BBC (2004d) Love train, 1 July 2005 (http://news.bbc.co.uk/1/hi/magazine/4640947.stm). Accessed 25 July 2005.

BBC (2005) US agrees climate deal with Asia, 28 July 2005 (http://news.bbc.co.uk/2/hi/science/nature/4723305.stm). Accessed 9 December 2005.

BBC (2006) Volunteers injured in A380 drill, 26 March 2006 (http://news.bbc.co.uk/2/hi/business/4847344.stm). Accessed 27 March 2006.

Beach, D.W. and Weinrich, M.T. (1989) Watching the whales. *Oceanus* 32 (1), 84–88.

Beard, J.G. and Ragheb, M.G. (1983) Measuring leisure motivation. *Journal of Leisure Research* 15 (3), 219–228.

Becken, S. (2002) Analysing international tourist flows to estimate energy use associated with air travel. *Journal of Sustainable Tourism* 10 (2), 114–131.

Becken, S. and Simmons, D.G. (2005) Tourism, fossil fuel consumption and the impact on the global climate. In C.M. Hall and J. Higham (eds) *Tourism, Recreation and Climate Change* (pp. 192–208). Clevedon: Channel View Publications.

Bejou, D. and Palmer, A. (1998) Service failure and loyalty: An exploratory empirical study of airline customers. Journal of Services Marketing 12 (1), 7–22.

Bell, P. and Cloke, P. (1990) *Deregulation and Transport: Market Forces in the Modern World*. London: Fulton.

Belobaba, P.P. and Wilson, J.L. (1997) Impacts of yield management in competitive airline markets. *Journal of Air Transport Management* 3 (1), 3–9.

Bennett, M.M. (1996) Airline marketing. In A.V. Seaton and M.M. Bennett *The Marketing of Tourism: Concepts, Issues and Cases* (pp. 377–398). London: International Thomson Business Press.

Beyhoff, S. (1995) Code-sharing: A summary of the German study. *Journal of Air Transport Management* 2 (2), 127–129.

Bhaumik, P.K. (2002) Regulating the domestic air travel in India: An umpires game. *Omega* 30, 33–44.

Bieger, T. and Laesser, C. (2001) The role of the railway with regard to mode choice in medium range travel. *Tourism Review* 56, 33–39.

Bieger, T. and Wittmer, A. (2006) Air transport and tourism – Perspectives and challenges for destinations, airlines and governments. *Journal of Air Transport Management* 12, 40–46.

Bitner, M.J. (1990) Evaluating service encounters: The effects of physical surroundings and employee responses. *Journal of Marketing* 54, 69–82.

Bitner, M.J., Fisk, R.P. and Brown, S.W. (1993) Tracking the evolution of the services marketing literature. *Journal of Retailing* 69 (1), 61–103.

Blane, J.M. and Jaakson, R. (1995) The impact of ecotourism boats on the St Lawrence beluga whales. *Environmental Conservation* 21 (3), 267–269.

Blenkey, N. (2005) New ships, new thinking. *Marine Log* 110 (2), 23–26.

Bloomberg News (2005) UK transport needs extra 50 bln pounds, business group says, 27 November 2005 (http://www.bloomberg.com/apps/news?pid=10000102&sid=a 3axm4YI54PQ&refer=uk#). Accessed 29 November 2005.

Boeing (2005) World demand for commercial airplanes, 2005 (http://www.boeing.com/commercial/cmo/). Accessed 30 December 2005.

Bonaire Department of Economic and Labour Affairs (2004) Cruise tourism statistics (http://www.bonaireeconomy.org/tourism_statistics/cruise_statistics.html). Accessed 24 May 2005.

The Bond Buyer (2004) House appropriations approves new spending package. 349 (31932), 23 July 2004.

Borenstein, S. (1989) Hubs and high fares: Dominance and market power in the US airline industry. *The RAND Journal of Economics* 20 (3), 344–365.

Botimer, T.C. (1996) Efficiency consideration in airline pricing and yield management. *Transportation Research Part A* 30 (4), 307–317.

Boyfield, K. (2003) Who owns airport slots? A market solution to a deepening dilemma. In K. Boyfield (ed) *A Market in Airport Slots* (pp. 21–50). London: The Institute of Economic Affairs.

Bowen, J. (2000) Airline hubs in Southeast Asia: National economic development and nodal accessibility. *Journal of Transport Geography* 8, 25–41.

Breugelmans, J.G., Zucs, P., Porten, K., Broll, S., Niedrig, M., Ammon, A. and Krause, G. (2004) SARA Transmission and commercial aircraft. Letter to editor. *Emerging Infectious Diseases* 10 (8), 1502–1503.

Brooker, P. (2003) Control workload, airspace capacity and future systems. *Human Factors and Aerospace Safety* 3 (1), 1–23.

Brueckner, J.K. (2001) The economics of international codesharing: An analysis of airline alliances. *International Journal of Industrial Organization* 19, 1475–1498.

Bruning, E.R. (1997) Country of origin, national loyalty and product choice: The case of international air travel. *International Marketing Review* 14 (1), 59–74.

BTS (2005a) Airline travel since 9/11. (http://www.bts.gov/publications/issue_briefs/number_13/html/entire.html). Accessed 30 December 2005.

BTS (2005b) Fourth quarter 2004 airline financial data: Regional passenger airlines only group to report domestic profit (http://www.bts.gov/press_releases/2005/bts022_05/html/bts022_05.html#table_01). Accessed 31 October 2005.

BTS (2005c) 2001 National Household Travel Survey (http://www.bts.gov/publications/highlights_of_the_2001_national_household_travel_survey/). Accessed 6 December 2005.

Buck, S. and Lei, Z. (2004) Charter airlines: Have they a future? *Tourism and Hospitality Research* 5 (1), 72–78.

Buhalis, D. (2004) eAirlines: Strategic and tactical use of ICTs in the airline industry. *Information and Management* 41, 805–825.

Buhalis, D. and Licata, M.C. (2002) The future of eTourism intermediaries. *Tourism Management* 23, 207–220.

Burger, J. and Leonard, J. (2000) Conflict resolution in coastal waters: The case of personal watercraft. *Marine Policy* 24, 61–67.

Butler, R. (1980) The concept of a tourist area cycle of evolution: Implications for management of resources. *Canadian Geographer* 24, 5–12.

Butler, R. (1997) Transportation innovations and island tourism. In D.G. Lockhard and D. Drakakis-Smith (eds) *Island Tourism: Trends and Prospects* (pp. 36–56). London: Pinter.

Butler, R., Hall, C.M. and Jenkins, J. (1998) *Tourism and Recreation in Rural Areas*. Chichester: John Wiley and Sons.

Button, K.J. (1993) *Transport Economics* (2nd edn). Aldershot: Edward Elgar.

Button, K.J. (2002) Debunking some common myths about airport hubs. *Journal of Air Transport Management* 8, 177–188.

Button, K., Clarke, A., Palubinskas, G., Stough, R., and Thibault, M. (2004) Conforming with ICAO safety oversight standards. *Journal of Air Transport Management* 10, 251–257.

Byran, D.L. and O'Kelly, M.E. (1999) Hub-and-spoke networks in air transportation: An analytical review. *Journal of Regional Science* 39 (2), 275–295.

CAA (2004) Introducing commercial allocation mechanisms: The UK Civil Aviation Authority's response to the European Commission's staff working paper on slot reform (http://www.caa.co.uk/docs/5/SlotReform-theUKCAAFinalResponse.pdf). Accessed 31 August 2005.

CAA (2005) (http://www.statistics.gov.uk/STATBASE/ssdataset.asp?vlnk=7819). Accessed 18 November 2005.

Calder, S. (2002) *No Frills: The Truth Behind the Low-cost Revolution in the Skies*. London: Virgin Books.

Cao, J.-M. and Kanafani, A. (2000) The value of runway time slots for airlines. *European Journal of Operational Research* 126, 491–500.

Carassava, A. (2005) Easy does it on the high seas. *Time*, 25 April 2005, 165 (17), A3.

Carey, H.C. (1858) *Principles of Social Science*. Philadelphia: J.B. Lippincott Company.

Carlsson, F. (2002) Environmental charges in airline markets. *Transportation Research Part D* 7, 137–153.

Cartwright, R. and Baird, C. (1999) *The Development and Growth of the Cruise Industry*. Oxford: Butterworth-Heinemann.

Caves, R. and Gosling, G. (1999) *Strategic Airport Planning*. Oxford: Pergamon.

Caves, R. and Pickard, C.D. (2001) The satisfaction of human needs in airport passenger terminals. *Transport* 147 (1), 9–15.

CBC (2006) Ferry uncertainty could cost northern BC, 20 April 2006 (http://www.cbc.ca/bc/story/bc_ferry-jobs20060420.html). Accessed 23 April 2006.

CE Delft (2005) Giving wings to emission trading: Inclusion of aviation under the European emission trading system (ETS): Design and impacts, reported commissioned by European Commission, DG Environment, Number 05.7789.20 (http://europa.eu.int/comm/environment/climat/pdf/aviation_et_study.pdf). Accessed 2 January 2006.

Central Otago Rail Trail (2004) (http://www.centralotagorailtrail.co.nz/index.htm). Accessed 9th May 2004.

Chapin, F.S. Jr (1966) The use of time budgets in the study of urban living patterns. *Research Previews* 13, 1–6.

Chapman, K. (1979) *People, Pattern and Process: An Introduction to Human Geography*. London: Edward Arnold.

Chen, F.-Y. and Chang, Y.-H. (2005) Examining airline service quality from a process perspective. *Journal of Air Transport Management* 11, 79–87.

Chin, A.T.H. (1997) Implications of liberalisation on airport development and strategy in the Asia Pacific. *Journal of Air Transport Management* 3 (3), 125–131.

Christaller, W. (1972) How I discovered the theory of central places: A report about the origin of central places. In P.W. English and R.C. Mayfield (eds) *Man Space and Environment* (pp. 601–610). London: Oxford University Press.

Christchurch International Airport (2005) Christchurch International Airport Annual Report. CIA Limited.

Clark, A. (2005) Frequent flyer miles soar above sterling. *The Guardian*, 8 January 2005 (http://www.guardian.co.uk/print/0,3858,5098772-103676,00.html). Accessed 9 January 2005.

CLIA (1995–2001) *The Cruise Industry: An Overview*. New York: CLIA.

CLIA (2005) Cruise news (http://www.cruising.org/CruiseNews/news.cfm?NID=174). Accessed 24 May 2005.

CNN (1998) NASA says space tourism is on its way but skeptics doubt it, 25 March 1998 (http://www.cnn.com/TECH/space/9803/25/space.tourism/#1). Accessed 25 March 1998.

Coase, R.H. (1960) The problem of social cost. *Journal of Law and Economics* 3, 1–44.

Cobin, J.M. (1999) *A Primer on Modern Themes in Free Market Economics*. Parkland, FL: Universal Publishers.

Cocks, C. (2001) *Doing the Town: The Rise of Urban Tourism in the United States, 1850–1915.* Berkeley: University of California Press.

Cohen, A.J. and Harris, N.G. (1998) Mode choice for VFR journeys. *Journal of Transport Geography* 6 (1), 43–51.

Cohen, E. (1974) Who is a tourist? A conceptual clarification. *Sociological Review* 22, 527–555.

Coles, T. (2004a) What makes a resort complex? Reflections on the production of tourism space in a Caribbean resort complex. In D.T. Duval (ed.) *Tourism in the Caribbean: Trends, Development, Prospects* (pp. 235–256). London: Routledge.

Coles, T. (2004b) Tourism and retail transactions: Lessons from the Porsche experience. *Journal of Vacation Marketing* 10 (4), 378–389.

Coles, T., Duval, D.T. and Hall, C.M. (2004) Tourism, mobility and global communities: New approaches to theorising tourism and tourist spaces. In W. Theobald (ed.) *Global Tourism: The Next Decade* (3rd edn) (pp. 463–481). Oxford: Butterworth-Heinemann.

Commission of the European Communities (1999) Communication from the Commission to the Council, the European Parliament, the Economic and Social Committee and the Committee of the Regions – Air transport and the environment: Towards meeting the challenges of sustainable development. Brussels, 1 December 1999 COM (1999) 640 final.

Commission of the European Communities (2004) Commission staff working document: Commercial slot allocation mechanisms in the context of a further revision of Council Regulation (EEC) 95/93 on common rules for the allocation of slots at Community airports (http://europa.eu.int/comm/transport/air/rules/competition2/doc/2004_09_17_consultation_paper_en.pdf). Accessed 31 August 2005.

Conroy, M. (2004) Upbeat developments in the state of the cruise industry. Business Briefings: Global Cruise 2004, May 2004 (http://www.bbriefings.com/cdps/cditem.cfm?NID=858&CID=6&CFID=3992128&CFTOKEN=96594492). Accessed 6 January 2005.

Cope, A., Cairns, S., Fox, K., Lawlor, D.A., Lockie, M., Lumsdon, L., Riddoch, C. and Rosen, P. (2003) The UK National Cycle Network: An assessment of the benefits of a sustainable transport infrastructure. *World Transport Policy & Practice* 9 (1), 6–17.

Copock, J.T. (ed.) (1977) *Second Homes: Curse or Blessing?* Oxford: Pergamon.

Coyle, J.J., Bardi, E.J. and Novack, R.A. (1994) *Transportation.* St Paul/Minneapolis: West Publishing Company.

Crabtree, R.M. (2000) A system dynamics model for visitors choice of transport mode to and from national parks. *Countryside Recreation* 8 (3), 2–5 (www.countrysiderecreation.org.uk/journal/journal2000.asp). Accessed 6 December 2004.

Crawford, G. and Melewar, T.C. (2003) The importance of impulse purchasing behaviour in the international airport environment. *Journal of Consumer Behaviour* 3 (1), 85–98.

Crenson, S.L. (2001) Cruise ship pollution under fire. *AP Online,* 7 April 2001.

Crockett, J. and Hounsell, N. (2005) Role of the travel factor convenience in rail travel and a framework for its assessment. *Transport Reviews* 25 (5), 535–555.

Crompton, J.L. (1979) Motivations for pleasure vacation. *Annals of Tourism Research* 6, 408–424.

Crouch, G.I. (2001) The market for space tourism: Early indications. *Journal of Travel Research* 40, 213–219.

CTO (2002) *Caribbean Tourism Statistical Report: 2000–2001 Edition.* St Michael: CTO.

Cycling Advocates Network (2004) DoC recreation opportunities review (http://www.can.org.nz/submissions/CAN-subm-040131-DoC-Review.pdf). Accessed 10 May 2004.

D'Angelo, A. (2005) Ghanian protectionism thwarts SAA, 25 October 2005 (http://www.businessreport.co.za/index.php?fArticleId=2964617). Accessed 9 November 2005.

Dann, G.M.S. (1981) Tourist motivation: An appraisal. *Annals of Tourism Research* 8 (2), 187–219.

Dann, G.M.S. (1994) Travel by train: Keeping nostalgia on track. In A.V. Seaton (ed.) *Tourism: The State of the Art* (pp. 775–782). Chichester: John Wiley and Sons.

Daskin, M.S. (1995) *Network and Discrete Location: Model, Algorithms, and Applications.* New York: John Wiley & Sons.

Department of Conservation (2004) The Otago Central Railway (http://www.doc.govt.nz/Conservation/Showcase-Areas/Otago-Central-Rail-Trail.asp). Accessed 11 May 2004.

Department for Transport (2004a) Transport statistics for Great Britain 2004 edition (http://www.dft.gov.uk/stellent/groups/dft_transstats/documents/page/dft_transstats_031999.hcsp). Accessed 21 October 2004).

Department for Transport (2004b) Transportation statistics bulletin: National travel survey 2002 (revised July 2004) (United Kingdom National Statistics) (http://www.dft.gov.uk/pgr/statistics/datatablespublications/personal/mainresults/nts2002/nationaltravelsurvey2002revised). Accessed 20 March 2007.

Department for Transport (2004c) UK response to European Commission's slot allocation consultation (http://www.dft.gov.uk/stellent/groups/dft_aviation/documents/pdf/dft_aviation_pdf_033272.pdf). Accessed 31 August 2005.

Department for Transport (2005) Transport trends 2005 (http://www.dft.gov.uk/stellent/groups/dft_transstats/documents/page/dft_transstats_026292.xls). Accessed 18 May 2006.

Department for Transport (2006) Transport statistics for Great Britain 2006 edition (http://www.dft.gov.uk/pgr/statistics/datatablepublications/tsgb/2006edition/transportstatisticsforgreatb1856). Accessed 11 March 2007.

DESA (Population Division) (2005a) World population ageing: 1950–2050 (http://www.un.org/esa/population/publications/worldageing19502050/regions.htm). Accessed 9 January 2006.

DESA (Population Division) (2005b) World population prospects: The 2004 revision and world urbanization prospects: The 2003 revision (http://esa.un.org/unpp). Accessed 9 January 2006.

Dickinson, J.E., Calver, S., Watters, K. and Wilkes K. (2004) Journeys to heritage attractions in the UK: A case study of National Trust property visitors in the south west. *Journal of Transport Geography* 12, 103–113.

Dijst, M., Lanzendorf, M., Barendregt, A. and Smit, L. (2005) Second homes in Germany and The Netherlands: Ownership and travel impact explained. *Tijdschrift voor Economische en Sociale Geografie* 96 (2), 139–152.

Dobruszkes, F. (2006) An analysis of European low-cost airlines and their networks. *Journal of Transport Geography* 14 (4), 249–264

Doganis, R. (1992) *The Airport Business*. London: Routledge.

Doganis, R. (2001) *The Airline Business in the 21st Century*. London: Routledge.

Doganis, R. (2002) *Flying Off Course: The Economics of International Airlines*. London: Routledge.

Domroes, M. (1999) Tourism in the Maldives: The resort-concept and tourist-related systems. *International Journal of Island Affairs* 8 (3), 7–14.

Douglas, N. and Douglas, N. (2001) The cruise experience. In N. Douglas, N. Douglas and R. Derrett (eds) *Special Interest Tourism: Context and Cases* (pp. 330–354). Brisbane: John Wiley & Sons Australia.

Douglas, N. and Douglas, N. (2004) Cruise ship passenger spending patterns in Pacific Island ports. *International Journal of Tourism Research* 6, 251–261.

Dowling, R. and Newsome, D. (eds) (2005) *Geotourism: Sustainability, Impacts and Management*. Oxford: Butterworth-Heinemann.

Dresner, M. (2006) Leisure versus business passengers: Similarities, differences, and implications. *Journal of Air Transport Management* 12, 28–32.

Dresner, M. and Windle, R. (1995) Are US air carriers to be feared? Implication of hubbing to North Atlantic competition. *Transport Policy* 2 (3), 195–202.

Driver, J.C. (1999) Developments in airline marketing practice. *Journal of Marketing Practice: Applied Marketing Science* 51 (5), 134–150.

DSEC (2007) Results of the vistor arrivals for February 2007 (http://www.dsec.gov.mo/e_index.html). Accessed 22 March 2007.

Duman, T. and Mattila, A.S. (2005) The role of affective factors on perceived cruise vacation value. *Tourism Management* 26, 311–323.

Dunlop, G. (2002) The European ferry industry – challenges and changes. *International Journal of Transport Management* 1, 115–116.

Duval, D.T. (2002) The return visit-return migration connection. In C.M. Hall and A.M. Williams (eds) *Tourism and Migration: New Relationships between Production and Consumption* (pp. 257–276). Dordrecht: Kluwer Academic Publishers.

Duval, D.T. (2003) When hosts become guests: Return visits and diasporic identities in a Commonwealth Eastern Caribbean community. *Current Issues in Tourism* 6 (4), 267–308.

Duval, D.T. (2004a) Trends and circumstances in Caribbean tourism. In D.T. Duval (ed.) *Tourism in the Caribbean: Trends, Development, Prospects* (pp. 3–22). London: Routledge.

Duval, D.T. (2004b) Future prospects for tourism in the Caribbean. In D.T. Duval (ed.) *Tourism in the Caribbean: Trends, Development, Prospects* (pp. 298–299). London: Routledge.

Duval, D.T. (2005a) Public/stakeholder perceptions of airline alliances: The New Zealand experience. *Journal of Air Transport Management* 11 (6), 355–462.

Duval, D.T. (2005b) Small steps, giant leaps: Space as the destination of the future. In M. Novelli (ed.) *Niche Tourism: Contemporary Issues and Trends* (pp. 213–222). Oxford: Butterworth-Heinemann.

Duval, D.T. (2005c) Tourism and air transport in Oceania. In C. Cooper and C.M. Hall (eds) *Oceania: A Tourism Handbook* (pp. 321–334). Clevedon: Channel View.

Duval, D.T. (2006) Coasian economics and the management of international aviation emissions. *International Journal of Innovation and Sustainable Development* 1 (3), 201–213.

Dwyer, L. (2004) Trends underpinning global tourism in the coming decade. In W. Theobald (ed.) *Global Tourism: The Next Decade* (3rd edn) (pp. 528–545). Oxford: Butterworth-Heinemann.

Dwyer, L. and Forsyth, P. (1998) Economic significance of cruise tourism. *Annals of Tourism Research* 25 (2), 393–415.

Eaton, B. and Holding, D. (1996) The evaluation of public transport alternatives to the car in British national parks. *Journal of Transport Geography* 4 (1), 55–65.

Ebersold, W.B. (2004) Cruise industry in figures. Business briefings: Global cruise 2004, (http://www.bbriefings.com/cdps/cditem.cfm?NID=858&CID=6&CFID=39921 28&CFTOKEN=96594492). Accessed 6 January 2005.

Eccles, G. and Costa, J. (1996) Perspectives on tourism development. *International Journal of Contemporary Hospitality Management* 8 (7), 44–51.

Economist (2005) Funny money. 377 (8458), 104–105.

Edinburgh News.com (2004) Chiefs fear Edinburgh tolls will drive away tourists, 6 April 2004 (http://edinburghnews.scotsman.com/index.cfm?id=390612004). Accessed 21 October 2004.

Edmondson, B. and Du, F. (1996) Who needs two cars? Automobile ownership statistics. *American Demographics*, December 1996 (http://www.findarticles.com/p/articles/mi_m4021/is_n12_v18/ai_18894247). Accessed 6 September 2004.

Edmonton Airports (2005) Edmonton Airport annual report 2004 (www.edmontonairports.com). Accessed ???.

Egli, R.A. (1991) Climate: Air-traffic emissions. *Environment* 33 (9), 2–5.

Ellett, T. (2003) Airport privatization after the Bush executive order (http://www.rppi.org/apr2003/airportprivatizationafter.html). Accessed 28 September 2005.

Elliot, C. and Silver, M. (2002) Cruise-a-palooza, *US News and World Report*, 18 March 2002, 72.

Ellis, C., Barrett, N. and Schmieman, S. (2005) Wilderness cruising: Turbulence, cruise ships, and benthic communities. *Tourism and Marine Environments* 2 (1), 1–12.

Essex, S. (1994) Tourism. In R. Gibb (ed.) *The Channel Tunnel: A Geographical Perspective* (pp. 79–100). Chichester: John Wiley & Sons.

Express Travel and Tourism (2005) Cruise industry: Future is bright, November 2005 (http://www.expresstravelandtourism.com/200511/lookout08.shtml). Accessed 15 April 2006.

European Commission (2001) White paper: European transport policy for 2010: Time to decide (http://europa.eu.int/comm/energy_transport/en/lb_en.html). Accessed 4 December 2005.

European Union Road Federation (2005) 2005 road statistics (http://www.erf.be/images/stat/ERF_stats.pdf). Accessed 18 May 2006.

Evening Times (2006) Airline cuts Sweden flights over tax row, 12 July 2006 (http://www.eveningtimes.co.uk/print/news/5054657.shtml). Accessed 3 August 2006.

Express (2004) New rail hell. 14 June 2004.

FAA (2004) Wildlife strikes to civil aircraft in the United States, 1990–2003, Serial Report Number 10, June 2004.

FAA (2006) Environmental impact statement scoping topics (http://ast.faa.gov/files/pdf/EIS%20Scoping%20Topics_Scoping%20Poster.pdf). Accessed 28 July 2006.

Fan, T. (2006) Improvements in intra-European inter-city flight connectivity: 1996–2004. *Journal of Transport Geography* 14 (4), 273–286.

Fellmann, J., Getis, A. and Getis, J. (1992) *Human Geography: Landscapes of Human Activities* (3rd edn). Dubuque, IA: Wm. C. Brown Publishers.

Field, D. (2005) Online charge. *Airline Business*, December 2005.

Firey, T. (2003) Nothing to fear from open skies with European Union. CATO Institute (http://www.cato.org/research/articles/firey-030924.html). Accessed 15 November 2005.

Fischer, M.M. (1993) Travel demand. In J. Polak and A. Heertje (eds) *European Transport Economics* (pp. 6–32). Oxford: Blackwell.

Florida–Caribbean Cruise Association (2001) *Cruise Industry's Economic Impact on the Caribbean*. Florida: PriceWaterhouseCoopers.

Forbes (2006) Ryanair slashes services to Sweden ahead of new passenger tax on Aug 1, 11 July 2006 (http://www.forbes.com/finance/feeds/afx/2006/07/11/afx2870654.html). Accessed 3 August 2006.

Forbes (2005) BA's Eddington says US protectionism props up failing airlines – report, 22 September 2005 (http://www.forbes.com/markets/feeds/afx/2005/09/22/afx2240404. html). Accessed 9 November 2005.

Forsyth, P. (2002) Privatisation and regulation of Australian and New Zealand airports. *Journal of Air Transport Management* 8, 19–28.

Forsyth, P. (2006) Martin Kunz memorial lecture. Tourism benefits and aviation policy. *Journal of Air Transport Management* 12, 3–13.

Francis, G., Humphreys, I. and Ison, S. (2004) Airports perspectives on the growth of low-cost airlines and the remodeling of the airport–airline relationship. *Tourism Management* 25, 507–514.

Freathy, P. and O'Connell, F. (1998) *European Airport Retailing: Growth Strategies for the New Millennium*. London: Macmillan Business.

Freathy, P. and O'Connell, F. (1999) A typology of European airport retailing. *Service Industries Journal* 19, 119–134.

GAO (2000) Aviation and the environment: Aviations effects on the global atmosphere are potentially significant and expected to grow. GAO/RCED-00-57.

Gauthier, H.L. (1970) Geography, transportation and regional development. *Economic Geography* 46 (4), 612–619.

Georggi, N.L. and Pendyala, R.M. (2001) Analysis of long-distance travel behaviour of the elderly and low income. In *Personal Travel: The Long and Short of It, Conference Proceedings* (28 June–1 July 1999), Transportation Research Circular E-C026. Washington, DC: Transportation Research Board.

Gerber, P. (2002) Success factors for the privatisation of airports – an airline perspective. *Journal of Air Transport Management* 8, 29–36.

Geurs, K.T. and van Wee, B. (2004) Accessibility evaluation of land-use and transport strategies: review and research directions. *Journal of Transport Geography* 12, 127–140.

Gibb, R. (1994) The Channel Tunnel project: origins and development. In R. Gibb (ed.) *The Channel Tunnel: A Geographical Perspective* (pp. 1–30). Chichester: John Wiley & Sons.

Gilbert, D.C. (1996) Relationship marketing and airline loyalty schemes. *Tourism Management* 17 (8), 575–582.

Gilbert, D. and Wong, R.K.C. (2003) Passenger expectations and airline services: A Hong Kong based study. *Tourism Management* 24, 519–532.

Giuliano, G. (1997a) Family structure and travel demand. *Journal of Transport Geography* 5 (1), 43.

Giuliano, G. (1997b) Age and trip-making. *Journal of Transport Geography* 5 (1), 44.

Glisson, L.M., Cunningham, W.A., Harris, J.R. and Di Lorenzo-Aiss, J. (1996) Airline industry strategic alliances: marketing and policy implications. *International Journal of Physical Distribution & Logistics Management* 26 (3), 26–34.

Goetz, A.R. (2002) Deregulation, competition, and antitrust implications in the US airline industry. *Journal of Transport Geography* 10, 1–19.

Golledge, R.G. (1981) Misconceptions, misinterpretations, and misrepresentations of behavioural approaches in human geography. *Environment and Planning A* 13, 1315–1344.

González-Savignat, M. (2004) Competition in air transport: The case of the high speed train. *Journal of Transport Economics and Policy* 38 (1), 77–108.

Goodman, J. (1998) Untitled. *Cruise News* (email newsletter), 12 June, 1.

Goodrich, J.N. (2002) September 11, 2001 attack on America: A record of the immediate impacts and reactions in the USA travel and tourism industry. *Tourism Management* 23 (6), 573–580.

Gössling, S. and Hall, C.M. (2006) Conclusion: Wake up…this is serious. In S. Gössling and C.M. Hall (eds) *Tourism and Global Environmental Change: Ecological, Social, Economic and Political Interrelationships* (pp. 305–320). London: Routledge.

Gössling. S., Hansson, C.B., Hörstmeier, O. and Saggel, S. (2002) Ecological footprint analysis as a tool to assess tourism sustainability. *Ecological Economics* 43, 199–211.

Gössling, S., Peeters, P., Ceron, J.-P., Dubois, G., Patterson, T. and Richardson, R.B. (2005) The eco-efficiency of tourism. *Ecological Economics* 54, 417–434.

Gow, D. (2004) BA outbid for Heathrow slots. *Guardian*, 21 January 2004 (http://www.guardian.co.uk/print/0,3858,4840673-103676,00.html). Accessed 7 March 2004.

Graβl, H., Kokott, J., Kulessa, M., Luther, J., Nuscheler, F., Sauerborn, R., Schellnhuber, H.-J., Schubert, R. and Schulze, E.-D. (2003) *Climate Protection Strategies for the First Century: Kyoto and Beyond.* Special Report. Berlin: WBGU.

Graham, A. (2001) *Managing Airports: An International Perspective.* Oxford: Butterworth-Heinemann.

Graham, B. (1995) *Geography and Air Transport.* Chichester: John Wiley & Sons.

Graham, B. (1997) Regional airline services in the liberalized European Union single aviation market. *Journal of Air Transport Management* 3 (4), 227–238.

Griswold, A. (2002) Carnival Cruise Lines beefs up campaign buy: Cooper & Hayes ads get new tagline, target, to increase exposure. *ADWEEK New England Edition*, 4 November 2002, 39 (44), 3.

Greene, D.L. and Wegener, M. (1997) Sustainable transport. *Journal of Transport Geography* 5 (3), 177–190.

Gubbins, E.J. (2004) *Managing Transport Operations* (3rd edn). London: Kogan Page Limited.

Gunderson, A. (2005) Cruise lines set their sights on Asia. *New York Times*, 2 October 2005, 5.3.

Gursoy, D., Chen, M.-H. and Kim, H.J. (2005) The US airlines relative positioning based on attributes of service quality. *Tourism Management* 26, 57–67.

Gutiérrez, J. (2001) Location, economic potential and daily accessibility: An analysis of the accessibility impact of the high-speed line Madrid–Barcelona–French border. *Journal of Transport Geography* 9, 229–242.

Gutiérrez, J., González, R., and Gómez, G. (1996) The European high-speed train network. *Journal of Transport Geography* 6, 227–238.

Guzhva, V.S. and Pagiavlas, N. (2004) US Commercial airline performance after September 11, 2001: Decomposing the effect of the terrorist attack from macroeconomic influences. *Journal of Air Transport Management* 10, 327–332.

Hägerstrand, T. (1967) *Innovation Diffusion as a Spatial Process*. Chicago: University of Chicago Press.

Hägerstrand, T. (1970) What about people in regional science? *Papers of the Regional Science Association* 24, 7–21.

Haggett, P., Cliff, A.D. and Frey, A. (1977) *Locational Analysis in Human Geography* (2nd edn). London: Arnold.

Hall, D.R. (1993) Transport implications of tourism development. In D.R. Hall (ed.) *Transport and Economic Development in the New Central and Eastern Europe* (pp. 206–225). London: Belhaven Press.

Hall, D.R. (1999) Conceptualising tourism transport: Inequality and externality issues. *Journal of Transport Geography* 7, 181–188.

Hall, D.R. (2004) Transport and tourism: Equity and sustainability issues. In L. Lumsdon and S.J. Page (eds) *Tourism and Transport: Issues and Agenda for the New Millennium* (pp. 45–55). Amsterdam: Elsevier.

Hall, C.M. (2005) *Tourism: Rethinking the Social Science of Mobility*. Harlow: Pearson Education/Prentice Hall.

Hall, C.M. and Müller, D.K. (2004) (eds) *Tourism, Mobility and Second Homes: Between Elite Landscape and Common Ground*. Clevedon: Channel View.

Hall, C.M., Timothy, D.J. and Duval, D.T. (2003) (eds) *Safety and Security in Tourism: Relationships, Management, and Marketing*. Binghamton, NY: Haworth Hospitality Press.

Halsall, D.A. (1992) Transport for tourism and recreation. In B.S. Hoyle and R.D. Knowles (eds) *Modern Transport Geography* (pp. 155–177). London: Belhaven.

Halsall, D.A. (2001) Railway heritage and the tourist gaze: Stoomtram Hoorn-Medemblik. *Journal of Transport Geography* 9, 151–160.

Halseth, G. (2004) The cottage priviledge: Increasingly elite landscapes of second homes in Canada. In C.M. Hall and D.K. Müller (eds) *Tourism, Mobility and Second Homes: Between Elite Landscape and Common Ground* (pp. 35–54). Clevedon: Channel View.

Hanley, N., Shogren, J.P. and White, B. (1997) *Environmental Economics: In Theory and Practice*. New York: Oxford University Press.

Hanlon, P. (1999) *Global Airlines* (2nd edn). Oxford: Butterworth-Heinemann.

Hanlon, P. (2001) *Managing Airports: An International Perspective*. Oxford: Butterworth-Heinemann.

Hardy, A. (2003) An investigation into the key factors necessary for the development of iconic touring routes. *Journal of Vacation Marketing* 9 (4), 314–330.

Hartman, D.E. and Lindgren, J.H. Jr (1993) Consumer evaluations of goods and services: Implications for services marketing. *Journal of Services Marketing* 7 (2), 4–15.

Harvey, D. (1990) Between space and time: Reflections on the geography of imagination. *Annals of the Association of American Geographers* 80, 418–434.

Hayashi, Y., Button, K. and Nijkamp, P. (eds) (1999) *The Environment and Transport*. Cheltenham: Edward Elgar Publishing.

Haynes, K.E. and Fotheringham, A.S. (1984) *Gravity and Spatial Interaction Models*. Beverly Hills, CA: Sage.

Haynes, K.E., Gifford, J.L. and Pelletiere, D. (2005) Sustainable transportation institutions and regional evolution: Global and local perspectives. *Journal of Transport Geography* 13 (3), 207–221.

Heggie, I.G. (1969) Are gravity and interactance models a valid technique for planning regional transport facilities? *Operational Research* 20 (1), 93–110.

Hennig-Thurau, T. and Hansen, U. (2000) *Relationship Marketing: Gaining Competitive Advantage Through Customer Satisfaction and Customer Retention*. Berlin: Springer-Verlag.

Hensher, D.A. (1997) A practical approach to identifying the market potential for high-speed rail: A case study in the Sydney–Canberra Corridor. *Transportation Research A* 31 (6), 431–436.

Hillman, A.L. (2003) *Public Finance and Public Policy: Responsibilities and Limitations of Government*. Cambridge: Cambridge University Press.

Hindustan Times (2005) Super luxury cruises, Indian tourism circuits latest fad, 1 December 2005 (http://www.hindustantimes.com/news/181_1561808,00110002001 3.htm). Accessed 27 December 2005.

Hodge, D. (1997) Accessibility-related issues. *Journal of Transport Geography* 5 (1), 33–34.

Holden, E. and Høyer, K.G. (2005) The ecological footprint of fuels. *Transportation Research Part D* 10, 395–403.

Holloway, S. (2003) *Straight and Level: Practical Airline Economics*. Aldershot: Ashgate.

Hooper, P. (2002) Privatisation of airports in Asia. *Journal of Air Transport Management* 8, 289–300.

Hoover's Company Profiles (2004) National Railroad Passenger Corporation, 5 August 2004; accessed via Factiva.

Høyer, K.G. (2000) Sustainable tourism or sustainable mobility? The Norwegian case. *Journal of Sustainable Tourism* 8 (2), 147–160.

Hoyle, B. and Knowles, R. (1998) Transport geography: An introduction. In B. Hoyle and R. Knowles (eds) *Modern Transport Geography* (2nd edn) (pp. 1–12). Chichester: John Wiley and Sons.

Hoyle, B. and Smith, J. (1998) Transport and development: Conceptual frameworks. In B. Hoyle and R. Knowles (eds) *Modern Transport Geography* (2nd edn) (pp. 13–40). Chichester: John Wiley and Sons.

Hubbard, P., Kitchin, R. and Valentine, G. (2004) *Key Thinkers on Space and Place*. London: Sage.

Huff, D.L. and Jenks, G.F. (1968) A graphic interpretation of the friction of distance in gravity models. *Annals of the Association of American Geographers* 58 (4), 814–824.

Hwang, Y.-H., Gretzel, U. and Fesenmaier, D. (2002) Multi-city pleasure trip patterns: An analysis of international travelers to the US. In K. Wöber (ed.) *City Tourism 2002: Proceedings of European Cities Tourisms International Conference in Vienna, Austria, 2002*. Wien: Springer Economics.

IATA (2004) World Air Transport Statistics (48th edn) (for 2003).

IATA (2005) Passenger traffic growth slows, freight remains flat, 30 November 2005 (http://www.iata.org/pressroom/pr/2005-11-30-02.htm). Accessed 31 December 2005.

ICCL (2003) ICCL Environmental Standard E-01-01 (Revision 2), 12 December 2003 (http://www.iccl.org/policies/environmentalstandards.pdf). Accessed 18 October 2005.

IEA (2002) IEA reports on ways to achieve sustainability in urban transport (http://www.iea.org/dbtw-wpd/Textbase/press/pressdetail.asp?year=%25&keyword=4121&Submit=Submit&PRESS_REL_ID=65). Accessed 8 December 2005.

Independent (2005) Bush must not stand in the way of new Kyoto deal (http://news.independent.co.uk/environment/article331973.ece). Accessed 9 December 2005.

Iso-Ahola, S. (1980) *The Social Psychology of Leisure and Recreation*. Dubuque, IA: Wm. C. Brown Company.

Iso-Ahola, S. (1982) Toward a social psychological theory of tourism motivation: Rejoinder. *Annals of Tourism Research* 9 (2), 256–262.

ITA (2005) World & US International Arrivals & Receipts 1984–2004p (http://tinet.ita.doc.gov/outreachpages/inbound.world_us_intl_arrivals.html). Accessed 19 November 2005.

Jackson, P. (1989) *Maps of Meaning*. London: Unwin Hyman.

Jamaica Gleaner (2005) Hoteliers call for Caribbean airline, 12 November 2005 (http://www.jamaica-gleaner.com/gleaner/20051112/business/business1.html). Accessed 13 November 2005.

Janelle, D.G. (1969) Spatial reorganization: A model and concept. *Annals of the Association of American Geographers* 59 (2), 348–364.

Janic, M. (2000) An assessment of risk and safety in civil aviation. *Journal of Air Transport Management* 6, 43–50.

Jarach, D. (2002) The digitalisation of market relationship in the airline business: The impact and prospects of e-business. *Journal of Air Transport Management* 8, 115–120.

Johnson, D. (2002) Environmentally sustainable cruise tourism: A reality check. *Marine Policy* 26, 261–270.

Johnston, J. (2006) Late-night train plan to forge Glasgow–Edinburgh link, 6 August 2006 (http://www.sundayherald.com/57001). Accessed 6 August 2006.

Kangis, P. and O'Reilly, M.D. (2003) Strategies in a dynamic marketplace: A case study in the airline industry. *Journal of Business Research* 56, 105–111.

Katz, R. (2006) Carnival plunges into cruise market. *China Daily*, 8 March, 11.

Kaul, R.N. (1985) *Dynamics of Tourism: A Trilogy (Volume III: Transport and Marketing)*. New Delhi: Sterling.

Kayton, M. and Fried, W. (1997) *Avionics Navigation Systems*. New York: John Wiley & Sons.

Kennedy, G. (1996) Airlines new buzzword cuts out tickets – and high costs. *National Business Review*, 5 July 1996.

Kenyon, T.A., Valway, S.E., Ihle, W.W., Onorato, I.M. and Castro, K.G. (1996) Transmission of multidrugresistant Mycobacterium tuberculosis during a long airplane flight. *New England Journal of Medicine* 334, 933–938.

Kerin, R.A. and Peterson, R.A. (1998) *Strategic Marketing Problems: Cases and Comments* (8th edn). Upper Saddle River, NJ: Prentice Hall.

Kidokoro, Y. (2004) Cost-benefit analysis for transport networks: Theory and application. *Journal of Transport Economics and Policy* 38 (2), 275–307.

Knowles, R. (1998) Passenger rail privatization in Great Britain and its implications, especially for urban areas. *Journal of Transport Geography* 6 (2), 117–133.

Knowles, R. and Hall, D. (1998) Transport deregulation and privatization. In B. Hoyle and R. Knowles (eds) *Modern Transport Geography* (2nd edn) (pp. 75–96). Chichester: John Wiley and Sons.

Kraft, R.M., Ballantine, J. and Garvey, D.E. (1994) Study abroad or international travel? The case of semester at sea. *Phi Beta Delta International Review* 4, 23–61.

Kremarik, F. (2002) A little place in the country: A profile of Canadians who own vacation property. *Canadian Social Trends* 65, 12–14.

Krippendorf, J. (1987) *The Holiday Makers: Understanding the Impact of Leisure and Travel*. London: Butterworth-Heinemann.

Lafferty, G. and van Fossen, A. (2001) Integrating the tourism industry: Problems and strategies. *Tourism Management* 22, 11–19.

Landon, M. (1997) *Cruise Ship Crews: The Real Truth About Cruise Ship Jobs*. London: Mark Landon.

Lanzendorf, M. (2000) Social change and leisure mobility. *World Transport Policy and Practice* 6 (3), 21–25.

Las Vegas Sun (2004) High-speed Vegas-LA train pitched to public, 22 June 2005 (http://www.lasvegassun.com/sunbin/stories/text/2004/jun/22/517058927.html). Accessed 25 June 2004.

Laws, E. (1997) *Managing Package Tourism.* London: International Business Press.

Laws, E. and Scott, N. (2003) Developing new tourism services: Dinosaurs, a new drive tourism resource for remote regions? *Journal of Vacation Marketing* 9 (4), 368–380.

Lester, J.-A. and Weeden, C. (2004) Stakeholders, the natural environment and the future of Caribbean cruise tourism. *International Journal of Tourism Research* 6, 39–50.

Levine, M.E. (1987) Airline competition in deregulated markets: Theory, firm strategy, and public policy. *Yale Journal on Regulation* 29, 393–494.

The Local (2006) EU gives negative signals over flight tax, 6 July 2006 (http://www.the-local.se/article.php?ID=4270&date=20060706). Accessed 3 August 2006.

Lois, P., Wang, J., Wall, A. and Ruxton, T. (2004) Formal safety assessment of cruise ships. *Tourism Management* 25, 93–109.

Long, M.M. and Schiffman, L.G. (2000) Consumption values and relationships: Segmenting the market for frequency programs. *Journal of Consumer Marketing* 17 (3), 214–232.

Long, M.M., Clark, S.D., Schiffman, L.G. and McMellon, C. (2003) In the air again: Frequent flyer relationship programmes and business travelers quality of life. *International Journal of Tourism Research* 5, 421–432.

Lösch, A. (1954) *The Economics of Location.* New Haven: Yale University Press.

Lowe, J.C. and Moryadas, S. (1975) *The Geography of Movement.* Atlanta: Houghton Mifflin Company.

Lozano, J.A. (2006) Residents concerned about environmental impact of spaceport, Star Telegram, 11 July 2006 (http://www.dfw.com/mld/dfw/news/state/15017006.htm). Accessed 27 July 2006.

Lumdson, L. (2000) Transport and tourism: Cycle tourism – A model for sustainable development? *Journal of Sustainable Tourism* 8 (5), 361–377.

Lumsdon, L. and Page, S.J. (2004) Progress in transport and tourism research: Reformulating the transport–tourism interface and future research agendas. In L. Lumsdon and S.J. Page (eds) *Tourism and Transport: Issues and Agenda for the New Millennium* (pp. 1–28). Amsterdam: Elsevier.

Lumdson, L. and Tolley, R. (2001) The National Cycle Strategy in the UK: To what extent have local authorities adopted its model strategy approach? *Journal of Transport Geography* 9, 293–301.

Lumdson, L., Downward, P. and Cope, A. (2004) Monitoring of cycle tourism on long distance trails: The North Sea Cycle Route. *Journal of Transport Geography* 12, 13–22.

McAdam, D. (1999) The value and scope of geographical information systems in tourism management. *Journal of Sustainable Tourism* 7 (1), 77–92.

McCartney, S. (2005) As airports try to add runways, many hurdles loom, *Wall Street Journal,* 31 August 2005 (http://www.post-gazette.com/pg/05243/563083.stm). Accessed 1 September 2005.

McHardy, J. and Trotter, S. (2006) Competition and deregulation: Do air passengers get the benefits? *Transport Research Part A* 40, 74–93.

McKercher, B. and Chon, K. (2004) The over-reaction to SARS and the collapse of Asian tourism. *Annals of Tourism Research* 31 (3), 716–719.

McKercher, B. and Lew, A. (2003) Distance decay and the impact of Effective Tourism Exclusion Zones on international travel flows. *Journal of Travel Research* 42, 159–165.

McKercher, B. and Lew, A. (2004) Tourist flows and the spatial distribution of tourists. In A. Lew, C.M. Hall and A.M. Williams (eds) *A Companion to Tourism* (pp. 36–48). Oxford: Blackwell.

Mahapatra, R. (2005) India must increase airport capacity. Associated Press, 18 October 2005. (www.usatoday.com/travel/news/2005-10-18-india-airports_x.htm). Accessed 1 December 2005.

Manorama Online (2005) Aviation 2005: Sky is not the limit, 27 December 2005 (http://www.manoramaonline.com/servlet/ContentServer?pagename=manorama/MmArticle/CommonFullStory&cid=1135260645119&c=MmArticle&p=1002194839100&count=10&colid=1002258272843&channel=News). Accessed 27 December 2005.

Marsh, J. and Staple, S. (1995) Cruise tourism in the Canadian Arctic and its implications. In C.M. Hall and M.E. Johnston (eds) *Polar Tourism: Tourism in the Arctic and Antarctic Regions* (pp. 64–72). Chichester: John Wiley & Sons.

Marti, B. (2005) Cruise line logo recognition. *Journal of Travel & Tourism Marketing* 18 (1), 25–31.

Martin, B.V., Memmott, F.W. and Bone, A.J. (1961) *Principles and Techniques of Predicting Future Demand for Urban Area Transportation*. Cambridge, MA: MIT Press.

Mason, K.J. (2000) The propensity of business travellers to use low cost airlines. *Journal of Transport Geography* 8, 107–119.

Mason, K.J. (2001) Marketing low-cost airline services to business travellers. *Journal of Air Transport Management* 7, 103–109.

Mason, P., Grabowski, P. and Du, W. (2005) Severe acute respiratory syndrome, tourism and the media. *International Journal of Tourism Research* 7, 11–21.

Meethan, K. (2001) *Tourism in Global Society: Place, Culture, Consumption*. Basingstoke: Palgrave.

Mercury News (2005) LA airport officials race to develop bird flu quarantine plan, 18 October 2005 (http://www.mercurynews.com/mld/mercurynews/news/local/states/california/northern_california/12930857.htm). Accessed 19 October 2005.

Middleton, V.T.C. and Clarke, J. (2001) *Marketing in Travel and Tourism* (3rd edn). Oxford: Butterworth-Heinemann.

Mill, R.C. and Morrison, A. (1985) *The Tourism System: An Introductory Text*. Englewood Cliffs, NJ: Prentice Hall.

Miller, A.R. and Grazer, W.F. (2002) The North American cruise market and Australian tourism. *Journal of Vacation Marketing* 8 (3), 221–234.

Millimet, D.L. and Slottje, D. (2002) Environmental compliance costs and the distribution of emissions in the US. *Journal of Regional Science* 42 (1), 87–105.

Milmo, D. (2006) Branson asks Brown to cut duty on fuel to power green trains, 16 October 2006 (http://business.guardian.co.uk/story/0,,1923153,00.html). Accessed 21 October 2006.

Minder, R. (2006) Airlines face peak-hour congestion charges, *Financial Times*, 5 April 2006 (http://news.ft.com/cms/s/9cfda466-c4c3-11da-b7c1-0000779e2340,_i_rss Page=1dffe558-c989-11d7-81c6-0820abe49a01.html). Accessed 6 April 2006.

Morgan, N. and Pritchard, A. (2001) *Advertising in Tourism and Leisure*. Oxford: Butterworth-Heinemann.

Morrison, S.A. and Winston, C. (1989) Airline deregulation and public policy. *Science (New Series)* 245 (4919), 707–711.

Mortishead, C. (2006) Carbon tax on airlines would never fly. Times Online (UK), 21 June 2006 (http://business.timesonline.co.uk/article/0,,13130-2234987,00.html). Accessed 6 July 2006.

Motevalli, V. and Stough, R. (2004) Aviation safety and security: Reaching beyond borders. *Journal of Air Transport Management* 10, 225–226.

Mowforth, M. and Munt, I. (1998) *Tourism and Sustainability: New Tourism in the Third World*. London: Routledge.

Müller, D.K. (2002a) Reinventing the countryside. German second home owners in southern Sweden. *Current Issues in Tourism* 5 (5), 426–446.

Müller, D.K. (2002b) Second home ownership and sustainable development in northern Sweden. *Tourism and Hospitality Research: The Surrey Quarterlt Review* 3 (4), 345–355.

Mulrine, A. (2002) A grand land cruise. *US News and World Report*, 21 October 2002, pd2.

National Journal (2004) Working on the railroad. 17 July 2004.

Nayar, B.R. (1995) Regimes, power, and international aviation. *International Organization* 49 (1), 139–170.

New Zealand Herald (2003) Cruise industry expects fall, 20 November 2003 (http://www.nzherald.co.nz/storyprint.cfm?storyID=3535087). Accessed 14 February 2004.

New Zealand Herald (2004) Air NZ plans San Francisco venture, 20 January 2004 (http://www.nzherald.co.nz/storyprint.cfm?storyID=3544434). Accessed 14 February 2004.

New Zealand Herald (2005) Airways profits dips slightly (http://www.nzherald.co.nz/section/story.cfm?c_id=3&ObjectID=10351386). Accessed 26 October 2005.

Niemeier, D., Redmond, L., Morey, J., Hicks, J., Hendren, P., Lin, J., Foresman, E. and Zheng, Y. (2001) Redefining conventional wisdom: Exploration of automobile ownership and travel behavior in the United States. Transportation Research Circular E-C026, 207–219 (http://wwwcf.fhwa.dot.gov/exit.cfm?link=http://trb.org/trb/publications/ec026/07_niemeier.pdf). Accessed 5 March 2004.

Nutley, S. (1998) Rural areas: The accessibility problem. In B.S. Hoyle and R.D. Knowles (eds) *Modern Transport Geography* (pp. 185–215). London: Belhaven.

Ocean Conservancy (2002) Cruise control: A report on how cruise ships affect the marine environment (http://www.oceanconservancy.org/site/DocServer/cruisecontrol.pdf?docID=141). Accessed 18 October 2005.

O'Connell, J.F. and Williams, G. (2005) Passengers perceptions of low cost airlines and full service carriers: A case study involving Ryanair, Aer Lingus, Air Asia and Malaysia Airlines. *Journal of Air Transport Management* 11, 259–272.

O'Connor, W.E. (1978) *An Introduction to Airline Economics*. New York: Preager Publishers.

OECD (1999) OECD workshop on regulatory reform in international air cargo transportation, Paris, 5–9 July 1999 (www.oecd.org/dataoecd/1/28/1821288.pdf). Accessed 27 November 2005.

OECD (2002) *International Mobility of the Highly Skilled*. Paris: OECD.

O'Kelly, M.E. (1998) A geographer's analysis of hub-and-spoke networks. *Journal of Transport Geography* 6 (3), 171–186.

Olsen, M. (2003) Themed tourism routes: A Queensland perspective. *Journal of Vacation Marketing* 9 (4), 331–341.

Olsthoorn, X. (2001) Carbon dioxide emissions from international aviation: 1950–2050. *Journal of Air Transport Management* 7, 87–93.

Onkvisit, S. and Shaw, J.J. (1991) Is services marketing really different? *Journal of Professional Services Marketing* 7 (2), 3–17.

Oppermann, M. (1995) A model of travel itineraries. *Journal of Travel Research* 33, 57–61.

Ortuzar, J. De Dios and Willumsen, L.G. (2001) *Modelling Transport* (3rd edn). Chichester: Wiley.

Otago Rail Trail (2004) Your accommodation and transport guide to the Rail Trail (http://www.otagorailtrail.co.nz/). Accessed 9th May 2004.

O'Toole, P. (2002) IT Trends Survey 2002. *Airline Business*, August 2002.

Oum, T.H. and Park, J.-H. (1997) Airline alliances: Current status, policy issues, and future directions. *Journal of Air Transport Management* 3 (3), 133–144.

Oum, T.H., Park, J.-H. and Zhang, A. (1996) The effects of airline codesharing agreements on firm conduct and international airfare. *Journal of Transport Economics and Policy* 30 (2), 187–202.

Oum, T.H., Yu, C. and Fu, X. (2003) A comparative analysis of productivity performance of the worlds major airports: Summary report of the ATRS global airport benchmarking research report 2002. *Journal of Air Transport Management* 9, 285–297.

Oum, T.H., Park, J.-H., Kim, K. and Yu, C. (2004) The effect of horizontal alliances on firm productivity and profitability: Evidence from the global airline industry. *Journal of Business Research* 57, 844–853.

Page, S.J. (1993) European rail travel. *Travel and Tourism Analyst* 1, 19–39.

Page, S.J. (1999) *Transport and Tourism*. Harlow: Prentice Hall.

Page, S.J. and Hall, C.M. (2003) *Managing Urban Tourism*. Harlow: Prentice Hall.

315

Palhares, G.L. (2003) The role of transport in tourism development: Nodal functions and management practices. *International Journal of Tourism Research* 5, 403–407.

Papatheodorou, A. (2002) Civil aviation regimes and leisure tourism in Europe. *Journal of Air Transport Management* 8, 381–388.

Parasuraman, A., Berry, L.L. and Zeithaml, V.A. (1991) Refinement and reassessment of the SERVQUAL scale. *Journal of Retailing* 67 (4), 420–450.

Parasuraman, A., Zeithaml, V.A. and Berry, L.L. (1985) A conceptual model of service quality and its implications for future research. *Journal of Marketing* 49 (4), 41–50.

Parasuraman, A., Zeithaml, V.A. and Berry, L.L. (1988) SERVQUAL: A multiple-item scale for measuring consumer perceptions of service quality. *Journal of Retailing* 64 (1), 12–40.

Park, J.-W., Robertson, R. and Wu, C.-L. (2004) 'The effect of airline service quality on passengers' behavioural intentions: A Korean case study. *Journal of Air Transport Management* 10, 435–439.

Parkes, D. and Wallis, W.D. (1978) Graph theory and the study of activity structure. In T. Carlstein, D. Parkes and N. Thrift (eds) *Human Activity and Time Geography* (pp. 75–99). London: Edward Arnold.

PATA (2005) *Annual Statistical Report 2004*.

Paylor, A. (2005) The slots game. *Air Transport World*, April 2005, 52.

PE.com (2005) Hawai'ian cruise industry booming, 3 December 2005 (http://www.pe.com/business/local/stories/PE_Biz_D_cruisebiz04.2921837.html). Accessed 12 December 2005.

Pearce, D.G. (2001a) Tourism and urban land use change: Assessing the impact of Christchurch's tourist tramway. *Tourism and Hospitality Research* 3 (2), 132–148.

Pearce, D.G. (2001b) Tourism, trams and local government policy-making in Christchurch, New Zealand. *Current Issues in Tourism* 4 (2–4), 331–354.

Pearson, C.S. (2000) *Economics and the Global Environment*. Cambridge: Cambridge University Press.

Peattie, K. and Peattie, S. (1996) Promotional competitions: A winning tool for tourism marketing. *Tourism Management* 17 (6), 433–442.

Peck, H., Payne, A., Christopher, M. and Clark, M. (1999) *Relationship Marketing: Strategy and Implementation*. Oxford: Butterworth-Heinemann.

Peisley, T. (1995) The North American cruise market. *Travel & Tourism Analysis*. London: Travel and Tourism Intelligence.

Pels, E., Nijkamp, P. and Rietveld, P. (2000) A note on the optimality of airline networks. *Economics Letters* 69, 429–434.

Pender, L. (1999) European aviation: The emergence of franchised airline operations. *Tourism Management* 20, 565–574.

People's Daily Online (2005) Number of Chinese airline passengers to be world's no. 2, 17 October 2005 (http://english.people.com.cn/200510/17/eng20051017_214812.html). Accessed 18 October 2005.

Pigou, A.C. (1920) *The Economics of Welfare*. London: Macmillan.

Plakhotnik, V.N., Onyshchenko, Ju. V. and Yaryshkina, L.A. (2005) The environmental impacts of railway transportation in the Ukraine. *Transportation Research Part D* 10, 263–268.

Plassard, F. (1992) Limpact territorial des transports a grande vitesse. In P.-H. Derycke (ed) *Espace et dinamiques territoriales* (pp. 243–261). Paris: Economia.

Preston, J. (2001) Integrating transport with socio-economic activity – A research agenda for the new millennium. *Journal of Transport Geography* 9, 13–24.

Prideaux, B. (2000a) The role of the transport system in tourism development. *Tourism Management* 21, 53–63.

Prideaux, B. (2000b) Links between transport and tourism – Past, present and future. In B. Faulkner, G. Moscardo and E. Laws (eds) *Tourism in the Twenty-first Century: Reflects on Experience* (pp. 91–109). London: Continuum.

Prideaux, B. (2004) Transport and destination development. In L. Lumsdon and S.J. Page (eds) *Tourism and Transport: Issues and Agenda for the New Millennium* (pp. 79–92). Amsterdam: Elsevier.

Prideaux, B. and Carson, D. (2003) A framework for increasing understanding of self-drive tourism markets. *Journal of Vacation Marketing* 9 (4), 307–313.

Priemus, H. and Konings, R. (2001) Light rail in urban regions: What Dutch policymakers could learn from experiences in France, Germany and Japan. *Journal of Transport Geography* 9, 187–198.

Qu, H. and Ping, E.W.Y. (1999) A service performance model of Hong Kong cruise travelers motivation factors and satisfaction. *Tourism Management* 20, 237–244.

Quinet, E. and Vickerman, R. (2004) *Principles of Transport Economics*. Cheltenham: Edward Elgar.

Rauscher, M. (1997) *International Trade, Factor Movements, and the Environment*. Oxford: Oxford University Press.

RCEP (2002) The environmental impacts of civil aircraft in flight (http://www.rcep.org.uk/avreport.htm). Accessed 12 July 2004.

Reuters (2005) Landing in Alaska? Fear moose collisions no more, 21 October 2005 (http://today.reuters.com/news/newsArticle.aspx?type=oddlyEnoughNews&storyID=2005-10-21T151442Z_01_ARM154815_RTRUKOC_0_US-MOOSE.xml&archived=False). Accessed 29 October 2005.

Reynolds-Feighan, A. (2001) Traffic distribution in low-cost and full-service carrier networks in the US air transportation market. *Journal of Air Transport Management* 7, 265–275.

Rhoades, D.L. and Waguespack Jr, B. (2000) Judging a book by its cover: The relationship between service and safety quality in US national and regional airlines. *Journal of Air Transport Management* 6, 87–94.

Rietveld, P. and Brons, M. (2001) Quality of hub-and-spoke networks: The effects of timetable co-ordination on waiting time and rescheduling time. *Journal of Air Transport Management* 7, 241–249.

Ritchie, B.W. (1998) Bicycle tourism in the South Island of New Zealand: Planning and management issues. *Tourism Management* 19 (6), 567–582.

Roberts, M.L. and Berger, P.D. (1999) *Direct Marketing Management.* Upper Saddle River, NJ: Prentice Hall, Inc.

Robinson, R. and Kearney, T. (1994) Database marketing for competitive advantage in the airline industry. *Journal of Travel and Tourism Marketing* 3 (1), 65–81.

Rodrigue, J.P. *et al.* (2006) The geography of transport systems, Hofstra University, Department of Economics and Geography (http://people.hofstra.edu/geotrans). Accessed 20 March 2007.

Ross, N. (1996) Otago central rail: Who's using it? Dissertation, Postgraduate Diploma in Tourism, University of Otago, Dunedin, New Zealand.

Ross, W. (2000) Mobility & accessibility: the yin & yang of planning. *World Transport Policy and Practice* 6 (2), 13–19.

Rowley, J. and Slack, F. (1999) The retail experience in airport departure lounges: Reaching for timelessness and placelessness. *International Marketing Review* 16 (45), 36–37.

RTE News (2005) Overturn data sharing law, says EU law officer, 22 November 2005 (http://www.rte.ie/news/2005/1122/eu.html?rss). Accessed 22 November 2005.

Ryan, C. and Trauer, B. (2004) Aging populations: Trends and the emergence of the nomad tourist. In W. Theobald (ed.) *Global Tourism: The Next Decade* (3rd edn) (pp. 510–528). Butterworth-Heinemann.

Ryan, C. and Trauer, B. (2003) Involvement in adventure tourism: Toward implementing a fuzzy set. *Tourism Review International* 7 (3–4), 143–152.

Sala, O.E., Chapin III, F.S., Armesto, J.J., Berlow, E., Bloomfield, J. *et al.* (2000) Global biodiversity scenarios for the year 2100. *Science* 287, 1770– 1774.

San Mateo County Times (2004) Bullet train system will lead the future. 28 January 2004.

Sarna, H. and Hannafin, M. (2003) *Frommer's Caribbean Cruises & Ports of Call.* Hoboken, NJ: Wiley Publishing, Inc.

Savage, I. and Scott, B. (2004) Deploying regional jets to add new spokes to a hub. *Journal of Air Transport Management* 10, 147–150.

Schulkin, A. (2002) Safe harbors: Crafting an international solution to cruise ship pollution. *Georgetown International Environmental Law Review* 15 (1), 105–132.

Scott, D.M., Novak, D.C., Aultman-Hall, L. and Guo, F. (2006) Network robustness index: A new method for identifying critical links and evaluating the performance of transportation networks. *Journal of Transport Geography* 14, 215–227.

Scull, T. (1996) Ship lives up to P&O's standards. *Travel Weekly*, 5 December, C3.

Seaton, A.V. and Bennett, M.M. (1996) *The Marketing of Tourism: Concepts, Issues and Cases.* London: International Thomson Business Press.

Seinfeld, J.H. (1998) Clouds, contrails and climate. *Nature* 391 (26), 837–838.

Sharpley, R. (2004) Tourism and the countryside. In A.A. Lew, C.M. Hall and A.M. Williams (eds) *A Companion to Tourism* (pp. 374–386). Oxford: Blackwell Publishing.

Shaw, J. and Farrington, J. (2003) A railway renaissance? In I. Docherty and J. Shaw (eds) *A New Deal for Transport?* (pp. 108–134). Oxford: Blackwell.

Shaw, S. (2004) *Airline Marketing and Management* (5th edn). Aldershot: Ashgate.

Sheth, J.N. (1975) A psychological model of travel mode selection. In *Advances in Consumer Research* (Volume 3) (pp. 425–430) (Proceedings of the Association of Consumer Research, Sixth Annual Conference).

Sinclair, M.T. and Stabler, M. (1997) *The Economics of Tourism.* London: Routledge.

SITA (2006) Airport IT trends survey (http://www.sita.aero/News_Centre/Airport_IT_Trends/webconference_launches_Airport_IT_Trends_Survey.htm). Accessed 29 July 2006.

Slater, S. and Basch, H. (1989) Carnival buys up Holland America. *Los Angeles Times*, 12 February, 7.

Song, J. (2006) A new Hawaiian airline sparks inter-island fare war with $39 one-way flights, *The Seattle Times*, 24 March 2006 (http://seattletimes.nwsource.com/html/traveloutdoors/2002887504_webhawaiiair24.html). Accessed 29 March 2006.

Smith, M. (2002) Caribbean mulls merging airlines. Associated Press Online, 12 November 2002.

Staniland, M. (1998) The vanishing national airline? *European Business Journal* 10 (2), 71–77.

Star Bulletin (Hawai'i) (2006) State funds obligate ferry to heed concerns (Editorial), 14 April 2006 (http://starbulletin.com/2006/04/14/editorial/editorial02.html). Accessed 24 April 2006.

Starkie, D. (1998) Allocating airport slots: A role for the market? *Journal of Air Transport Management* 4, 111–116.

Starkie, D. (2002) Airport regulation and competition. *Journal of Air Transport Management* 8, 63–72.

Starkie, D. (2003) The economics of secondary markets for airport slots. In K. Boyfield (ed.) *A Market in Airport Slots* (pp. 51–79). London: The Institute of Economic Affairs.

Statistics New Zealand (2002) Tourism and Migration 2001, Table 3.01 (http://www.stats.govt.nz/domino/external/web/prod_serv.nsf/htmldocs/Tourism+and+Migration+2001). Accessed 25 April 2003.

Statistics Norway (2005) Substantial growth in water transport (http://www.ssb.no/english/subjects/10/12/stranskom_en/). Accessed 18 May 2006.

Stevens, A. (2005) Air corridor: Mozambique's channel to competition. *Airways*, August 2005, 28–33.

St Paul Pioneer Press (2005) Loss of airline hub can be a boon for travelers, 6 November 2005 (http://www.twincities.com/mld/twincities/business/13084956.htm). Accessed 7 November 2005.

Suzuki, Y., and Tyworth, J.E. (1998) A theoretical framework for modeling sales-service relationships in the transportation industry. *Transportation Research Part E* 34, 87–100.

Suzuki, Y., Tyworth, J.E. and Novack, R.A. (2001) Airline market share and customer service quality: A reference-dependent model. *Transportation Research Part A* 35, 773–788.

Svenson, S. (2004) The cottage and the city: An interpretation of the Canadian second home experience. In C.M. Hall and D.K. Müller (eds) *Tourism, Mobility and Second Homes: Between Elite Landscape and Common Ground* (pp. 55–74). Clevedon: Channel View.

Sykes, L. (1997) Watch a whale. *Geographical*, September 1997.

Szodruch, J. (2001) Aircraft drag reductions as an answer to global challenges. *Air & Space Europe* 3 (3/4), 93–97.

Taaffe, E.J., Gauthier, H.L. and O'Kelly, M.E. (1996) *Geography of Transportation* (2nd edn). Upper Saddle River, NJ: Prentice Hall.

Tagami, K. (2005) Whats it mean for the airport?, 13 September 2005 (www.ajc.com/news/content/business/delta/0905/deltaharts.html). Accessed 28 September 2005.

Tam, M.-L. and Lam, W.H.K. (2004) Determination of service levels for passenger orientation in Hong Kong International Airport. *Journal of Air Transport Management* 10, 181–189.

Taneja, N.K. (2003) *Airline Survival Kit: Breaking Out of the Zero Profit Game*. Aldershot: Ashgate.

Tarry, C. (2004) The difficult part is yet to come: Profit rather than traffic alone remains they key to airline prosperity. *Tourism and Hospitality Research* 5 (1), 79–83.

Teye, V. and Leclerc, D. (1998) Produce and service delivery satisfaction among North American cruise passengers. *Tourism Management* 19 (2), 153–160.

Times Record (2004) Fate of excursion train remains to be decided, 2 September 2004 (http://www.swtimes.com/archive/2004/September/02/business/fate.html). Accessed 6 September 2004.

This Is Money (2005) Budget airline growth continues, 10 October 2005 (http://www.thisismoney.co.uk/money-savers/article.html?in_article_id=404245&in_page_id=5). Accessed 11 October 2005.

Thomas, C.D., Cameron, A., Green, R.E., Bakkenes, M., Beaumont, L.J. *et al.* (2004) Extinction risk from climate change. *Nature* 427, 145–148.

Thorton, P.R., Shaw, G. and Williams, A.M. (1997) Tourist group holiday decision-making and behaviour: The influence of children. *Tourism Management* 18 (5), 287–298.

Times Online (2005) Cornwall flights cut by Ryanair, 31 August 2005 (http://travel.times-online.co.uk/article/0,,10295-1758709,00.html)

Tomkins, R. (1994) Six airlines agree not to use 'price fixing system'. *Financial Times*, 19 March.

Tourism Scotland (2003) Tourism in Scotland 2003 (www.scotexchange.net/tourism_in_scotland_2003.pdf). Accessed 6 December 2005.

Tran, M. (2004) Q&A: Eurotunnel. *Guardian*, 7 April 2004 (http://www.guardian.co.uk/print/0,3858,4897368-103630,00.html). Accessed 22 April 2004.

Transport Canada (2004) *Creating a Transportation Blueprint for the Next Decade and Beyond: Defining the Challenges* (http://www.tc.gc.ca/aboutus/straightahead/challenges/menu.htm). Accessed 14 October 2004.

Transport Canada (2001) Transportation in Canada, Annual Report 2001 (http://www.tc.gc.ca/pol/en/Report/anre2001/tc0100ae.htm). Accessed 6 December 2005.

Travel and Tourism Analyst (2004) Cruises: North America and the Caribbean, No. 9, June 2004.

Travel Daily News (2005) The skies open up over India (http://www.traveldailynews.com/makeof.asp?central_id=894&permanent_id=33). Accessed 9 November 2005.

Travel Trade Gazetta Europa (1998) Agents must by ready for urgent action over fees. 9 April 1998, 9.

Tretheway, M. and Mak, D. (2006) Emerging tourism markets: Ageing and developing economies. *Journal of Air Transport Management* 12 (1), 21–27.

Tretheway, M.W. and Oum, T.H. (1992) *Airline Economics: Foundation for Strategy and Policy*. University of British Columbia: The Centre for Transportation Studies.

Tretheway, M.W. and Waters, W.G. II (1998) Reregulation of the airline industry: Could price cap regulation play a role? *Journal of Air Transport Management* 4, 47–53.

Tripp, C. and Drea, J.T. (2002) Selecting and promoting service encounter elements in passenger rail transportation. *Journal of Services Marketing* 16 (5), 432–442.

Trivedi, C. (2003) Mixed views on Heathrow growth, BBC News, 16 December 2003 (http://news.bbc.co.uk/go/pr/fr/-/1/hi/england/london/3324710.htm). Accessed 10 September 2005.

Tsaur, S.-H., Chang, T.-Y. and Yen, C.-H. (2002) The evaluation of airline service quality by fuzzy MCDM. *Tourism Management* 23, 107–115.

Tucker, H. (2002) Welcome to Flintstones-Land: Contesting place and identity in Goreme, Central Turkey. In S. Coleman and M. Crang (eds) *Tourism: Between Place and Performance* (pp. 143–159). New York: Berghahn Books.

Turner, R.K., Pearce, D. and Bateman, I. (1993) *Environmental Economics: An Elementary Introduction*. Baltimore: The Johns Hopkins University Press.

Turnock, D. (2001) Railways and economic development in Romania before 1918. *Journal of Transport Geography* 9, 137–150.

Turton, B. (2004) Airlines and tourism development: The case of Zimbabwe. In L. Lumsdon and S.J. Page (eds) *Tourism and Transport: Issues and Agenda for the New Millennium* (pp. 69–78). Amsterdam: Elsevier.

Turton, B. and Black, W.R. (1998) Inter-urban transport. In B. Hoyle and R. Knowles (eds) *Modern Transport Geography* (2nd edn). Chichester: John Wiley and Sons.

Turton, B.J. and Mutambirwa, C.C. (1996) Air transport services and the expansion of international tourism in Zimbabwe. *Tourism Management* 17 (6), 453–462.

UK National Statistics (2005) (www.statistics.gov.uk/STATBASE/Expodata/Spread sheets/D8066.xls). Accessed 12 November 2005.

United States DoT (2003) Strategic Plan, 2003–2008: Safer, simpler, smarter transportation solutions (http://www.dot.gov/stratplan2008/strategic_plan.htm). Accessed 3 September 2004.

Unnrikrishnan, M. (2005) Indias aviation market surges on strong domestic and international demand, *Aviation Daily*, 3 November 2005 (http://aviationnow.com/avnow/news/channel_aviationdaily_story.jsp?id=news/INDINT.xml). Accessed 21 December 2005.

UNWTO (2005a) *Tourism Highlights, 2005 Edition*. World Tourism Organization.

UNWTO (2005b) High oil prices yet to impact on tourism says WTO, 14 November 2005 (http://www.world-tourism.org/newsroom/Releases/2005/november/oil.htm). Accessed 14 November 2005.

UNWTO (2006) International tourism up 4.5% in the first four months of 2006, 28 June 2006 (http://www.unwto.org/newsroom/Releases/2006/june/barometer.html). Accessed 29 July 2006.

Urquhart, C. (2005) Why I don't like Ryan Air. *The Times Online*, 17 September 2005 (http://business.timesonline.co.uk/article/0,,9077-1782464,00.html). Accessed 17 September 2005.

Urry, J. (1995) *Consuming Places*. London: Routledge.

van der Knaap, W. (1999) GIS oriented analysis of tourist time–space patterns to support sustainable tourism development. *Tourism Geographies* 1 (1), 56–69.

Van der Pool, L. (2004) Royal Caribbean battles winter blues. *ADWEEK Southeast*, 28 December 2004.

van Wee, B., Hagoort, M. and Annema, J.A. (2001) Accessibility measures with competition. *Journal of Transport Geography* 9, 199–208.

Vlek, S. and Vogels, M. (2000) AERO – Aviation emissions and evaluation of reduction options. *Air & Space Europe* 2 (3), 41–44.

Vowles, T.M. (2000) The effect of low fare air carriers on airfares in the US. *Journal of Transport Geography* 8, 121–128.

Wagner, J.E. (1997) Estimating the economic impacts of tourism. *Annals of Tourism Research* 24, 592–606.

Walters, T. (2005) Tourism has real potential in 2006, 19 December 2005 (http://www.theroyalgazette.com/apps/pbcs.dll/article?AID=/20051220/NEWS/112200174). Accessed 27 December 2005.

Walton, W. (2003) Roads and traffic congestion policies: One step forward, two steps back. In I. Docherty and J. Shaw (eds) *A New Deal for Transport?* (pp. 75–107). Oxford: Blackwell.

Wang, D. (2001) Impacts of institutional policies on individuals participation in non-work activities. *Journal of Transport Geography* 9, 61–74.

Ward, D. (2005) *Ocean Cruising and Cruise Ships 2005* (15th edn). London: Berlitz Publishing.

Wardell, J. (2005) Virgin spaceport to be built in NM, 13 December 2005 (http://news.yahoo.com/s/ap/20051213/ap_on_sc/britain_space_tourism&printer=1;_ylt=A9FJqZP8fp9Ds4UBGhRxieAA;_ylu=X3oDMTA3MXN1bHE0BHNlYwN0bWE-). Accessed 14 December 2005.

Washington Times (2004a) Traveling the rails to picturesque sites, 2 September 2004 (http://www.washingtontimes.com/weekend/20040901-100727-2446r.htm). Accessed 6 September 2004.

Washington Times (2004b) Russian railways plans for future, 31 May 2004 (http://www.washtimes.com/upi-breaking/20040531-094033-5911r.htm). Accessed 4 June 2004.

Wastnage, J. (2004) EC considers airlines and airports in new emissions trading scheme. *Flight International* 9–15 November: 10.

Weaver, A. (2005) The McDonalization thesis and cruise tourism. *Annals of Tourism Research* 32 (2), 346–366.

Weaver, D. (2005) The distinctive dynamics of exurban tourism. *International Journal of Tourism Research* 7, 23–33.

Weaver, A. and Duval, D.T. (in press) International and transnational aspects of the global cruise industry. In C.M. Hall and T.E. Coles (eds) *Tourism and International Business*. London: Routledge.

Webster, B. (2005) Aircraft emissions to double by 2030 despite hi-tech jets, 21 June 2005 (http://www.timesonline.co.uk/article/0,,2-1662662,00.html). Accessed 15 November 2005.

Webwire (2005) VIA Rail exceeds Kyoto targets, 12 April 2005 (http://webwire.com/ViewPressRel.asp?SESSIONID=&aId=6440). Accessed 6 December 2005.

Weinstein, A. (1987) *Market Segmentation: Using Niche Marketing to Exploit New Markets.* Chicago: Probus Publishing Company.

Welsby, J. and Nichols, A. (1999) The privatisation of Britain's railways. *Journal of Transport Economics and Policy* 33 (1), 55–76.

Werner, C. (1985) *Spatial Transportation Modeling.* Beverly Hills, CA: Sage.

Wilkinson, P. (1999) Caribbean cruise tourism: Delusion? Illusion? *Tourism Geographies* 1 (3), 261–282.

Wolfe, R. (1952) Wasaga Beach: The divorce from the geographic environment. *Canadian Geographer* 1 (2), 57–65.

Wolfe, R. (1966) Recreational travel: The new migration. *Canadian Geographer* 10 (1), 1–14.

Wood, R. (2000) Caribbean cruise tourism: Globalization at sea. *Annals of Tourism Research* 27 (2), 345–370.

Wood, R. (2004a) Global currents: Cruise ships in the Caribbean Sea. In D.T. Duval (ed.) *Tourism in the Caribbean: Trends, Development, Prospects* (pp. 152–171). London: Routledge.

Wood, R. (2004b) Cruise ships: Deterritorialised destinations. In L. Lumsdon and S.J. Page (eds) *Tourism and Transport: Issues and Agenda for the New Millennium* (pp. 133–145). Amsterdam: Elsevier.

World Bank (1996) *Sustainable Transport: Priorities for Policy Reform.* World Bank.

WTO (1991) *Resolutions of International Conference Travel and Tourism, Ottawa.* Madrid: WTO.

WTO (2003a) *Worldwide Cruise Ship Activity.* Madrid: WTO.

WTO (2003b) Climate Change and Tourism: Proceedings of the 1st International Conference on Climate Change and Tourism Djerba, Tunisia, 9–11 April 2003 (http://www.world-tourism.org/sustainable/climate/brochure.htm). Accessed 8 December 2005.

WTTC (2003) *Special SARS Analysis: Impact of Travel and Tourism.* London: WTTC.

Wynne, C., Berthon, P., Pitt, L., Ewing, M. and Napoli, J. (2001) The impact of the Internet on the distribution value chain: The case of the South African tourism industry. *International Marketing Review* 18 (4), 420–431.

Yang, J.-Y. and Lui, A. (2003) Frequent flyer program: A case study of China airlines marketing initiative – Dynasty Flyer Program. *Tourism Management* 24, 587–595.

INDEX